The Technological Imperative in Canada

R. Douglas Francis

The Technological Imperative in Canada
An Intellectual History

UBCPress · Vancouver · Toronto

20 19 18 17 16 15 14 13 12 11 10 09   5 4 3 2 1

Printed in Canada with vegetable-based inks on FSC-certified ancient-forest-free paper (100% post-consumer recycled) that is processed chlorine- and acid-free.

Library and Archives Canada Cataloguing in Publication

Francis, R.D. (R. Douglas), 1944-
The technological imperative in Canada : an intellectual history / by R. Douglas Francis.

Includes bibliographical references and index.
ISBN 978-0-7748-1650-2 (bound)
ISBN 978-0-7748-1651-9 (pbk.)
ISBN 978-0-7748-1652-6 (e-book)

1. Technology – Philosophy – History. 2. Technology – Social and ethical aspects – Canada – History. 3. Technology – Social aspects – Canada – History. 4. Technology and civilization – History. 5. Intellectuals – Canada – History. I. Title.

T14.F73 2009   601   C2008-907951-5

Canadä

UBC Press gratefully acknowledges the financial support for our publishing program of the Government of Canada through the Book Publishing Industry Development Program (BPIDP), and of the Canada Council for the Arts, and the British Columbia Arts Council.

This book has been published with the help of a grant from the Canadian Federation for the Humanities and Social Sciences, through the Aid to Scholarly Publications Programme, using funds provided by the Social Sciences and Humanities Research Council of Canada.

Printed and bound in Canada by Marquis Book Printing
Set in Stone by Artegraphica Design Co. Ltd.
Copy editor: Stephanie VanderMeulen
Proofreader: Dallas Harrison
Indexer: Christine Jacobs

UBC Press
The University of British Columbia
2029 West Mall
Vancouver, BC V6T 1Z2
604-822-5959 / Fax: 604-822-6083
www.ubcpress.ca

I dedicate this book to Murray Fraser (1937-97)

for his encouragement and support

while president of the University of Calgary

Contents

Acknowledgments

A number of individuals have been of great help and have provided encouragement in the writing and publishing of this book. I want to thank the staff at the various archives and libraries listed in my bibliography where I did research; special thanks goes to the staff at Library and Archives Canada. While working on this project, I received moral support from the following colleagues: Sarah Carter, Betsy Jameson, and Don Smith. The following individuals generously gave of their time from their busy schedules to read part or all of the manuscript and to provide constructive criticism: William Christian, Ramsay Cook, Herman Ganzevoort, Anthony Rasporich, Keith Walden, and Bill Westfall. Susan Gingell of the Department of English at the University of Saskatchewan suggested literary references to technology and read an early draft of the sections of the manuscript pertaining to Canadian poets and novelists. Charlotte Gray shared her research on Alexander Graham Bell with me. Joyce Woods word-processed chapters as they went through revisions. To these colleagues and friends, I offer sincere thanks.

The two anonymous readers for UBC Press provided thoughtful and helpful critiques of the manuscript. Ramsay Derry read an earlier draft and provided valuable editorial assistance, reminding me in particular of the importance of making ideas accessible to the general reader. At UBC Press, Director Peter Milroy has been a loyal supporter of the project. A special note of thanks goes to Melissa Pitts. She has been an exemplary editor, consistently offering encouragement and helping to bring clarity and coherence to the book. Thanks too to Ann Macklem as production editor; she has been most helpful at guiding my manuscript through the production process. I also want to acknowledge the following: Stephanie VanderMeulen as copy editor, Dallas Harrison as proofreader, and Christine Jacobs as indexer. I alone am responsible for any limitations, errors, or shortcomings in judgment that remain.

I want to thank the University Research Grants Committee at the University of Calgary for a research grant to assist in archival research. Thanks also to

Research Services at the University of Calgary and to the Killam Foundation for a Killam Resident Fellowship, which gave me a term free from teaching and administrative responsibilities to work on the project. This book has been published with the help of a grant from the Canadian Federation for the Humanities and Social Sciences, through the Aid to Scholarly Publications Program, using funds provided by the Social Sciences and Humanities Research Council of Canada.

I offer warm thanks to my wife Barbara and my children and their spouses – Marc (Jennifer), Myla (Eric), and Michael, and my grandson, Blake – for their unwavering support and their encouragement for a project that has gone on for some time. Their good sense of humour and their affection were healthy reminders that there is life beyond Canadian history.

I want to thank the following publishers for permission to quote from their publications.

The Heidegger quotation on pp. 21-22 is reproduced by permission of HarperCollins Publishers from Martin Heidegger, *The Question Concerning Technology and Other Essays,* trans. William Lovitt (New York: Harper and Row, 1977). © 1977 by Harper and Row Publishers, Inc.

The University of Toronto Press has given me permission to quote Archibald Lampman's poetry on pp. 137-40 from *The Poems of Archibald Lampman (including 'At the Long Sault'),* ed. Margaret Coulby Whitridge (Toronto: University of Toronto Press, 1973). © 1973 by University of Toronto Press.

I acknowledge the permission of Oxford University Press for the quotations on pp. 227-28 from Northrop Frye, *The Modern Century*. The Whidden Lectures (Oxford: Oxford University Press, 1967). © 1967 by Oxford University Press.

I greatly appreciate the permission of Anansi Press for quotations from the following publications: the quotations on pp. 207-8 from *Essential McLuhan,* ed. Eric McLuhan and Frank Zingrone (Toronto: Anansi, 1995). © 1995 by Anansi Press; the Dennis Lee poetry on pp. 256-58 from Dennis Lee, *Civil Elegies and Other Poems* (Toronto: Anansi Press, 1972). © 1972 by Anansi Press; the quotations on pp. 252-54 from George Grant, *Technology and Empire: Perspectives on North America* (Toronto: Anansi Press, 1969). © 1969 by Anansi Press; the quotations on pp. 261-65 from George Grant, *Technology and Justice* (Toronto: Anansi Press, 1986). © 1986 by Anansi Press; and the quotations on pp. 211-12 from David Cayley, *Northrop Frye in Conversation* (Toronto: House of Anansi Press, 1992). © 1992 by Anansi Press.

The quotations on pp. 201-4 from Marshall McLuhan, *Understanding Media: The Extensions of Man* (New York: McGraw Hill Book Company, 1965) have been reproduced by kind permission of the Estate of Herbert Marshall McLuhan and Gingko Press.

The Technological Imperative

Introduction

'Technology is the metaphysics of our age; it is the way being appears to us.'[1] The distinguished Canadian philosopher George Grant, author of the quotation and one of the intellectuals whose ideas on technology are examined in this book, came to this conclusion after a sustained period of thinking about the nature and meaning of technology. His observation also capped over a century of thought on technology by Anglo Canadian thinkers who analyzed the nature of technology and contributed to an understanding of how technology came to be, as Grant notes, 'the way being appears to us.' This book examines that thought.

In the process of tracing the evolution of Anglo Canadian thought on technology, I realized that these thinkers saw technology as the most pervasive and dominant force in the modern world; it became for them an imperative – what I call the 'technological imperative.' This technological imperative, they believed, created a mindset that was itself technological, shaped by the very technology the mind was attempting to comprehend. In analyzing their perspective on this mindset, and in noting the importance they gave to the dominating influence of technology, I realized – and this is a central argument of my book – that these Canadian intellectuals were the *makers* of the technological imperative. This book examines the unfolding of the Canadian perspective on that technological imperative.

Ironically, this was not what these Canadian thinkers intended; it was quite the opposite. Having so closely read the ideas of Canadian thinkers on technology over a long period of time, I realized that they were moralists who were attempting to retain or salvage a moral order – a moral imperative – that they believed the technological imperative either enhanced or else threatened. Their solution was to attempt to reconcile the two imperatives or at least to make Canadians aware of the benefits or dangers that the technological imperative posed to the moral imperative. Those Canadian thinkers who favoured technology maintained that the technological imperative would complement the moral imperative by instilling moral values essential for the advancement of society and Western civilization. Those who saw technology as a threat feared that the technological imperative would undermine the moral imperative by breaking down communal ties that were important for the well-being of society and by undermining moral and spiritual values that had been the underpinning of Western civilization. Yet these latter intellectuals realized they could not simply dismiss the technological imperative. They had to confront it and seek a balanced perspective. In their attempts at reconciliation, or at least coming to terms with these two imperatives, these Canadian thinkers created and magnified a tension between the two imperatives that remained a constant in Canadian thought from the mid-nineteenth century to the beginning of the twenty-first century. The end result was that neither imperative became dominant or absolute. It is possibly another example of the Canadian tendency to compromise – a compromise, however, that failed to satisfy any of the Canadian thinkers examined in this book.

Analysts of the moral imperative in Canada have attributed its decline to the emergence of critical thought, Darwinian science, and higher criticism. In *A Disciplined Intelligence: Critical Inquiry and Canadian Thought in the Victorian Era,* historian Brian McKillop focuses on the role critical inquiry played in challenging and, by the 1890s, shattering the moral imperative that was founded on constraint and dominated by a myth of concern that was largely closed.[2] He sees a new attempt at reconciling the moral imperative with critical thought in the emergence of the philosophy of idealism that underlay the social gospel movement at the turn of the twentieth century. In *The Regenerators: Social Criticism in Late Victorian English Canada,* distinguished Canadian historian Ramsay Cook picks up where McKillop left off to argue that, in their attempt to reconcile the sacred and the secular, social gospellers in Canada actually contributed to the secularization of society.[3] They abandoned their traditional role as religious leaders of society to focus on secular concerns in the hopes of creating 'the Kingdom of God on Earth.' Cook blames Darwinian science, along with its offshoot Social Darwinism, and higher criticism, the challenge to the infallibility of the Bible as a sacred text, as the major causes of the undermining of the moral imperative.

I contend that it was the emergence of the technological imperative in the late nineteenth and twentieth centuries that challenged the moral imperative and weakened it as an absolute in Canadian thought. Equally, however, the continuous presence of the moral imperative during the same period of time prevented the technological imperative from becoming dominant. One important purpose of this book is to show the playing out of this rivalry in Canadian thought and the resulting tension it generated.

Another way that the perspective on technology among these Canadian thinkers had a 'Canadian twist' to it (besides the Canadian attempt at compromising the technological and moral imperatives) was in their association of technology with civilization within the context of the Canadian identity. 'Civilization' is identified in two ways: as Western civilization with Britain as its centre in the nineteenth century, and as American civilization, a 'bastardized' form of Western civilization, according to some Canadian theorists, that by the twentieth century had become pervasive. In the nineteenth century Canadian theorists of technology saw technology as the means by which Canadians could partake of the virtues of Western civilization by physically *and intellectually* linking their country to Britain. In so doing, technology enabled Canadians to be world citizens, able to rise above their parochial existence on the North American continent. Technology also became associated with freedom for these nineteenth-century Canadian thinkers, a means by which, through their technological association with Britain and Western civilization, Canadians could be independent of the United States.

In the twentieth century, technology had a negative identity when associated with the United States and American imperialism. Technology was seen as instilling American values into Canadian society that were antithetical to traditional British Canadian morality. Technology was also seen as a source of power that had enabled the United States to dominate Canada and, through American imperialism, to control the entire world. Thus, the theme of technology as freedom versus power, which has its roots in Greek thought as a theme in Aeschylus's play *Prometheus Bound* and in early Christian thought in Saint Augustine's *The City of God,* appears in Canadian thought as part of the ongoing debate over Canadian identity.

I had originally intended to include both Anglo Canadian and French Canadian thinkers on technology in this book. Certainly, the theme of technological imperative versus moral imperative runs through French Canadian thought, although the moral imperative in French Canadian Roman Catholicism is nuanced differently than in Anglo Canadian Protestantism. In addition, French Canadian intellectuals looked at technology from the perspective of national identity and in the context of survival, or, in the case of French Canadian thinkers, *la survivance,* as their English Canadian counterparts did, although the 'nation' for French Canadian thinkers usually centred on Quebec. However, as I further pursued French Canadian thought on technology, I realized the plethora of intellectuals and the richness of perspectives to be analyzed. I came to the conclusion that to provide the same depth of analysis for French Canadian thinkers on technology that I have tried to provide for Anglo Canadian thinkers in one study was simply impossible. To attempt to do so would make French Canadian thought on technology appear to be an 'appendage,' or in addition, to that of Anglo Canadian thought rather than as a study in its own right.

The issue of gender arises in dealing with the subject of technology. A number of recent feminist studies of technology have noted the many ways that technology is gendered in their disfavour. As well, it has been noted that until recently the vast majority of individuals writing about technology have been males. Certainly, this study bears that out. All of the Canadian thinkers whose ideas I analyze in this book are Anglo Canadian males, with one exception: Adelaide Hoodless. I had hoped to find more Canadian women in the past who reflected at length on the meaning of technology, especially from a female perspective, but Hoodless was the only one. In her case, she looked at domestic science as a form of technical education, a topic I address in Chapter 3. In the current period, Ursula Franklin, a professor of engineering at the University of Toronto, and Heather Menzies, an adjunct professor at Carleton University in Ottawa, have addressed the problem of a lack of female analysts of technology and offered their own perspectives as to why this is the case and what impact this deficiency has had on our understanding of technology. I discuss their views in my Conclusion. I had considered examining in what ways Canadian theorists of technology have gendered the technological imperative, providing in essence a postmodernist approach to the subject. However, I concluded that, since no one has looked at technological thought in Canada, the first step is to examine what these thinkers had to say about technology and leave it to others to analyze the views of Canadian theorists of technology from the perspective of gender and even from the perspectives of race and class as well. I am aware that all of the individuals whose ideas I discuss in this book are of Anglo-Celtic upbringing and of a privileged class, most of them with an academic background. Again, why this is so and what impact race and class might have on their perspective could be another important and related topic of study in and of itself.

One popular theme in postmodernist studies is power, particularly relating to the questions of who holds power, who is powerless, and how those in positions of power utilize it to keep the powerless under their control. This theme arises in my study because technology has always been a source of power, a theme that I note, if only in passing, when important to an understanding of the ideas of the individuals discussed. The exception is my chapter on Harold Innis and Eric Havelock: the theme of power was front and centre in their understanding of technology.

The challenge in writing a history of this kind is to find a theoretical model or form of classification that is appropriate to a subject as vast as technology. What we usually think of when we hear the word 'technology' is objects and machines. But one does not have to venture very far into the literature on technology or into the ideas of Canadian theorists to realize that technology is much more than objects and machines. Carl Mitcham argues in his article 'Philosophy of Technology' that technology can be classified into four broad categories: technology as object, technology as knowledge, technology as process, and technology as volition. Under technology as object,

Mitcham notes that the common-sense view is to associate technology with 'tools, machines, electronic devices, consumer products and the like,' and then to classify 'technological objects into various types and ultimately the articulation of an ontology of artifacts.'[4] Technology as knowledge, Mitcham's second category, is chiefly concerned with laws and how they relate to human nature. As Mitcham notes, 'To view technology as a kind of knowledge not only invites epistemological analysis, it transforms technology from an extension of man into an inherent constituent of human nature.'[5] Mitcham's third classification, technology as process, assumes that what is important about technology is the process of 'making and using' rather than how things are made and used.[6] The former – 'making' – became the domain of the engineer, while the latter – 'using' – has become the concern of the social scientist. Mitcham's last category, technology as volition, is concerned with the 'aims, intentions, desires, and choices' of those utilizing technology.[7] Increasingly, the debate focuses on whether the aims, intentions, desires, and choices are human ones or whether technology has a will of its own that dictates the choices humans make. I have found Mitcham's classification useful as an organizing principle for the Canadian thinkers in this book, as I will show in a moment. But I realized there is another category, at least with regard to Canadian analysts of technology: technology as imperative. Canadian thinkers may have differed as to whether they identified technology as object, knowledge, process, or volition, but they were united in their belief that technology was an imperative.

I begin my work with a chapter on the ideas of major international thinkers on technology. It is the evolution of their thinking that provides the historical and intellectual contexts in which to place the Canadian thinkers in this book. Historically, technology went from being seen first as objects or machines, then as a form of knowledge, as a process, and finally as volition. This was the case among the Canadian intellectuals that I discuss as well. Thus, by examining the ideas of the major international analysts of technology, I provide an intellectual context and backdrop for my discussion of the ideas of Anglo Canadians. As well, the issues raised by these international analysts of technology are ones that I too address in my analysis of Anglo Canadian theorists of technology.

My historical examination of Anglo Canadian thought on technology begins in the mid-nineteenth century with the writings of Thomas Coltrin Keefer and Thomas Chandler Haliburton on railways. They both saw railways, or more precisely railway locomotives, as objects or machines, the first level of identification of technology. These mighty juggernauts were so powerful, and the influence of railways so pervasive, Keefer and Haliburton argued, as to inaugurate a new modern era in Canada. To begin with Keefer and Haliburton on railways is not to suggest that no significant technological inventions preceded locomotives or that no Canadian thinkers before Keefer noted the importance of earlier technological inventions. However, I do contend that no technological invention prior to the locomotive had as significant an impact to cause sustained reflections on its significance, and no Anglo Canadian thinkers prior to Keefer and Haliburton reflected at length on the impact of technology on Canadian thought. Only with the advent of railways did Canadian thinkers begin to think systematically and deeply about the nature, meaning, and significance of technology as opposed to just using it. There was a sense of wonder, excitement, and awe about railways as a form of technology that marks the railway era and Keefer's and Haliburton's writings as new and distinct. They claimed that railways were inaugurating a new world in which technology would be the dominant force, an imperative.

In the late nineteenth and early twentieth centuries, as Canada underwent its industrial revolution, a number of Canadian educational theorists saw technology as knowledge, the second phase of identification. Borrowing Francis Bacon's dictum that 'knowledge is power,' these educational theorists argued for the importance of technical education not only for the material advancement of Canada but also, and more importantly, for the moral and spiritual advancement of the country. I analyze their ideas in Chapter 3.

During the First World War and in the interwar years when large-scale industrialism and mechanization took hold, Canadian analysts of technology saw technology as process, the third way of classifying technology. During the war years, technology was defined as a process of war. In Chapter 4, I have used the ideas of the distinguished philosopher-psychologist George Sidney Brett as my focal point for analyzing the ideas of technology as a process of war. In the immediate postwar era, William Lyon Mackenzie King wrote of technology as a process of industrialism in *Industry and Humanity* (1918), as did the noted novelist Frederick Philip Grove in his novel *Master of the Mill* (1944). Their ideas are analyzed in Chapter 5. In *The Unsolved Riddle of Social Justice* (1920), Stephen Leacock examined technology as a process of mechanization. The same theme is evident in a number of poems by Archibald Lampman, one of the 'Confederation poets.' I discuss their ideas in Chapter 6.

In the post–Second World War era, theorists of technology realized that technology was much more than tools, machines, mechanization, or mode of production: it represented a pervasive value system. Canadian analysts of technology confronted the larger issue of the power of technology to control human thought. For these intellectuals, the issue of technology as volition loomed large, the final form of classification of technology. In Chapter 7, I trace the theme of technology as power in the context of volition in the writings of noted Canadian economist, historian, and communication theorist Harold Innis, as well as in the writings of Eric Havelock, Innis's colleague in the classics department at the University of Toronto in the 1930s and early 1940s. The ideas on technology, especially electronic technology, of Canadian guru of communication technology Marshall McLuhan are analyzed in Chapter 8. In Chapter 9, I explore the association of technology with mythology in the writings of Northrop Frye, world-renowned scholar of mythology, and discuss the poetry of E.J. Pratt, Frye's teacher and later colleague, in which so many of the Canadian myths on technology have taken form. In Chapter 10, I trace the evolution of the ideas on technology of George Grant, one of the most extensive analysts of technology, and that of the poet Dennis Lee. I focus particularly on Grant's perspective of technology as 'being' and on the implications of this perspective on the concept of technology as volition. Grant came to see technology as so pervasive that the only meaningful response was one of silence. Dennis Lee, moved by Grant's emotional appeal to Canadians to fight American technological dominance, also used the theme of silence in his reflections on technology in his epic poem *Civil Elegies*.

In tracing Anglo Canadian thought on technology through the four perspectives of technology – as object, knowledge, process, and volition – I also show how technology goes from being seen as machines, external to human beings but having an impact on them, even on their perspective on the world, to being seen as a pervasive force that shapes our very essence as human beings, including the values and ideals by which we live. Equally, technology goes from being 'out there,' external objects or processes that humans can react to and possibly control, to being 'in here,' a force within the human mind that controls our ways of thinking. In all cases, Canadian theorists of technology have come to see it as a mindset that is itself shaped by the very technology that humans are attempting to comprehend and control – hence, an imperative.

One final issue needs to be addressed. How does this study fit into the existing English Canadian historiography on technology and Canadian thought? What is surprising is how little has been written on this topic despite its pervasiveness. To date, no one has looked at the evolution of Anglo Canadian thought on technology as this book does. While most of the individuals whose perspectives on technology form the basis of this book are well known to Canadian intellectual historians, most have not been looked at from the perspective of their views on technology. The exceptions are Harold Innis, Marshall McLuhan, and George Grant. Arthur Kroker, an analyst of technology, did a study of the ideas of these three theorists in *Technology and the Canadian Mind: Innis/McLuhan/Grant*. While insightful in terms of the views of these intellectuals and how they interrelate, the discussion occurs in a vacuum since Kroker does not trace the roots of their thinking on technology, nor does he put their ideas into a historical context. The other important study that includes Innis, McLuhan, and Grant, as well as Northrop Frye, and that deals with the subject of technology is Robert E.

Babe's *Canadian Communication Thought: Ten Foundational Writers*. Babe does provide brief biographical sketches of these individuals, as does this book, to reveal the roots of their thinking. However, the focus of his study is not on their views of technology per se but rather, as his title indicates, on their communication thought. While technology is certainly an important component of these intellectuals' views on communication, it does not get adequate treatment. Also, Babe, like Kroker, fails to place the ideas of these communication theorists in a broader historical context. These criticisms of Kroker and Babe apply as well to other analysts of Innis, McLuhan, Frye, and Grant. The tendency is to focus on a theme other than technology. For the few analysts who do dwell on the subject of technology in the thought of these Canadian intellectuals, the approach is to look at the topic of technology in selective writings only, such as in Innis's communication studies, rather than as a theme throughout all their major writings. Finally, no one has identified technology as an important topic in English Canadian intellectual life, that is, as an imperative – the technological imperative – and then shown how it has interacted with the other dominant imperative in English Canadian thought since the mid-nineteenth century, namely, the moral imperative, as I have done in this book.

The evolution of the idea of technology as metaphysics within Anglo Canadian thought has been long and multifaceted. I highlight the peaks of that intellectual journey by focusing on the ideas of the major Anglo Canadian theorists of technology from the mid-nineteenth century to the present. In so doing, I show precisely how these Canadian theorists came to shape a technological imperative that continually came up against a moral imperative in a way that accentuated the tension between these two dominant modes of thought. It was this tension, and the feeling among some Canadian intellectuals, like George Grant, that the technological imperative invariably dominated over the moral imperative, that caused him to reluctantly conclude that 'technology is the metaphysics of our age; it is the way being appears to us.'

1
Perspectives on Technology

International analysts of technology have grappled with the nature, meaning, and significance of technology since its emergence as a dominant force in Western civilization. Their ideas enrich our understanding of the multifaceted ways of seeing technology. They also offer a theoretical lens through which to frame Canadian thought. More importantly, however, the ideas of the international thinkers on technology provide an intellectual backdrop for the ideas of Canadian thinkers on technology. In discussing the issues these international theorists raised and the perspectives they offered, one becomes aware to what extent Canadian theorists of technology were in tune with the thinking of their times; they were not writing in an intellectual vacuum. While Canadian thinkers may not have been cognizant of the ideas on technology being put forward by intellectuals elsewhere, they did nevertheless share a Zeitgeist of the times in which they wrote. They also raised questions and grappled with issues from their own Canadian perspectives that were age-old ones relating to technology. Identifying these issues and noting the changing currents of thought on technology among international theorists of technology thus provide the intellectual framework needed to enrich our understanding and enhance our appreciation of the depth of insights on technology offered by Canadian thinkers.

Lewis Mumford, distinguished and insightful writer on the history of technological thought, has traced the historical and intellectual evolution of technology, first perceived as objects and then, by the advent of the Industrial Revolution, as machines, from prehistoric times to the twentieth century. He shows how technology began as an idea in the minds of primitive human beings that grew to keep pace with the numerous technological inventions over the centuries. He argues that over time machines came to shape a mindset that was itself 'mechanical,' that in essence the idea of technology consumed itself. As early as the sixteenth century, technology had come to be seen as a form of knowledge. The provocative Renaissance thinker Francis Bacon, identified as the first philosopher of technology, or what he called the mechanical arts, saw technology as a way of thinking that was superior to scholastic philosophy and that could create a utopian world if it were to become the dominant paradigm of thought. To him, knowledge was power, and technical knowledge was the greatest source of intellectual power. William Leiss, a Canadian analyst of theories of technical education, explores the ideas underlying the perspective of technology as knowledge from Francis Bacon's time to the present in *Under Technology's Thumb*. I also note the importance of John Kenneth Galbraith's work on technical education as 'technostructure.' Karl Marx, writing in the mid-nineteenth century, became the first analyst of technology to identify the multiplicity of ways in which the process of industrialism and its offshoot, mechanization, affected all aspects of society, especially the lifestyle of the working class. What Marx did for the nineteenth century, the noted analyst Siegfried Giedion did for the twentieth century in *Mechanization Takes Command,* showing that the process of mechanization had even 'invaded' the private spaces of the home and of the mind. Martin Heidegger, distinguished German philosopher of technology, provided the first significant analysis of technology as volition, noting in his seminal essay 'The Question Concerning Technology' the ways in which it shapes the values and beliefs of the modern age. French theorist Jacques Ellul explored the subject further, especially the idea of technology as a *mentalité* that was itself technologically induced, in his important study entitled *The Technological Society*. System designers, particularly advocates of cybernetics, have explored ways of using communication data to create patterns of thought that can be applied to solving problems. Norbert Weiner presents this perspective on technology in *Cybernetics: Control and Communication in the Animal and the Machine*.

What follows is an overview of the key ideas of these major international thinkers on technology as representative of the perspective of technology from one of the four broad categories of machines, knowledge, process, and volition. These major international thinkers on technology also provide an intellectual and historical context for analyzing the ideas of Canadian theorists of technology that form the essence of this work.

Analyst Lewis Mumford went the furthest in studying technology as objects or machines.[1] In his monumental works, beginning with *Technics and Civilization* (1934) and including *Art and Technics* (1952), *The Transformations of Man* (1956), and his two-volume study *The Myth of the Machine: Technics and Human Development* (1966) and *The Myth of the Machine: The Pentagon of Power* (1970), Mumford explored all aspects of the world of technology as 'the machine.' He differentiated between machines as specific objects, such as the printing press or the power loom, and 'the machine' as a 'shorthand reference to the entire technological complex.' Concerning the latter, he noted, 'This will embrace the knowledge and skills and arts derived from industry or implicated in the new technics, and will include various forms of tool, instrument, apparatus and utility as well as machines proper.'[2]

He traced the historical evolution of technology from prehistoric tools to machines run by the power of wind, water, animals, and humans in the 'ecotechnic' phase (from roughly 1000 to 1750 AD), by coal and steam in the 'paleotechnic' phase (1750 to 1900 AD), and by electricity in the 'neotechnic' phase (1900 on). What is significant about his study, besides the extensive period covered in his analysis of 'the machine,' is threefold. First of all, he includes as 'machines,' and therefore within his definition of technology, artifacts not usually identified as machines, such as utensils (baskets, tables, and chairs), apparatus (dye vats and brick kilns), utilities (reservoirs, aqueducts, and roads), works of art, and even human beings themselves. Indeed, he argued that the greatest machine – the 'megamachine' – has been collective human power, initially used to build the Egyptian pyramids, for example, and later used as large-scale armies. Second, he maintained that machines have both shaped the culture of the society from which they developed (far more than has been recognized by the people within that society and analysts since) and, more importantly, were themselves shaped by humans through a cultural context of the time. During the paleotechnic phase, for example, the technology of the Industrial Revolution, particularly related to the production of coal and the introduction of the steam engine, created new institutions such as capitalism and modern armies, which in turn resulted in the creation of 'a new civilization.'[3]

Mumford argued that machines came to dominate human life to a greater extent than ever before or since, making humans quantifiable entities valued only for their productivity. He believed that in the recent neotechnic phase and the advent of electricity, the imbalance was corrected, and machines served human needs and were patterned on organic life. A redeeming feature of this age was the shift from quantitative to qualitative standards through automation, which, he claimed in *Technics and Civilization* (1934), would liberate man from inhuman work. However, Mumford came to doubt this claim by the time he wrote *The Myth of the Machine* in the 1960s.

Closely related to his second point on the interaction of technology and culture is his third point on seeing all technology as machines. Even his term 'the machine' to refer to items not usually seen as machines, or to sources of power in the ecotechnic phase, and even to a way of perceiving the world – a mindset – indicates the pervasiveness of his image and definition of technology as machines. This image not only weakens his analysis of technology in the twentieth century, when the image distorts more than it explains, but also limits his analysis of technology by seeing the technological mindset as only 'mechanical' in nature, thus eliminating aspects of this technological mindset that for some theorists of technology go well beyond what might be imagined by the term 'the machine.' While Mumford certainly suggests some of these wider implications,[4] his insistence on using the term 'the machine' for this wider mindset clearly links technology with machines only in his perspective on the subject of technology. It also makes humans out to be 'mechanical beings,' lacking in moral and spiritual values. Note for example Mumford's discussion of the 'new scientific method' as an aspect of the mindset of Western civilization that underlay our modern technological age and the extent to which he sees that scientific method as associated with machinery:

> Machines – and machines alone – completely met the requirements of the new scientific method and point of view: they fulfilled the definition of 'reality' far more perfectly than living organisms. And once the mechanical world-picture was established, machines could thrive and multiply and dominate existence: their competitors had been exterminated or had been consigned to a penumbral universe in which only artists and lovers and breeders of animals dared to believe ... By renouncing a large part of his humanity, a man could achieve godhood: he dawned on this second chaos and created the machine in his own image: the image of power, but power ripped loose from his flesh and isolated from his humanity.[5]

The cultural aspect of technology clearly interested Mumford. He believed that the greatest impact of new machines and new energy sources was qualitative rather than quantitative and thus more accessible to the cultural sensitized than the statistician or the scientist. Yet even the 'cultural sensitized' would miss the qualitative impact of new technology, Mumford maintained, if they saw it only as a one-way interaction – the machine on society. Mumford refused to see machines as having ultimate sway and autonomous power over humans. From the beginning, he noted, human spirituality and creativity, through dreams and the imagination, created the ideas and human energy needed to create the machines. He wrote, 'His [man's] first task was not to shape tools for controlling the environment, but to shape instruments even more powerful and compelling in order to control himself, above all, his unconscious. The invention and perfection of these instruments – rituals, symbols, words, images, standard modes of behavior (mores) – was, I hope to establish, the principal occupation of early man, more necessary to survival than tool-making, and far more essential to his later development.'[6] Thus, man was a thinker and a creator of ideas before he was a tool-making animal or later a worker or technician. That is why, Mumford explained, the earliest great machines in the West that were precursors of the Industrial Revolution – the clock, watermill, horse-powered treadmill, and windmill – were creations of the monastery. Christianity provided the essential intellectual milieu for such inventions by providing a spiritual motivating force for work – that is, the belief that in doing good works one was serving God – and incentive – that is, faith in the creative ability of humans as 'children of God.' Mumford contended reluctantly that once in place, however, the mechanical prevailed over the spiritual component of technology, providing both the rationale and the means for humans to advance.

Because of Mumford's persistence in seeing technology and culture as an interactive dynamic, he remained essentially optimistic about the ability of humans to control technology. While there were periods when it appeared as though humans were the slaves of the machine, such as during the Industrial Revolution of the paleotechnic phase, ultimately the imbalance corrected itself, and humans emerged dominant over the machine. Only as Mumford reached the twentieth century and the current age did he become pessimistic. He feared that, in the pursuit of perfection, humans had come to trust in the machine to change them for the better. As a result, society in general was allowing technology to come under the control of an elite intent on creating a utopia that would be authoritarian and uniform rather than liberating. Mumford noted with concern that we are moving toward the age of 'megatechnics' when 'the dominant minority will create a uniform, all-enveloping, super-planetary structure, designed for automatic operation. Instead of functioning actively as an autonomous personality, man will become a passive, purposeless, machine-conditioned animal, whose proper functions ... will either be fed into the machine or strictly limited and controlled for the benefit of depersonalized, collective organizations.'[7] Mumford believed that what contributed to this 'misdirection' was the myth of prehistoric man as predominantly a tool-making animal and modern man as essentially a worker and a technician. Ultimately, technology had to be seen not as an autonomous entity but as a human creation, existing at man's will for human ends.

The perspective of technology as objects or machines was followed by the perspective of technology as knowledge. This view dates back to the seventeenth century, when Francis Bacon declared in *The New Organon; Or, True Directions Concerning the Interpretations of Nature* (1620) that 'Knowledge is Power.' Bacon, who has been called the first philosopher of industrial science,[8] believed that technical knowledge, what was then called the 'mechanical arts,' was the most powerful and useful form of knowledge, superior in every way to scholastic discourse. Technical knowledge held the possibility of conquering nature by eliminating its vagaries and unpredictability and overcoming the devastation of human life through flood, famine, disease, and pestilence. Bacon maintained that scholastic knowledge, by comparison, saw physical nature as shaping human nature, thus becoming bogged down on issues of morality that he believed had no place in dealing with technology. Technical knowledge held the potential to create a utopian state, a state he sketched out in his novel *The New Atlantis* (1624).

Bacon's faith in technology as knowledge and power continues into the present. It can be found in the teachings of schools of the practical arts and schools of engineering and in our belief that with each new invention and the accumulation and integration of new technical knowledge the world comes closer to a state of perfection. Technical knowledge is seen as the key to the future, and 'from this standpoint,' William Leiss, the Canadian analyst of technology, notes, 'technologies are essentially a crystallized form of human knowledge.'[9]

Unlike the Canadian thinkers who follow in this study, who are theoreticians of technology, Leiss analyzes in his major work *Under Technology's Thumb* the ideas of those thinkers over time who have perceived technology as knowledge. He argues that the popularity of technical knowledge arose out of the political theory of classical liberalism in the early modern period of the West. It assumed that social progress would occur on the basis of rational or educated self-interest, as opposed to emotional appeal to the social conscience of society as a whole. As John Locke reasoned, the rational person was the 'industrious' being who accumulated property. In the self-regulating and highly competitive marketplace of the laissez-faire economy, knowledge, especially technical knowledge, became a commodity, a form of 'capital,' and a factor in success. Locke pointed out that a knowledgeable worker was more productive than an untrained one. It was Karl Marx, however, who pointed out the direct link between technical knowledge and the productivity of labour; what he condemned was the application of this knowledge and resulting productivity for private ends (as private property) and one class only (the bourgeoisie) rather than for the good of society as a whole and in particular for the working class that contributed substantially to that productivity. By the nineteenth century, Leiss notes, 'many writers regarded scientific and technical knowledge as the cornerstone of the truly revolutionary changes in production made through the industrial system.'[10] As Alfred Whitehead observed, a process of invention was the greatest invention of the nineteenth century, and that process was the by-product of scientific and technological knowledge. From that starting point, technology as knowledge evolved to the point at which it has now become one of the sacred and unquestioned truths of our modern value system, so much so, in fact, that it has created its own idols – what Leiss calls 'the idols of technology' – comparable to the idols that Francis Bacon fought against to get the 'mechanical arts' or technology accepted in his day.

In *The New Industrial State* (1967), John Kenneth Galbraith, an economist by training but a social activist by conviction,[11] argued that technology in the late twentieth century was so sophisticated as to require megacorporations with an array of highly educated specialists – researchers, designers, lawyers, accountants, economists, engineers, personnel specialists, public relations agents, among others – to ensure maximum technological efficiency.[12] He called this technical knowledge 'technostructure' and noted that it had replaced capital as the crucial factor of production. This technostructure, created and sustained by technology, has its own imperative that may appear less threatening than power structures in the past, since power resides in many experts in a corporation rather than in one individual. However, these techno-corporations with their collective technical expertise, Galbraith pointed out, wield infinitely more power than any individual could have hoped to wield in the past. Worse still, these corporations operate purely on the profit motive: to produce more consumer goods at huge profits for the financial benefit of the shareholders in the corporation; they lack a social conscience. Hence Galbraith's concern that these techno-corporations, rather than politicians or the public at large, are driving the economies of the world, and especially the largest of those economies, that of the United States. He also noted the difficulty in the modern world of controlling the technical will to power because it is invisible, complex, and diverse. Nevertheless, he warned, it is no less dangerous to the good of society: it is all the more so because it provides the goods that consumers want and thus appears to be a 'friend' or at least 'benign,' until we are awakened to the motives behind its drive to power and the negative implications that result from it. Galbraith argued that 'we are becoming the servants in thought, as in action, of the machine we have created to serve us.' Society is allowing economic goals to drive social policy rather than vice versa. This needs to be corrected, he emphasized, so that what counts 'is not the quantity of our goods but the quality of life.'[13]

Daniel Bell, an American cultural analyst, has also noted the importance of technical knowledge in the modern age. In *The Coming of Post-Industrial Society,* he argued that 'broadly speaking, if industrial society is based on machine technology, post-industrial society is shaped by an intellectual technology. And if capital and labor are the major structural features of industrial society, information and knowledge are those of the post-industrial society.'[14] Technical knowledge, Bell noted, has replaced machinery and land as the key resource in society, and the possession of technical know-how has become the route to power, thus fulfilling Bacon's prediction that knowledge is power.

Such hyperbole, Leiss points out, creates the danger of elevating technical knowledge, and therefore technology itself, to the status of 'new god' of the modern world. It also poses the danger of seeing technology as a mammoth and autonomous entity – a somewhat friendly Frankenstein – over which humans have little control and technical knowledge as the only worthy kind of knowledge to acquire. The danger becomes what Leiss calls 'a despotic marriage of knowledge and power.'[15] The answer is to realize that technical knowledge is only one form of knowledge and not necessarily the best at coming to terms with complex social problems arising out of the very technology that needs to be studied and addressed. Ultimately, Leiss points out, we must come to realize that the motivating factor for change in society today is not technology but the social values that lie behind the technology. These are the cultural and intellectual assumptions on which we operate, the moral values for which we strive. In that search, the 'technological imperative' can be a positive factor but should not be the ultimate goal. In the end, then, Leiss remains optimistic that technical knowledge can be a positive factor for social change so long as it is seen as only one – and not the primal – factor in coping with problems and moving toward an enlightened future. As he notes in his tempered conclusion on the subject, 'We shall need every ounce of technological ingenuity and scientific understanding we can muster to pull us back from the abyss of irremediable environmental disaster. But there is no hope of healing so long as the illusion persists that those instruments themselves can bring about the harmonization of human interests.'[16]

Implicit in Leiss's analysis of technology as knowledge is a corollary of this perspective on technology, namely, that technology – technical education – can be a powerful agent for social reform. Francis Bacon himself envisioned technical knowledge being used in this way in his promise that the benefits of applying technical knowledge – the mechanical arts – to nature and society far outweighed its dangers and limitations. From Bacon onward, a host of social reformers and utopianists alike have seen in technology the panacea to the multiplicity of ills besetting society, many of which are, ironically, the direct results of the very technology that has become the object of faith.

Machines seldom operate as self-contained entities. They form part of a larger complex that we call industrialism and a process that is described as mechanization. This became the third way of perceiving technology: as process. Karl Marx has often been seen as the first to analyze the dynamics and implications of the process of mechanization in the factory system that became the core of industrialization. In his section called 'Machinery and Large-Scale Industry' in *Capital: A Critique of Political Economy,* Marx began by distinguishing between tools and machines: the former being the instruments of humans, whereas the latter were implements of their own mechanization. Once set in motion, Marx pointed out, these mechanical instruments 'perform with [their] tools the same operations as the worker formerly did with similar tools.'[17] Hence machines began to replace workers, as opposed to tools that were used by workers. When these machines began to operate as a system, as in a factory, then they constituted a 'vast automaton.' Marx described such mechanized systems as 'a mechanical monster whose body fills whole factories, and whose demonic power, at first hidden by the slow and measured motions of its gigantic members, finally bursts forth in the fast and feverish whirl of its countless working organs.' Here then, Marx noted, was full-scale mechanization.

What followed were 'the abolition of the old handcraft and manufacturing systems in the spheres of production,'[18] and the division of labour, along with the exploitation of workers, including the women and children of the male workers, as mere 'commodities' for the benefit of capitalists. Marx noted the chain reaction set in place by the mechanization of the mode of production in terms of its impact on the working class:

> Partly by placing at the capitalists' disposal new strata of the working class previously inaccessible to him, partly by setting free the workers it supplants, machinery produces a surplus working population, which is compelled to submit to the dictates of capital. Hence that remarkable phenomenon in the history of modern industry, that machinery sweeps away every moral and natural restriction on the length of the working day. Hence too the economic paradox that the most powerful instrument for reducing labour-time suffers a dialectical inversion and becomes the most unfailing means for turning the whole lifetime of the worker and his family into labour-time at capital's disposal for its own valorization.[19]

Mechanization took command, Marx noted, and workers were forced to keep pace with the relentless motion of the machines to the point of nervous exhaustion. As well, workers became slaves to the machines. They reacted by fighting the machines, by railing against the system that created the machines, through riots and strikes, and by denouncing the capitalists who controlled and benefited from the system. What is impressive about Marx's analysis is his ability to show the multifaceted ways that machine technology in the form of a mechanized industrial system had an impact on all aspects of society, especially on the working class. However, he failed to address the moral and spiritual impacts of technology on the working class. In the end, Marx believed there were potential economic benefits associated with technology for the working class but that such benefits could not accrue until control was taken out of the hands of the bourgeoisie and placed in the hands of the proletariat. In Marx's mind, industrialism and the accompanying process of mechanization were value-neutral; as to the outcome, it depended in whose hands the power of technology resided.

A more recent major study of technology as mechanization is Siegfried Giedion's *Mechanization Takes Command: A Contribution to Anonymous History* (1948).[20] Surprisingly, Giedion did not define mechanization but rather showed its multifaceted nature in the material, natural, organic, and human realms. Its roots lay in the rationalistic view of the world of the eighteenth century, with its faith in progress and human perfectibility. In the nineteenth century, mechanization was seen as the means to achieve these ideals. Mechanizing production became the ultimate goal, but this could only come about when the guilds were abolished, Giedion argued. This in turn required a change in perspective from what Giedion called 'the miraculous to the utilitarian.'[21] Thus, Giedion also maintained, like Mumford, that technology in the form of mechanization is the by-product of a particular mindset that emerged in the modern world – a mindset that saw the world in mechanistic terms, that measured everything in quantitative as opposed to qualitative terms, that put a premium on utilitarian over spiritual values, and that had as its ultimate goal the rational, systematic, and calculated mechanization of the organic and inorganic, natural and human worlds. This ultimately required the severing of thoughts from feeling, and the dehumanizing of human beings through an alteration of human nature. Although noting this disjuncture between the practical and moral aspects of the new technological mindset, Giedion did not explore its implications. Instead he simply noted by way of introduction to his study: 'At the origin of the inquiry stood the desire to understand the effects of mechanization upon the human being; to discern how far mechanization corresponds with and to what extent it contradicts the unalterable laws of human nature.'[22]

Giedion argued that the ultimate form of mechanization was the assembly line: 'It aims at an uninterrupted production process. This is achieved by organizing and integrating the various operations. Its ultimate goal is to mould the manufactory into a single tool wherein all the phases of production, all the machines, become one great unit. The time factor plays an important part; for the machines must be regulated to one another.'[23] The assembly line was 'an American institution,' just as mechanization was most notably 'an American phenomenon.' Underlying the assembly line and giving it its rationality or purpose was scientific management, another American phenomenon and the brainchild of Frederick W. Taylor. Both reflected the initial optimism and euphoria with which mechanization was greeted. Giedion maintained that mechanization reached its extreme form in the interwar years, when it penetrated the private sphere: the household through the kitchen and the bathroom; nutrition through food processing; and leisure through the automobile. On the latter, he noted, 'The automobile is a personal appurtenance which comes to be understood as a movable part of the household ... The automobile is a harbinger of full mechanization ... Walking, relaxation for its own sake, because the body demands it, or because the brain requires a pause in which to recuperate, is increasingly eliminated by the motor-car.'[24] Even the senses come under the sway of mechanization: the eye by the silent cinema, the ear by the radio, and both senses by the television. Heralding Marshall McLuhan's later theories on communication technology, Giedion argued that new mediums create new values and new modes of imagination.

For Giedion, change is the one 'constant' in the modern world of technology. He began his study with the concept of movement, which, he claimed, 'underlies all mechanization.'[25] Our modern concept of motion has its roots, he argued, in the Judeo-Christian tradition, the belief that the world was created at a point in time – *ex nihilo* – 'and set in motion by an act of will.'[26] Out of the search for the principles and first causes that underlay God's action came the question of change and thus an interest in the nature of movement, especially although not exclusively in the realm of astronomy. By the eighteenth century, movement in all its forms was of interest, Giedion pointed out, best embodied in Étienne Jules Marcy's popular book *Le Mouvement* (1894). By the nineteenth century, movement forward became associated with progress. And progress was measured in terms of technology, which became translated into a faith in production as an end in itself. According to Giedion, this was the basis of a change in orientation of thought from 'the miraculous to the utilitarian,'[27] and it expressed itself in every sphere of life. Synchronous movement – constant change – became synonymous with mechanization and thus associated with technology, in contrast to the erratic and spontaneous movements of organized life. The former was predictable and therefore considered the ideal; humans had to fit the ideal, and thus they become like the machine: mechanized, regular, and predictable. As William Kuhns notes in his analysis of Giedion's ideas, 'Man increasingly moves less by the measure of his own body and mind than by that of the machine.'[28]

What was Giedion's evaluation of technology as mechanization? He claimed in his evaluative summation of his study that he refrained 'from taking a positive stand for or against mechanization. We cannot simply approve or disapprove. One must discriminate between those spheres that are fit for mechanization and those that are not.'[29] For Giedion, these were moral questions that had no place in evaluating the impact of technology as mechanization. However, Giedion still had a desire to believe, like Mumford, that technology is an external entity that may inform and shape the human mind to certain desired ends but can ultimately be controlled and used by humans 'to protect [themselves] against its inherent perils.'[30] Yet the whole thrust of his argument and study is that mechanization is so pervasive, insidious, anonymous (see his 'Anonymous History') as to be beyond human control, a 'Frankenstein' that humans have created and are unable to contain. Indeed, Giedion's description of mechanization reminds one of Mary Shelley's *Frankenstein:* 'Because mechanization sprang entirely from the mind of man, it is more dangerous to him. Being less easily controlled than natural forces, mechanization reacts on the senses and on the mind of its creator.'[31] As for the future judgment of history on the age of mechanization, Giedion was not optimistic from his perspective in 1948, having experienced the Second World War: 'Never has mankind possessed so many instruments for abolishing slavery. But the promises of a better life have not been kept ... Future generations will perhaps designate this period as one of mechanized barbarism, the most repulsive barbarism of all.'[32]

Technology as volition is the fourth and final perspective. Volition refers to 'the aims, intentions, desires, and choices'[33] that humans see in and bring to technology. The term assumes that technology in itself is neutral; its value depends on its uses. However, there is also a counter-belief, one that we have already seen: that technology is beyond human control and indeed controls humankind, shaping and dictating the very values that society sees as of utmost importance. Indeed, implicit in the concept of technology, as opposed to science, is the belief that the ends or intentions of these two disciplines are different: science aims at knowing the world, while technology aims at controlling or manipulating it. The question then becomes: Who or what is controlling or manipulating the world? Do humans control technology, or does technology itself control by shaping and dictating the values, ideals, and aims humans bring to it? From the perspective of technology as volition, it appears that technology takes command, dictating our views of it, including the belief that technology is controlling us by dictating how we think.

Martin Heidegger provided the most extensive philosophical study of technology as volition or 'willing' in his *Being and Time* (1927) and especially in his later essay 'The Question Concerning Technology' (1954).[34] In *Being and Time,* he explored the implications of the concept of *Dasein* (literally 'to be here'), a being-in-the-world. Part of this 'being' has been of a practical concern for 'manipulating things and putting them to use.' These things were identified initially and fundamentally by their use – a hammer, for example, 'for hammering.' Thus, Heidegger believed that practical knowledge was the fundamental form of knowledge from which other forms, like abstract knowledge, derived. With regard to human beings, then, they were tool makers before they were abstract thinkers, the exact opposite perspective from that of Lewis Mumford.

In 'The Question Concerning Technology,' Heidegger went further, arguing that technology is 'by no means anything technological'[35] – that is, to do with tools, machines, industrialism, or mechanization – 'or, in Latin, an *instrumentum.*'[36] Rather it is 'a kind of truth, a kind of revealing or disclosing of what is.' However, the truth that is 'revealed' or 'disclosed' is different in the modern world of technology from anything that preceded it. Heidegger pointed out that ancient technology revealed by means of 'bringing-forth' from nature through art and poetry; modern technology (beginning with the Industrial Revolution) reveals by 'challenging,' a 'setting upon,' nature.

The implication here is that ancient technology created artifacts in cooperation with nature, whereas modern technology imposes on nature, 'forcing it to yield up materials and energies that are not otherwise to be found.'[37] Furthermore, the objects of nature that were used in ancient technology continued to have an inherent value of their own independent of their use by humans, whereas objects in modern technology have no inherent value apart from human use.

Heidegger also argued that nature too became an object of manipulation (his 'setting upon' in the modern technological age). He called this perspective *Gestell* ('enframing'), which is the technological attitude toward the world and thus the essence of the modern mindset. This perspective or modern mindset, Heidegger noted, had its origins in Descartes, *ego cognito* [ergo] *sum,* in which humans found their self-certainty *within themselves* rather than within the world over against themselves. The world – nature – became a *representation* of reality formed within the human mind, set as object against humans as subjects, to be understood within the human mind and thus humanly manipulated and controlled. Heidegger believed that his altered philosophical perspective lay at the root of modern science. As William Lovitt points out in his Introduction to *The Question Concerning Technology and Other Essays,* for Heidegger 'the modern scientist does not let things presence [sic] as they are in themselves. He arrests them, objectifies them, sets them over against himself, precisely by representing them to himself in a particular way.'[38] Such a mindset was an essential – indeed the *central* – component of the modern technological world, because everything is seen over against humans to be controlled by them.

However, *Gestell,* or 'setting upon' and 'challenging,' occurs not only toward nature but also toward humans themselves. 'The essence of modern technology,' Heidegger wrote, 'starts man upon the way of that revealing through which the real everywhere, more or less distinctly, becomes *Bestand* ["stock," "standing-reserve," things "in supply"].' It is a way of thinking or perceiving the world that humans as technological beings, or 'tool makers,' are destined or fated to uphold. This way of thinking or perceiving is 'the modern volitional stance towards the world.' Because it is itself 'technologically based,' this mindset is extremely difficult, if not impossible, to get outside of to view objectively. Hence, Heidegger referred to it as 'the supreme danger.' He explained why:

> This danger attests itself to us in two ways. As soon as what is unconcealed no longer concerns man even as object, but does so, rather, exclusively as standing-reserve, and man in the midst of objectlessness is nothing but the orderer of the standing-reserve, then he comes to the very brink of a precipitous fall; that is, he comes to the point where he himself will have to be taken as standing-reserve. Meanwhile man, precisely as the one so threatened,

> exalts himself to the posture of lord of the earth. In this way the impression comes to prevail that everything man encounters exists only insofar as it is his construct. This illusion gives rise in turn to one final delusion: It seems as though man everywhere and always encounters only himself ... *In truth, however, precisely nowhere does man today any longer encounter himself, i.e., his essence.* Man stands so decisively in attendance on the challenging-forth of Enframing that he does not apprehend Enframing as a claim, that he fails to see himself as the one spoken to, and hence also fails in every way to hear in what respect he ek-sists [sic], from out of his essence, in the realm of an exhortation or address, and thus can never encounter only himself.
>
> But Enframing does not simply endanger man in his relationship to himself and to everything that is. As a destining [willing], it banishes man into that kind of revealing which is an ordering. Where this ordering holds sway, it drives out every other possibility of revealing.[39]

As Gregory Bruce Smith notes, for Heidegger, 'modern technology is no mere instrumentality that man can consciously and rationally control by imposing "values" upon it. We stand within its mode of revealing and cannot stand outside it.'[40] While Heidegger presented this all-encompassing technological mindset within which human beings view the world and to which they look for understanding, guidance, and meaning, he refused to pass moral judgment on its value. He also resisted passing judgment on whether this technological mindset had been beneficial or destructive to modern society.

As well, Heidegger refused to see the modern technological perspective as 'fated' to exist. He believed that in the very awareness of danger comes what he calls the 'saving power' of technology, and reasoned, 'Enframing, as a destining of revealing, is indeed the essence of technology, but never in the sense of genus and *essentia*. If we pay heed to this, something astounding strikes us: It is technology itself that makes the demand on us to think in another way what is usually understood by "essence."'[41] That other way is to think of technology as an 'Idea' whose truth can be known to man and, in the very act of 'human reflection,' free man. Heidegger argued that this view of technology as an 'Idea' is evident in the other Greek meaning for *techne:* 'the bringing-forth of the true into the beautiful,'[42] an idea captured in the Greek word *poiesis,* which refers to both *techne* and art. Art, he noted, was both 'akin to the essence of technology' and yet 'fundamentally different from it.'[43] Thus, in coming to terms with the meaning of art, we come closer to understanding the essence of technology and, paradoxically, better able to rise above it to see it for what it is – its essence.

For Heidegger, then, technology is a way of perceiving – a mindset – that is itself 'technologically induced' in that it orders a way of perceiving the world that is in essence 'technological.' Only by coming to terms with this technological perspective are humans able to see the very essence of technology that is the basis from which humankind can 'liberate' itself from the mindset. This is no facile act of will or a shallow way of thinking, Heidegger cautioned, but a coming to terms with technology as a supreme act of faith, a volition in the highest sense of the term.

Jacques Ellul questioned the implications of seeing technology as volition in his ground-breaking book *The Technological Society* (1964).[44] Ellul argued that, due to the power of technology, the modern world is radically different from anything that went before it. Ellul is not clear as to when this modern 'technological civilization' began, although his study concentrated mainly on the post–Industrial Revolution (Ellul called it the 'technical revolution') period. Indeed, one of Ellul's arguments is that the modern world of technology has developed slowly and unconsciously, making it a cumulative rather than sudden phenomenon.

According to Ellul, to use the term 'technology' in speaking of the modern world is to misunderstand its pervasive nature, because the term has come to be associated with machines, whereas the world of technology is much more than machines. This is the exact opposite perspective from that of Mumford, who saw technology in all guises only as machines. It is a mindset, a way of perceiving the world, that incorporates the machine and all it symbolizes in its perspective but is evident in every sphere of life: economics, politics, law, education, religion, eating habits, work, and recreation. To distinguish this mindset from machines, Ellul chose to use the term 'technique' rather than 'technology.' He provided a definition of technique in a 'Note to the Reader,' a section in the revised edition of his text, published in 1967: 'The term *technique,* as I use it, does not mean machines, technology, or this or that procedure for attaining an end. In our technological society, *technique* is the *totality of methods rationally arrived at and having absolute efficiency* (for a given stage of development) in *every* field of human activity. Its characteristics are new; the technique of the present has no common measure with that of the past.'[45]

Such a definition incorporates three concepts that Ellul argued are integral to the modern world of technology: totality, rationality, efficiency. Technology, or technique, is total in that it 'integrates everything,' including humans themselves. He showed the totality of technique by contrasting it with the machine: 'As long as technique was represented exclusively by the machine, it was possible to speak of "man *and* the machine." The machine remained an external object, and man ... remained none the less independent. He was in a position to assert himself apart from the machine; he was able to adopt a position with respect to it. But when technique enters into every area of life, including the human, it ceases to be external to man and becomes his very substance. It is no longer face to face with man but is integrated with him, and it progressively absorbs him.'[46]

Ellul described rationality, the second concept, as a process by which 'mechanics is brought to bear on all that is spontaneous or irrational.' Examples of rationality, he noted, are 'systematization, division of labor, creation of standards, [and] production norms.' All are characterized by 'the reduction of method to its logical dimension alone. Every intervention of technique is, in effect, a reduction of facts, forces, phenomena, means, and instruments to the schema of logic.'[47]

The final common characteristic of the modern world of technology, according to Ellul, is efficiency. '"The one best way": so runs the formula to which our technique corresponds,' he pointed out. 'When everything has been measured and calculated mathematically so that the method which has been decided upon is satisfactory from the rational point of view, and when, from the practical point of view, the method is manifestly the most efficient of all those hitherto employed or those in competition with it, then the technical movement becomes self-directing. I call this process *automatism*.'[48]

Automatism, in turn, is one of five characteristics that made the modern world of technology totally different from any previous world orders. He identified the other four characteristics as self-augmentation, monism, technical universalism, and autonomy. Self-augmentation is the ability of technique to pursue its own course without the decisive intervention by human beings, because of an unquestioning faith in technical progress as inherently good. Monism is the fact that the 'technical phenomenon, embracing all the separate techniques, forms a whole.'[49] Technical universalism is the ability of technique to pervade the whole world and equally to master *all* the qualitative elements of civilization, including art, literature, and religion. '*Technical civilization*,' Ellul argued, 'means that our civilization is constructed *by* technique (makes a part of civilization only what belongs to technique), *for* technique (in that everything in this civilization must serve a technical end), and *is* exclusively technique (in that it excludes whatever is not technique or reduces it to technical form).'[50] The fourth characteristic is autonomy, by which technique becomes an end in itself to be achieved by its own means. In such a technological world, humans only begin the operation without participating in it, a perspective analogous to the concept of *deus ex machina*, by which God started the world that then ran on its own accord. Furthermore, autonomy of technique is premised on the elimination of all human activity, because the latter is inferior due to variability and elasticity.

Ellul argued that technique has not only created a modern age that is totally different from any previous age but has also created 'a new man,' who must fit into a world that is not of his own making and that runs counter to his 'human nature.' Ellul noted, for example, 'He was made to go six kilometers an hour, and he goes a thousand. He was made to eat when he was hungry and to sleep when he was sleepy; instead, he obeys a clock. He was made to have contact with living things, and he lives in a world of stone. He was created with a certain essential unity, and he is fragmented by all the forces of the modern world.'[51]

This technological world, Ellul argued, is beyond human control. It operates according to its own laws that are beyond our comprehension. While technology has liberated humankind from nature's oppression, it has in the process ironically subjected humans to a greater oppression, 'the forces of the artificial necessity of the technical society which has come to dominate our lives.'[52] In such a world, it is meaningless to talk in terms of human volition and freedom of choice. According to Ellul, 'technique can never engender freedom.' For him, then (in keeping with the metaphor of technology as Frankensteinian), technology is, as William Kuhns perceptively notes, 'best symbolized by Dr. Frankenstein's monster, which, once alive, cannot be killed, coped with, related to, or compromised ... The monster also has the supreme advantage of invisibility. And not seeing him, people refuse to believe that he roams their world.'[53]

While Ellul and Heidegger were pessimistic about the ability of humans to control and use technology for beneficial ends – to see technology as being within the purview of human will – there have been others who believe that it is within human power not only to change a present system through technology but also to create a totally new system that can be applied to specific problems. They are systems designers, and the most sophisticated of these with regard to the rise of technology is Norbert Wiener in his science of cybernetics.[54] Wiener noted that cybernetics is derived from the Greek word *kubernetes,* or 'steersman,' implying the ability of humans to control their actions through technology. Indeed, the subtitle of Wiener's book *Cybernetics* (1948) makes the perspective implicit: *Control and Communication in the Animal and the Machine*.

The assumption underlying cybernetics is that 'society can only be understood through a study of the messages and the communication facilities which belong to it; and that in the future development of these messages and communication facilities, messages between man and machines, between machines and man, and between machine and machine, are destined to play an ever-increasing part.'[55] The importance of messages lies in the process of communication rather than in the message itself, a reflection of Marshall McLuhan's famous dictum 'the medium is the message.' Once that process is understood, it can be classified and controlled. Weiner called the process of understanding 'information.' Such information, based on past decisions and patterns of action, becomes 'feedback' to predict future action. These decisions and patterns are characteristic of modern machines, like computers, as much as of human beings, Wiener argued – hence his belief that the same laws apply to humans as to machines. One such law, coming out of the study of thermodynamics, is that action within machines tends toward entropy – that is, to dissipate energy and run down, or to go from a state that is highly organized, differentiated, and less probable to one that is more probable, undifferentiated, and chaotic. Wiener believed this tendency to entropy is true of the universe in general, eventually resulting in the self-destruction of the world. However, within the increasingly entropic world are pockets of decreasing entropy, of which humans are one such pocket. As a result of their faith in progress, they are able to continue, through faith in the future, to generate new energy and new sources of information. Modern information technology has the same ability to resist entropy. Such realizations can be used to create a better world so long as the information is used for positive ends. Wiener was both optimistic and pessimistic about human potential. He believed that cybernetics offered a way of channelling human knowledge and energy for positive ends. Equally, like all forms of technology, it had the possibility to destroy the world if humans were not clear as to what end technology should serve. Wiener distinguished between technological 'know-how' and human 'know-what.' Technology can provide the 'know-how' to accomplish certain ends or purposes, but humans needed to provide the 'know-what' – that is, the values and ends to which technology should be directed. Without the 'know-what,' the 'know-how' of technology would take command and control humankind. Wiener's warning goes to the heart of the concept of technology as volition:

> Any machine constructed for the purpose of making decisions, if it does not possess the power of learning, will be completely literal-minded. Woe to us if we let it decide our conduct, unless we have previously examined the laws of its action, and know fully that its conduct will be carried out on principles acceptable to us! On the other hand, the machine like the djinnee, which can learn and can make decisions on the basis of its learning, will in no way be obliged to make such decisions as we should have made, or will be acceptable to us. For the man who is not aware of this, to throw the problem of his responsibility on the machine, whether it can learn or not, is to cast his responsibility to the winds, and to find it coming back seated on the whirlwind.[56]

Wiener pointed out that the 'machine' he was referring to may be made of 'brass and iron' or 'flesh and blood.' 'When human atoms are knit into an organization in which they are used, not in their full right as responsible human beings, but as cogs and levers and rods,'[57] they are machines, equally as dangerous as man-made machines. Only free human agency – human volition – can control the power of machines, be they mechanical, electrical, or human 'machines.' Wiener cautioned that, in all cases, the technology to fear is the closed mindset that tends to be endemic to all forms of technology. Thus, for Wiener, the analogy of technology was not, as it was for

Ellul, to Mary Shelley's *Frankenstein* but rather to Samuel Butler's *Erewhon,* where machines were better able to cope with the changing environment than humans and therefore outdistance and control humankind. Wiener feared that this possibility could become a reality unless human beings resisted the danger of technology encroaching on their freedom and assumed their own responsibility.

The historical and intellectual evolution of international theorists' perspectives on technology from objects, tools, and machines to knowledge, process, and volition parallels the evolution of Canadian thought on technology from the mid-nineteenth century to the present. Thus, Lewis Mumford's historical overview of the evolution of the idea of technology as machines illuminates our understanding of the ideas of T.C. Keefer, T.C. Haliburton, and Sandford Fleming, who saw railways, especially the locomotive, as mighty machines that were transforming the physical and intellectual landscape in the mid-to-late nineteenth century. Alexander Graham Bell saw communication technology, especially the telephone, as shaping a new world in which technology would be supreme. William Leiss's exploration of the idea of technology as knowledge offers insight into the ideas of the Canadian advocates of technical education who put forward similar ideas to their European counterparts to get technical education accepted at the elementary, secondary, and university levels of education. Canadian thinkers in the interwar years, most notably George Sidney Brett, William Lyon Mackenzie King, and Stephen Leacock, put forward perspectives on technology as process of war, industrialism, and mechanization respectively that were very much in keeping with the views of Karl Marx and Siegfried Giedion. The ideas of Martin Heidegger, Jacques Ellul, and Norbert Wiener provide an intellectual context for illuminating the ideas of Harold Innis, Marshall McLuhan, Northrop Frye, and George Grant writing in the same period and equally concerned about the implications of the idea of technology as volition.

Part 1
Approaching the Imperative

2
T.C. Keefer, T.C. Haliburton, Sandford Fleming, and Alexander Graham Bell: Technology as Railways, Communication Media, and Time

The nineteenth century was the great age of mechanical inventions. In an article entitled 'Signs of the Times' (1829), with its biblical reference, noted British literary critic Thomas Carlyle attempted to find the descriptive word that encapsulated the new age he saw emerging. He dismissed such terms as 'Heroical, Devotional, Philosophical, or Moral' and chose instead the term 'Mechanical.' 'It is the Age of Machinery, in every outward and inward sense of that word,'[1] he wrote. He noted the ways in which machines were shaping society outwardly but also, and more importantly, inwardly. He pointed out that the introduction of machinery resulted in 'a preoccupation with the external arrangement of human affairs as against their inner meaning and consequences.'[2] Even more striking, according to Carlyle, was the way that machines were regulating 'not our modes of action alone, but our modes of thought and feeling.' Humans have 'grown mechanical in head and in heart, as well as in hand,'[3] he wrote. Carlyle was suggesting that machines appeared to be taking over, dictating even, what it meant to be human. They were forcing the adoption of a new mindset, a new way of thinking in keeping with the new mechanical age.

Railways, especially the locomotive – the 'iron horse' – were seen as the greatest technological invention of the mechanical age. 'Nothing else in the nineteenth century seemed as vivid and dramatic a sign of modernity as the railroad,' Wolfgang Schivelbusch wrote in *The Railway Journey*. 'Scientists and statesmen joined capitalists in promoting the locomotive as the engine of "progress," a promise of imminent Utopia.'[4] A number of intellectuals in Britain and the United States attempted to capture the significance of this new technology for their time. In England, novelist William Thackeray noted the difference his generation of Victorians experienced having grown up in the pre-railway era and then in the railway age. It was like living in two entirely different worlds:

> It was only yesterday; but what a gulf between now and then! *Then* was the old world. Stage-coaches, more or less swift, riding-horses, pack-horses, highway-men, knights in armour, Norman invaders, Roman legions, Druids, Ancient Britons painted blue, and so forth – all these belong to the old period ... But your railroad starts the new era, and we of a certain age belong to the new times and the old one. We are of the time of chivalry as well as the Black Prince of Sir Walter Manny. We are of the age of steam.[5]

When Matthew Arnold saw the first trains pass through the Rugby countryside, he remarked, 'Feudality is gone for ever.'[6]

In the United States, Charles Francis Adams, of the famous Adams family that produced presidents and literary critics, noted the impact of railways on society and culture in an article entitled 'The Railroad System' (1868):

> Though the material or financial aspect of the [railroad] system is almost invariably dwelt upon, it is by no means the most interesting one. Here is an enormous, an incalculable force practically let loose suddenly upon mankind; exercising all sorts of influences, social, moral, and political; precipitating upon us novel problems which demand immediate solution; banishing the old before the new is half matured to replace it; bringing the nations into close contact before yet the antipathies of race have begun to be eradicated; giving us a history full of changing fortunes and rich in dramatic episodes. Yet, with the curious hardness of a material age, we rarely regard this new power otherwise than as a money-getting and time-saving machine ... Not many ever stop to think of it as ... the most tremendous and far-reaching engine of social change which has ever either blessed or cursed mankind.[7]

Canada also had its prophets for the new technological age. Three of the most insightful with regard to railways were Thomas Coltrin Keefer, Thomas Chandler Haliburton, and Sandford Fleming. Alexander Graham Bell would join these three in providing insights into the new mechanical age by discussing the importance of the telephone in revolutionizing communications. All three theorists of railways went beyond the obvious identification of railways as mighty machines that lessened distance and time as limiting factors to trade and communication to reflect on the impact of railways internally – on the mind – as Carlyle did. These thinkers wanted to understand the impact of railways in shaping values and culture. Keefer and Haliburton wrote in the mid-nineteenth century when the British North American colonies were about to embark on an age of railway building. Their views expressed the exuberance for the new technology of railways, an exuberance that is characteristic whenever a new form of technology is introduced into society. They were also aware that they were experiencing a technological invention that was truly revolutionary, one that was about to inaugurate a new and modern world of technology. Sandford Fleming also wrote in the railway age from his perspective as an engineer involved in building for the greatest railways in British North America/Canada. By the end of his life, he was involved in other great technological feats as well, some of them the by-products of railways, like the invention of standard time, others the result of the discovery of electricity, like the laying of the Pacific Cable. Bell saw the telephone as the beginning of a revolution in communication technology, even foreseeing the possibility of computers.

Thomas Keefer and Sandford Fleming were engineers and thus were intimately involved in and knowledgeable about technology. This is equally true of Alexander Graham Bell as an inventor. They knew first-hand of which they spoke. In that respect, they had insights into technology beyond that of the average Canadian and even beyond that of later philosophically inclined Canadian theorists of technology who lacked technical expertise and therefore came to the subject as an observer of rather than an active agent in the technological process. These early thinkers were the 'makers of technology,' to use one of the two terms identified by Mitcham when discussing technology as a process – that is, 'making and using.' They reflected on technology while they were creating it. They could also be described as 'mechanical prophets,' showing the way in which the technological imperative could lead to the fulfillment of the moral imperative.

Keefer, Haliburton, Fleming, and Bell each had his own unique perspective on the nature, meaning, and significance of the new mechanical inventions of their time. All four, however, would have agreed that these inventions were shaping a new world with a new technological mindset, which they believed was compatible with the moral imperative that up to this point in time had been the underpinning of Western civilization. Fundamental to that moral imperative was the belief that the world had been created by God. Within the world order existed a hierarchy with humans at the top since they were created in the image of God, followed by the animal world and then the inanimate world of nature. All were part of a divine plan, which, when understood, would reveal the mind of God. The material world, especially the array of new and marvellous inventions, reflected human ingenuity in the same way the natural world reflected God's creativity. While these early Canadian theorists of technology still talked about human ingenuity as a gift from on high, they increasingly saw it as an innate human trait, especially in those with the gift of invention.

Keefer, Haliburton, Fleming, and Bell also believed that technology would be the means to unlock the mysteries of the natural world, the human mind, and even the mind of God. They were coming to believe that the world of physical nature and human nature were not 'natural' at all but rather 'mechanical,' ruled by the same mechanical laws that regulated machines. Even creativity became defined in mechanical as opposed to intellectual terms, reinforcing the belief, as Lewis Mumford noted (although he lamented), that humans were most active as 'tool makers' rather than as 'imaginative thinkers.' And machines gave humans the ability – and power – to control nature and cause it to do their will. In this way, technology held the potential to elevate humans to Godhead and maybe even immortality. And just as humans were believed to be created in God's image, so too machines were seen as being created in the image of human beings. Machines reflected the greatness and ingenuity of their creators. In essence, then, machines were becoming the 'new God' for the new mechanical age. Little wonder that the reflections of Keefer, Haliburton, Fleming, and Bell on technology were often infused with religious and moral terminology. Seeing machines as acts of creation, these Canadian theorists of technology borrowed the phraseology of Christianity, previously reserved only for God and humans, to describe machines. Technology dethroned God and even humans as the supreme source of power and as the criterion by which to judge greatness. God became the *deus ex machina,* the 'engineer' who started the world moving, but once in motion, it operated of its own volition. Humans created the machines that also took on lives of their own, independent of the will of their masters. Indeed, machines became the standard by which to judge perfection. Machines with their reliability, predictability, and efficiency were preferred over human beings who were unreliable, unpredictable, and inefficient. Machines dictated the values by which to judge success and progress. Those values were productivity, autonomy, and industriousness. These values also underlay the moral imperative. What changed was the belief now that technology best served the greater good of society by promoting those moral values. Progress in this world, not salvation in the afterlife, became the new aspiration of humans, societies, and Western civilization. And progress was measured in mechanical terms. This was the essence of the new technological mindset that was taking shape in the mid-to-late nineteenth century.

What these early theorists of technology most feared even more than negativity was the threat of lassitude and complacency on the part of the Canadian public in the face of the need for technological advancement in transportation and communication if Canada was going to progress. Keefer talked about the 'apathy' of the residents of his fictitious 'Sleepy Hollow'; Haliburton lamented the indifference of his fellow Nova Scotians (Bluenosers) to the potential within their own colony; Fleming worried about the failure of Canadians to be receptive to new forms of technology that held the potential of making Canada greater than the United States and an equal to Britain in terms of power and influence; Bell implied that Canadians lacked an inherent drive to use technology to improve the world. He once described an inventor as 'a man who looks around the world and is not contented with things as they are. He wants to improve whatever he sees, he wants to benefit the world; he is haunted by an idea.'[8] Was this the reason Bell turned to the United States to patent his telephone invention? For all four men, technology held out the promise of enabling Canada to withstand the threat of American absorption; their moral duty as they saw it was to sound the alarm and to show how the technological imperative was the means to ensure that the moral values inherited from the mother country – the moral imperative – would ultimately prevail.

Thomas Coltrin Keefer, a civil engineer from a prominent Upper Canadian family of engineers, made his mark in history not by any great engineering feat but by writing a little pamphlet entitled *The Philosophy of Railroads* (1849). In it he reflected on the nature and importance of railways. What distinguishes his writing from others of his day is his ability to go beyond the obvious qualities of the new technology of railways as machines with incredible power and speed that were capable of bringing about economic prosperity to reflect on their *moral* and *intellectual* significance as symbols of modernity and agents of civilization. For Keefer, railways were machines whose real power and importance were *not* physical but *mental,* capable of expanding the mind – indeed, of transforming the mind – by creating a new way of thinking that would lead to a higher form of civilization. That is why Keefer described the locomotive as 'the iron civilizer' and emphasized its 'moral influence.' As he noted, 'The civilizing tendency of the locomotive is one of the modern anomalies, which however inexplicable it may appear to some, is yet so fortunately patent to all, that it is admitted as readily as the action of steam, though the substance be invisible and its secret ways unknown to man.'[9]

Keefer saw a magic in railways reminiscent of the magic of the alchemists of the Middle Ages in their search for the elixir of life. Lewis Mumford has noted the link between magic and technology: 'The conception of new kinds of power-machines, which could be put together and made to work without magical hocus-pocus, fascinated various minds from the thirteenth century on ... These fantasies were no doubt incited by such rudimentary machines as were already in operation: for there must have been a moment when the first windmill or the first automaton moving by clockwork seemed as marvellous as the first dynamo or the first "talking machine" less than a century ago.'[10] Mumford goes on to note that the locomotive transcended anything previous in terms of its magical quality. Keefer's American contemporary Ralph Waldo Emerson described railroad iron as 'a magician's rod in its power to evoke the sleeping energies of land and water,'[11] although later he came to realize the danger of allowing machines to cast too magic a spell on human beings, dulling their critical faculties and lulling them into a false sense of security. It was this magical quality of railways to transform nature and society, and even to alter human nature and the mind, that excited Keefer.

It is this excitement, characteristic of enthusiasts of any new form of technology, combined with a reflective perspective that makes Keefer's writings so insightful and engaging, if somewhat one dimensional.

What immediately strikes the reader of *The Philosophy of Railroads* is that Keefer felt he had to convince his fellow Canadians to accept railways. We assume today that these mighty machines, symbols of power, progress, and prosperity, were enthusiastically accepted as great technological inventions of unquestionable value; yet this was not the case. While Keefer was clearly a railway enthusiast, not all Canadians were. Many were skeptical and suspicious of the new technology. They feared the impact of the 'iron horse' on their daily lives, their society, and their culture. Living in an age that prided itself on being 'cultured' and 'civilized,' Keefer realized the need to pitch his argument for railways beyond the economic and material gains to show their value as 'civilizing agents' or, more precisely, to show how the material and economic gains from railways could be the means to ultimately reach the higher goal of intellectual and spiritual rejuvenation.

How could railways serve this purpose? Keefer argued that railways had a dual purpose. First of all, they provided a means for Canadians to remain independent of the United States by enabling them to be economically self-sufficient. Keefer's concluding remarks in *The Philosophy of Railroads* speak to this concern:

> We are placed beside a restless, early-rising, 'go-a-head' people – a people who are following the sun Westward, as if to obtain a greater portion of daylight; we cannot hold back – we must tighten our own traces or be overrun – we must use what we have or lose what we already possess – capital, commerce, friends and children will abandon us for better furnished lands unless we *at once* arouse from our lethargy; we can no longer afford to loiter away our winter months, or slumber through the morning hours. Every year of delay but increases our inequality, and will prolong the time and aggravate the labour of what, through our inertness, has already become a sufficiently arduous rivalry.[12]

Secondly, and more importantly, railways served as lifelines that radiated outward, linking isolated communities to larger centres in British North America and beyond the border to the North American continent as a whole and even, along with steamships, to the mother country, the centre of Western civilization. Thus, for Keefer, railways were not only or mainly tracks of trade with the Yankees but also lines of communication that linked British North Americans intellectually and spiritually. Railways conveyed not only goods but also, and more importantly, culture. In this way, railways had the potential to provide British North Americans with intellectual stimulation and moral advancement that could liberate them from their parochial attitudes that resulted from their isolated geographical location. Railways could bring British North Americans into the modern age. This is why Keefer describes the locomotive as 'the iron civilizer' and entitles his essay *The Philosophy of Railroads*.

Thomas Coltrin Keefer was well qualified to undertake such reflections. The sixth son of George Keefer, first president of the Welland Canal Company, Thomas was educated at the finest schools of his day: Grantham Academy in St. Catharines, and Upper Canada College in Toronto.[13] Upon graduating from seventh form, he worked as an engineer apprentice for two years on the Erie Canal, first begun by American engineers in 1825 to link the Great Lakes to the Atlantic by connecting Buffalo on Lake Erie with the port of New York. In 1840, he returned to Upper Canada, where he served as an assistant engineer on the Welland Canal for five years, and then, thanks to the influence of his brother Samuel, the chief engineer of the Department of Public Works, he worked as the engineer in charge of the timber slide and river improvements at Bytown (Ottawa). This position lasted only three years, at which time Keefer was dismissed, allegedly for economic reasons, but more likely because he was rumoured to have been exposing expenditures that were politically necessary but professionally questionable.

Keefer used his period of unemployment to write two essays: *The Philosophy of Railroads*, and *The Canals of Canada* (1850). The former essay resulted from a request by a group of Montreal merchants who wanted to promote western rail connections, while the latter became Keefer's submission to an essay competition sponsored by Lord Elgin, governor of the United Canadas, to celebrate the long-awaited completion of the St. Lawrence canal system. *The Canals of Canada* celebrated the construction of three major canals: the Welland, the Lachine, and the Rideau, the latter noted for its new technology in the form of series of impressive, albeit expensive, locks. But the dominant age of canals was passing, and no one was more conscious of that fact than the engineers, like Keefer, who had been involved in building and promoting them. Canals still had a role to play in Canada's transportation network, but only if they were integrated with railroads, whose age had just begun. Thus, Keefer's *The Philosophy of Railroads* was a hymn of praise to the new machine and industrial age, while his essay on the canals of Canada was a eulogy to the commercial age that was passing.

Keefer began *The Philosophy of Railroads* with a subject very much on the minds of British North Americans at the time (and Canadians ever since): the weather. 'Old Winter is once more upon us,' he reminded his audience, 'and our inland seas are "dreary and inhospitable wastes" to the merchant and to the traveller; – our rivers are sealed fountains – and an embargo which no human power can remove is laid on all our ports.'[14] British North Americans could identify immediately with the crippling effect of winter weather on commerce; it was one of the perennial arguments made at the time for the disadvantage of Montreal, British North America's greatest port, over its American rival, New York, with its year-round, ice-free port. Keefer accentuated the idleness of Montreal in wintertime by using active words generally associated with machines – 'the splashing wheels' and 'the roar of steam' – during the busy shipping season juxtaposed with words denoting inactivity, such as 'silenced' and 'hushed,' thus reinforcing the association of machinery with activity and nature with idleness. To make the contrast even more poignant, he compared the silence of Montreal with the bustle of New York: 'Far away to the South is heard the daily scream of the steam-whistle,' a reminder that winter weather did not have the same debilitating effect on commerce in the United States. Here Keefer was pointing out that the new technology of railways had the power to control the weather. 'Old Winter' may be an embargo that no 'human power' could remove, but it was not beyond the power of technology to control. Railways held the promise of achieving one of the dreams of the Enlightenment – that is, liberating humanity from the age-old restraints of nature. For British North Americans, railroads could eliminate the weather as a limiting factor to their creative ability.

Keefer introduced the subject of weather for another reason. Not only did 'Old Winter' debilitate commerce, but it also incapacitated human beings by creating an attitude of apathy, what might appropriately be called 'frozen minds.' In the very sentence in which he points out that winter weather had no detrimental effect on commerce south of the border, Keefer noted that in British North America 'there is no escape: blockaded and imprisoned by Ice and Apathy.' Both crippled communication, creative action, and commerce. To make the connection of cold weather and human limitation stronger, Keefer used the metaphor of blood in describing the debilitating effect of cold weather on British North American commerce. 'The animation of business is suspended,' he wrote of Montreal in winter, 'the life blood of commerce is curdled and stagnant in the St. Lawrence – the great aorta of the North.' Railways, then, could also overcome apathy. Thus, these marvellous machines had the ability not only to overcome the vicissitudes of physical nature but also the limitations of human nature.

Keefer attributed the apathy of Canadians to their inferiority complex next to Americans. While Americans were moving ahead, confident in the technology of railways, Canadians were still hesitant and uncertain. To dismiss the perennial argument used to account for the success of the United States over the British North American colonies, namely, that the United States was bigger and more populated, Keefer pointed out that even American states much smaller in size and population and with no greater financial resources than that of the colony of the Canadas had embraced railways. They realized what Canadians failed to realize: that railways were the wave of the future, symbols of progress. To reinforce his point, Keefer highlighted the progress already made in transportation, of which railways were the latest and finest achievement: 'The cart road is succeeded by the turnpike, this again by the macadam or plank roads, and these last by the Railway. The latter is the perfected system and admits of no competition – and this characteristic pre-eminently marks it out as the most desirable object for investment in the midst of an enterprising and increasing population.'[15] What Canadians needed, Keefer argued, was 'faith' in the new technology of railways as a prerequisite for a sense of self-confidence. For Keefer, railways had become the new god that, once accepted, could transform British North Americans into new beings. Keefer saw a spiritual dynamic to railways, with a miraculous power and energy that seemed unparalleled. It was the faith in railways – a new spiritual perspective of this new technology – more than mere money that Canadians required to ensure their success – and their survival independent of the United States – on the North American continent.

To prove his point, Keefer took the example of a well-cultivated, prosperous little village 'situated twenty or fifty miles from the chief towns upon our great highway, the St. Lawrence, and without navigable water communication with it.'[16] He conveniently called his imaginary village 'Sleepy Hollow,' reflecting in its name the apathy of British North Americans but also associating it with the popular legendary hamlet in the American story about Rip van Winkle, an amicable but unimaginative and unenterprising individual, in Washington Irving's *Sketch Book,* who falls asleep and awakes 100 years later to a transformed Sleepy Hollow that was anything but 'sleepy.' As well, by using the generic name 'Sleepy Hollow,' as opposed to the actual name of a real village, Keefer was reminding his fellow Canadians that there were all kinds of 'Sleepy Hollows' in British North America that lacked contact with the wider world and thus remained 'unprogressive,' subject to the vagaries of nature in the wilderness.

However, Keefer's Sleepy Hollow is not a 'straw village,' an easy target to knock down and ridicule. It is a prosperous and contented community with a subsistence economy that appears to have all the attributes of success, except a railway. In fact, Keefer described the community in such an appealing way that the reader wonders why its inhabitants would want or need a railway. 'Each farmer has his comfortable house, his well stored barn, variety of stock, his meadows and his woodland; he cultivates only as much as he finds convenient, and his slight surplus is exchanged for his modest wants ... To themselves, to the superficial observer, their district has attained the limit of improvement.' Note, however, the sense of self-deception that governs the occupants' outlook and the uncritical eye of the 'superficial observer' who judges a place by its outward appearance only. What keeps the inhabitants of Sleepy Hollow deceived and smug is, ironically, their lack of knowledge, their ignorance of the world beyond their immediate community. What they require is 'information,' which can only come from contact with the outside world by means of a railway. In their isolated world, Sleepy Hollowers are self-righteous and complacent, 'grateful that their farms have not been disfigured by canals or railroads, or the spirits of their sires troubled by the hideous screech of the steam-whistle.'

Then suddenly 'the iron bond of railway' enters their rustic village. The inhabitants initially fear the new technology. They imagine their farms being '"cut up" or covered over ... of rifled gardens and orchards, of plundered poultry yards and abducted pigs.' Mothers fear that their children will be '"drawn and quartered" on the rail by the terrible locomotive,' and a 'whole hecatomb of cattle, pigs and sheep' will be offered up as a sacrifice to 'this insatiate Juggernaut,' this metal god.[17] Note that the evils are more imagined than real, the by-products of a distrust of or lack of faith in the new technology.

What alleviates the fears and uncertainties of the villagers and converts them to railways is the arrival of 'a handsome Rodman' – a knight-in-armour, with an explicit sexual connotation – who falls in love with 'Eliza Ann, the rival spokesman's daughter.' The romance of the railway is personified in the romance between the Rodman, representing the masculine 'Juggernaut' (and reminiscent of Emerson's depiction of rail iron as 'a magician's rod'), who woos and subdues the local rural damsel, who represents Mother Earth. The union promises a new beginning and produces a new offspring that is a fusion of technology and nature. The implication is that, rather than raping or desecrating nature, railways would fulfill the potential of technology through a blissful union of the machine with the land. The railway would not destroy rural society or obliterate the rural piety and moral values of its inhabitants but rather enable the rural way of life to achieve its full potential.

When the inhabitants of Sleepy Hollow accept the railway as an act of faith rather than resist it out of fear, a miraculous transformation takes place in Sleepy Hollow. The first changes appear in the material and economic spheres: more work for local youth; increased demand for farmer's goods; better prices for local commodities; rising land values; and the beginning of manufacturing as city investors take advantage of cheaper rent, a surplus population, and ample foodstuffs in the countryside to set up local industries. Suddenly, the pace of life begins to quicken, and apathy dissipates. Keefer not only described the change but evoked and enacted it in the active verbs he used to describe the new pace and in the very rhythm of his sentences that quicken to keep pace with the new technology he described: 'the patient click of the loom, the rushing of the shuttle, the busy hum of the spindle, the thundering of the trip-hammer, and the roaring of steam'[18] – all mechanical devices with familiar mechanical sounds.

However, for Keefer, more magical than the material transformation of Sleepy Hollow was 'the moral influence of the iron civilizer.' Culture comes to the little hamlet in 'well-dressed and rich-looking people,' some of whom are sons and daughters of local inhabitants who have gone off to the city for education and training and then have returned home transformed: a neighbour's son as 'a first class passenger with all the prestige of broadcloth, gold chains, rings, gloves, and a travelled reputation'; or damsels with 'gay bonnets, visites [sic] etc., of that superior class of beings who are flying (like angels) over the country.'

Most important is a new 'spirit' that comes to Sleepy Hollow, a new way of looking at the world, a new mindset. It is most evident, Keefer maintained, 'where it is most needed – in the improved character it gives to the exercise of the franchise.' It might appear strange that Keefer would emphasize the positive role that railways could play on politics. But he had numerous disagreements with Upper Canadian politicians who, he believed, lacked the vision for railways that he had and only wanted to use them for self-serving purposes. So the importance of politicians with noble intentions and good moral scruples was uppermost in his mind. For Keefer and his age, the quality and public-spirited nature of the democratically elected political leaders were a reflection of the cultural sophistication, the civility, of its populace. Prior to the coming of a railway, those who exercised the franchise or ran for public office, in Keefer's estimation, were people with selfish interests or who were uninformed, which meant that 'the greatest talker is elected, and an improved judicature, instead of an improved country, is the result.'[19] With the coming of a railway, discernment in judgment comes into play, and people expect more from their elected representatives, because the voting public is more informed thanks to knowledge from the outside world. Keefer admitted that how exactly this transformation occurs is as mysterious, magical, and paradoxical as the machines that cause it to happen. It is a case of matter over mind, material improvements leading to spiritual well-being. He poignantly noted, 'The civilizing tendency of the locomotive is one of the modern anomalies, which however inexplicable it may appear to some, is yet so fortunately patent to all, that it is admitted as readily as the action of steam, though the substance be invisible and its secret ways unknown to man.'

Once a railway is accepted, then its 'invisible power which has waged successful war with the material elements, will,' Keefer proclaimed, 'assuredly overcome the prejudices of mental weakness or the designs of mental tyrants.' It will, through its magical power, dispel 'poverty, indifference, the bigotry or jealousy of religious denominations, local dissensions or political demagogueism.' For Keefer, the mental power of the mighty locomotive in altering attitudes and overcoming mental frailties was as great as its physical power to overcome natural obstacles in its way. Both acts are performed by the Juggernaut 'with a restless, rushing, roaring assiduity.' Railways, then, had the ability to satisfy both material and spiritual needs at the same time. People need not forsake their spiritual needs in pursuing their material wants or vice versa. Keefer wrote, 'While ministering to the material wants, and appealing to the covetousness of the multitude, it [a railway] unconsciously, irresistibly, impels them to a more intimate union with their fellow men.'[20]

In a sentence that foreshadows the later thinking of Harold Innis and Marshall McLuhan, Keefer compares the influence of the steam engine to that of the printing press on the mind. Both have 'brought us out of the mental wilderness of the dark and middle ages,' he claimed. Together they have enabled humans to reach time immemorial by eliminating the restraints of time: 'Ideas are exchanged by lightning – readers and their books travel together but little behind their thoughts – while actors, materials, scenes and scenery are shifted with the rapidity and variety of the kaleidoscope.'[21] For Keefer, the real magic of the technology of railways was in the expansion of the human mind – indeed, the creation of a new mindset by bringing the power of information and knowledge within easy reach of every human being. The mind becomes one with itself, capable of thinking great thoughts based on informed judgment without the individual having to move outside of his or her physical space. Railways would bring the world to each and every individual's doorstep. For British North Americans, confined to the distant and isolated space of the northern half of the North American continent, railways would broaden their horizons, expand their minds, and make them civilized. They were lifelines of communication; in that respect, the ribbon of steel was 'the iron civilizer.'

Having provided the vision of a community intellectually, morally, and spiritually transformed by the railway, Keefer then proceeded in *The Philosophy of Railroads* to deal with the more practical values of railways. But the loftier vision is never far from his mind. He knew, as he noted in his introductory remarks, that British North Americans were a practical and skeptical people and that he had to 'sell them' on the material and economic benefits of this new technology if they were to 'buy into' his greater vision of railways. Equally, he realized that the economic and material value of the new technology was an essential prerequisite for its greater intellectual and moral benefits. Thus he proceeded to note the value of railways in terms of increased speed, especially for transporting perishable goods. As well, time was money, and it became more economical to ship goods at faster speeds over railways than the slower pace of ships or by horse and wagon. Add to speed the lower costs of an 'iron horse' compared to the natural horse in terms of the carrying capacity, not to mention feed, disease, and fatigue, and the economic benefits of the railway were even greater. Regularity was another benefit, especially for delivering mail and passengers, once again improving communication. In terms of safety, too, railways had fewer dangers than steamboats and no greater danger than transportation by carriage. Finally, railways were also convenient and efficient: able to be used year round; to traverse the most rugged terrain; and to 'clamber over mountains and penetrate the most remote corner of the land.' What Keefer discovered and admired in railways were some of the attributes that Jacques Ellul would later note (as discussed in Chapter 1) as being some of the dangers of the modern technological mindset: its admiration of speed and its emphasis on the qualities of efficiency, reliability, convenience, and predictability.

With such great possibilities, how could Montrealers delay any longer in building a railway to connect Montreal with Hamilton, its inland 'principal Western town'? Never mind the cost, Keefer proclaimed, for in the end it would pay for itself. Anyway, he noted paradoxically, British North Americans 'are too poor to do without it.' Thus, he concluded that 'as a people we may as well ... attempt to live without books or newspapers, as without Railroads.'[22] It was an appropriate analogy given Keefer's belief that railways, like books and newspapers, were civilizing agents that dispelled ignorance and broadened the knowledge of all who utilized them. Furthermore, railways were first and foremost an idea, a vision, that had to be conceived and accepted in the mind before they could become a physical reality. In keeping with his theme of apathy as the reason why British North Americans had not embraced railways, he believed that it was neither physical nor financial obstacles that prevented British North Americans from embracing the new technology of railroads but 'the barriers of indifference, prejudice and ignorance.'[23] Until the idea, the vision, of railways took hold, Keefer argued, the actuality of achieving them would never occur. Railways had to be in the mind as a dream or vision (as Mumford noted concerning new technology) before they could become a reality. There had to exist a 'railway mentality,' a way of thinking that put railways front and centre, if railways were to come to fruition. This is what British North Americans lacked and Americans and others in the Western world had. They had already adjusted their thinking to realize the potential to be actualized by these wondrous machines. Unless British North Americans did the same, Keefer maintained, they would forever remain in a state of apathy, ignorance, and inertia, incapable of advancing to a higher level of civilization. They needed to realize the power of 'the iron civilizer.'

The positive reception afforded *The Philosophy of Railroads* and the winning of the Governor General's prize for *The Canals of Canada* essay enabled Keefer to enjoy a brief vogue in Montreal. He was invited to give a series of lectures at the Montreal Mechanics' Institute, the first of which he entitled simply 'Montreal.' In characteristic fashion, Keefer pitched his message, both in style and content, to his audience, in this case the artisans, working men, and their wives. 'It is usual, I know, on these occasions,' he began with deference, 'to make the subject a scientific one, to take up some of the -isms or the -ologies, and expound them. If I am guilty of any innovation in meddling with the domestic affairs of this City, my apology is that having been honored with a request to address you, I am more anxious to benefit than to amuse you. We are a practical people, and we live in an eminently practical age, and what more edifying, what more profitable subject can the Mechanics of this City discuss, than the causes which favor or which threaten the prosperity of Montreal.'[24] What followed was anything but a 'practical' speech. In characteristic fashion, Keefer took a practical question – in this case the economic problems and geographical limitations facing the city of Montreal at this critical time in its history – and elevated it into a visionary quest of what the city could become if it adopted his recommendation to build more railways. A great orator, a master visionary, and a genuinely concerned citizen of Montreal, Keefer offered his audience that night another hymn of praise to technology – to railways – as the elixir for the city's ills.

He began by noting that it was precisely railways that at the moment put New York ahead of Montreal in the competitive race between the two port cities to monopolize the trade of the continent. Earlier, British mercantile trade, with its 'protected demand for our products in Britain,' had given the British North American colonies a commercial advantage, but after the advent of British free trade in the late 1840s this was no longer the case. Now economic victory went to the city best able to take advantage of its natural assets. Here New York had been more successful by building railways to overcome the limitations of mountains, difficult streams to traverse, and five months of winter to link Great Lakes trade with their city.

Once again, as in the *Philosophy of Railroads,* Keefer reminds his audience that railways were more than vehicles of trade; they were also avenues of communication. In a wonderful metaphor, he compared the canal to the street and the railway to the sidewalk. Just as the street was the avenue of commerce, the sidewalk was the medium of communication. Canals were still vital to Montreal's trade (all the more so if complemented by railways), but railways were the *sine qua non* of communication and travel. In addition, Keefer noted, 'nothing is more certain than that travel follows trade ... Now, if trade and travel are inseparable, we cannot expect to enjoy much of either until our travelling facilities are improved.'[25]

At this point, Keefer presented the key point of his address: not that Montreal must build more railroads, but that it must build *the right ones*. He noted that, in some respects, Montreal was too eager to build railways and then built them to the wrong destinations. Instead of looking to Portland and New England, Montrealers should have been looking 'to the West and

North behind her.' The Ottawa valley and the peninsulas at Niagara, Detroit, and Sault Ste. Marie all offered promise of trade for Montreal if connected by rail. Even more exciting would be the possibilities such lines would afford in opening up trade for Montreal with the American Midwest and eventually the British North West. Once this inland trade was secured, Montreal needed to envision a Halifax and Quebec railroad to the sea. In essence, Keefer was presenting as convincing an argument as found anywhere for the east-west axis of Canadian trade and communication. In presenting justification for the existence of Montreal as a key link in continental and oceanic trade, he was by implication justifying the separate existence of the British North American colonies and the western hinterland that only two decades later would constitute Canada.

Having presented a compelling argument for how railways could satisfy Montreal's economic needs, Keefer then turned to the question of how railways could assist in the city's 'moral wants.' First of all, the city needed a 'Public Library' and an 'Alms House.' As a public-spirited individual, Keefer appealed to fellow Montrealers to be concerned about the poor and unfortunate. Then, too, the city needed parks and gardens 'to let in the light and air of heaven amongst our thickening streets.' He pointed out that the financial returns the city would enjoy from railways could be used to finance these social needs. Keefer the engineer was also the social reformer, a combination that is rare but not absent in Canadian history.

Having focused solely on Montreal in the body of his lecture, Keefer used his conclusion to show his audience how his vision (and hopefully theirs too) for the greatness of Montreal through technology – chiefly railways but also canals – formed part of a larger vision:

> Is there not a marked change in the general appreciation of what are called public improvements? Is not the English tongue rapidly girdling the earth? California and Australia – and who is not interested in them – who has not friends there – having in the duly appointed time revealed their hidden treasures, America has opened up the Isthmus of Darien while England is breaking through that of Suez. America is agitating a Railway from the Atlantic to the Pacific – England one from the British Channel to the Ganges, from Calais to Calcutta, passing through Constantinople and the valley of the Euphrates, with a station at Antioch and a junction to Jerusalem. In the Ohio basin, in the Mississippi valley, on the Atlantic slope of the Alleghanies [sic], throughout Western Canada, from the Saguenay to Panama, from Halifax to San Francisco – everywhere one subject, the making of Railways, rules the public mind.[26]

Thanks to railways, ours is the greatest of all civilizations, Keefer reminded his audience, greater than even the civilizations of Greece and Rome. The

Greeks had their great scientists, 'the fathers of Astronomy, of Mathematics and Sculpture: in Euclid, in Archimedes, they had their Bacons and Newtons.' The Greeks, however, lacked the great inventors: 'their Watts and their Arkwrights.'[27] Rome had 'perfect roads,' what Keefer calls 'one great civilizing engine.' It was over these roads (harkening back to the analogy of streets to railways), he pointed out to his audience, that Christianity, 'the first great moral revolution applied to the earth,' spread its word throughout Europe. What even greater potential for disseminating knowledge existed now with our modern means of communication! 'May not the vast, the almost incredible extension of the Railway system, the Electric Telegraph, and the Ocean Steamer over all the Christian Earth,' Keefer asked rhetorically, 'be a forerunner – a necessary and an indispensable forerunner – to that second great moral revolution, the Millennium'? The association of the wondrous new technology of the nineteenth century – railways, telegraphs, and ocean steamers – with Christianity, the miracle of the first millennium, and with the 'moral revolution' that he predicted would come about in the next millennium, thanks to the technological inventions of their age, was a mental link Canadians could readily make.

The distinguishing features of this new and promising civilization would be peace, Keefer declared, and the absence of ignorance and prejudice. All this through the wonders of modern technology! Keefer could hardly contain his enthusiasm for railways and his vision of what these 'iron civilizers' could do for humankind. 'Wherever a railway breaks in upon the gloom of a depressed and secluded district,' he proclaimed with the conviction of a religious convert, 'new life and vigour are infused into the native torpor ... The pulpit will have then its grateful listeners, the school its well filled benches – the stubborn opponents of wordy philosophy will then surrender to a practical one the truth of which they have experienced.'[28] His promise of renewed vigour, of a reawakening in the new 'Railway Age' from the slumber, apathy, and torpor that existed before the advent of railroads, harkens back to his analogy five years earlier in *The Philosophy of Railroads* of the conditions that existed in 'Sleepy Hollow' before and after the coming of the railway. Before, everything seemed frozen, motionless, and therefore uncivilized; afterward, all was alive, active, and civilized. He made the same point in 1854: 'Every new manufacture, every new machine, every mile of railway built is not only of more practical benefit, but is a more efficient civilizer, a more speedy and certain reformer, than years of declamation, agitation, or moral legislation.'[29] It was as though for Keefer with every turn of the mighty wheels of the locomotive through the North American wilderness, humankind was getting closer to the 'Promised Land.' Indeed, he spoke as though railways could reform society so as to return it to a state of grace akin to the Garden of Eden before the expulsion of Adam and Eve. For Keefer, there were no limits to the power of the technology of railways.

Keefer then proclaimed that the real prophets of the new age, the ones preparing the (rail)way to the Promised Land, the harbingers of the New Civilization, were not the great philosophers, or the great inventors, or even the great engineers, but the lowly mechanics. As they went about their busy daily activities – 'as you ply the busy hammer or wield the heavier sledge,' Keefer reminded the mechanics in his audience – they could see their work as

> fast driving nails into the coffin of prejudice, of ignorance, of superstition and national animosities; that as you turn down the bearings or guide the unerring steel over all the 500 parts of a locomotive engine, fancy will picture you cutting deep, and smooth, and true, into obstacles which have so long separated one district, one family, one people from another – and that you may exult in the reflection that those huge drivers will yet tread out the last smouldering embers of discord, that those swift revolving wheels – by practically annihilating time and space and by re-uniting the scattered members of many a happy family – will smooth the hitherto rugged path, fill up the dividing gulf, break through the intervening ridge, overcome or elude the ups and downs of life's chequered journey, and speed the unwearied traveller upon his now rejoicing way.[30]

Thomas Keefer, like the majority of Victorians, found the increased speed of life exhilarating and exciting. (It would be this constant motion, the frantic pace of life in the modern world of technology, that later generations would find weary, tedious, and even frightening, one of the negative repercussions of the age of technology.) To Keefer and his age, the technology of railways, with their penchant for change, their sense of novelty, their love of motion – 'movement,' which Siegfried Giedion identified as the one 'constant' in a world of constant change in the new technological age – was liberating, freeing Victorians from the restraints of time, space, ignorance, and apathy. In Keefer's mind, railways were transforming British North Americans into more sophisticated and cultured beings, especially those individuals who resided in the 'backwaters' of British North America. Unquestionably, the railway locomotive for Keefer was 'the iron civilizer.'

Keefer's *Philosophy of Railroads* and his series of lectures for the Mechanics Institute in the mid-1850s were the high-water mark in his career, giving him a reputation as an outstanding engineer and a visionary. Nevertheless, his crowning achievements met with disdain and came under attack from politicians, promoters, and even some fellow engineers, in a series of letters and editorials designed to malign his character. Keefer collected and published these letters along with his rebuttals in a little-known pamphlet entitled *A Sequel to the Philosophy of Railroads*.[31] The title is clearly satirical, for the pamphlet was more a repudiation or at least a questioning of Keefer's optimistic and buoyant view of railroads in *The Philosophy of Railroads* rather than 'a sequel' to it, although his doubt lay not with railways per se but with those who were controlling them.

The first essay contains extracts from lectures on civil engineering, given at McGill University in the winter of 1855-56, in which Keefer set out the noble qualities of the engineering profession to which his students were entering, doing so by comparing the morally upright engineer to the morally depraved politicians in Canada. Indeed, he envisioned the ideal government as a technocracy, a government ruled by engineers, or at least one in which engineers acted as consultants and experts. But he doubted this would happen, since engineers were too 'professional' and 'morally upright' to lower themselves to the guileful actions of politicians. Better prospects existed for engineers as managers since he believed technical knowledge was becoming, or should become, a requisite for employment. But before engineers became too involved in other endeavours, Keefer cautioned, they needed to establish their own professional standards, otherwise they would become the puppets and innocent victims of 'thieving politicians' or else would end up being dishonest themselves. As he reminded the students in his audience, '"Knowledge is power," and the strong in railway knowledge prey upon the weak.'[32] Clearly, Keefer had a noble vision of what the engineer should be: a technocrat who would use his expertise nobly for the public good. Keefer the moralist was concerned that engineers use their technological expertise for the good of society rather than for selfish or amoral political ends.

The feud that broke out between Keefer and members of the Legislative Assembly involved in railway building, especially the most ambitious Grand Trunk Railway, need not concern us. Both sides accused the other of incompetence, low morals, and corruption. However, Keefer got the last word in an article entitled 'Travel and Transportation' in *Eighty Years' Progress of British North America* (1863), a book designed, as the title indicates, to celebrate progress in British North America with particular focus on the contribution of railways.

Keefer began his article by reminding Canadians once again, as he had in *The Philosophy of Railroads,* that they were a skeptical people when it came to supporting railways. Yet despite that initial skepticism, the British North American colonies had by 1863 laid more miles of track 'than Scotland, or Ireland, or any of the New England States, and [was] only exceeded in this respect by five States in America.'[33] Then he turned to a discussion of the building of the latest and most ambitious railway, the Grand Trunk Railway, and his tone changed. While he agreed to the need for such a railway, he condemned the corruption behind its financing on the part of British financiers and Canadian politicians. Thus, he reminded his readers that the fault for the failure of the Grand Trunk Railway to achieve solvency lay not with the technology of railways itself but with the individuals who used the technology for selfish ends rather than for the public good.

However, Keefer failed to realize that the wonderful power of the steam engine to overcome the limitations of nature and even to transform human nature for the good could also be used by individuals in positions of power to suppress those who stood in the way of its relentless drive. Keefer saw railways as benign machines if not technological wonders, capable of turning inertia into action, potential into actuality, parochial enterprises into national edifices, a wilderness into a civilization. Others, however, saw the potential of the new technology for power and domination. In the end, those in positions of political power used railways to promote their own interests at the expense of Keefer, the very railways in which he held such faith.

Nevertheless, Keefer's faith in technology remained undaunted. In his presidential address to the Canadian Society of Civil Engineers in January 1880, he waxed eloquent on the most recent railway enterprise, the building of the Canadian Pacific Railway. He described it as 'our last and greatest effort, and is not only the most important road in Canada, but, in some respects, in the world,' applying the latest in engineering know-how, utilizing the most up-to-date equipment, and displaying the way engineers and public officials could work together for the public good. But by 1880, Keefer realized that railway engineering was only one classification of engineering along with hydraulic, civic, mechanical, mining, and electrical. All promised a better future. No wonder Keefer could conclude that, thanks to engineers and the wonders of technology that lay at their fingertips, Canada stood at the beginning of a new and wondrous age, one of unlimited possibilities and pure potentiality.

Thomas Chandler Haliburton, creator of the legendary character Sam Slick in his popular *Clockmaker* series, was the Maritime equivalent of Thomas Keefer in the mid-nineteenth century in terms of his enthusiasm for railways. Born in 1796 into a family that prided itself on its British ancestry and believed that the Haliburtons numbered Sir Walter Scott among their clan, Thomas Haliburton maintained an abiding faith in *everything* British. Schooled at the finest educational institutions in Nova Scotia, his education reinforced the belief in the importance of the moral imperative: the need for moral standards to live by and to guide society. He was active in colonial politics until elevated to the bench as a judge in 1829. His public career reflected his belief in the moral responsibility of public officials to serve their fellow human beings, a sense of noblesse oblige, although he was not opposed to using public positions for personal gains too. But his real passion and fame came from his writing career, especially his *Clockmaker* series. Fred Cogswell notes that 'the *clockmaker* can be regarded as a series of moral essays.'[34]

One of the 'moral' lessons was for Nova Scotians to continue their abiding faith in everything British and to always remain faithful members of the British Empire. Railways became the means, both physically and intellectually, to maintain this tie between the colonies and the mother country. For Haliburton, railways were lifelines that radiated outward from his native colony of Nova Scotia to the other British North American colonies, the North American continent, and, when linked to steamships, to Britain, the heart of Western civilization. He believed that through a series of railways, British North Americans could rise above their parochial existence on the northern half of the North American continent, maximize the natural resources within their own colonies, and enjoy the finest of culture that association with Britain provided. To convince his fellow Nova Scotians that this was their destiny and railways were the means to achieve that destiny became his reason for writing his *Clockmaker* series. As he noted in his Preface to *Sam Slick's Wise Saws and Modern Instances: On What He Said, Did, and Invented* (1853), 'The original design in writing the sketches known as "The Sayings and Doings of the Clockmaker" was to awaken Nova Scotians to the vast resources and capabilities of their native land. To stimulate their energy and enterprise, to strengthen the bond of union between the colonies and parent state, and by occasional reference to the institutions and governments of other countries to induce them to form a just estimate, and place a proper value on their own.'[35]

Like Thomas Keefer, who lamented the apathy of Canadians, Haliburton complained about the indifference of Nova Scotians to the potential in their own colony.[36] His fictional character Sam Slick notes, 'I never seed or heard tell of a country that had so many natural privileges as this ... They have iron, coal, slate, grindstone, lime, firestone, gypsum, freestone, and a list as long as an auctioneer's catalogue. But they are either asleep, or stone blind to them ... The folks of Halifax take it all out in talking – they talk of steam boats, whalers and rail roads – but they all end where they began – in talk.'[37] In contrast, Sam Slick is depicted as an enterprising American clockmaker who doles out advice on how to succeed. Literary critic George L. Parker points out that clock manufacturing was 'the first industry to pioneer successful mass production technology – however crude – and in doing so, to demonstrate the economic returns to technological change.'[38] Clocks were also one of the great technological inventions of the fourteenth century that derived their importance from regulating and regimenting workers during the Industrial Revolution and since that time. They were also indispensable for efficient train travel. Thus, it seemed logical that Haliburton's legendary clockmaker should give advice on railways.

While the problems besetting Nova Scotia may have been many, the solution, for Haliburton, was simple and one dimensional: build railways. Haliburton saw railways as a great technological invention able to unite a people, foster enterprise, and advance their cultural sophistication. Using a metaphor similar to Keefer's analogy of a canal to a street and a railway to a sidewalk (the former fostering trade while the latter advanced communication), Haliburton compared a railway to a combination of a bridge, river, thoroughfare, and canal in terms of its ability to foster enterprise and communication so essential for the success of a town. He noted, 'A Bridge makes a town, a river makes a town, a thoroughfare makes a town, a canal makes a town, but a rail-road is bridge, river, thoroughfare, and canal all in one.'[39] 'The only thing that will make or break Halifax,' Haliburton went on, 'is a rail-road across the country to the Bay of Fundy.'[40] If put into operation, he predicted, 'the activity it will inspire into business, the new life it will give the place, will surprise you ... This here rail-road will not perhaps beget other rail-roads, but it will beget a spirit of enterprise that will beget other useful improvements.'[41] Like Keefer, who believed that British North Americans needed a vision of railways before they could come to fruition, Haliburton maintained that Nova Scotians needed a faith in railways as wondrous machines capable of transforming their society by cultivating an enterprising spirit before this new technology could perform its magic. In the minds of both Keefer and Haliburton, the *idea* of railways necessarily preceded and ultimately dictated their success. Once envisioned for the greatness they beheld, railways held unlimited potentiality; they could be the new god, raising humans to godhead.

Haliburton envisioned the rail line from Halifax to Windsor on the Bay of Fundy as only the first of three rail lines necessary to link Nova Scotians to the wider world. He advocated building a rail link between the Maritimes and the colonies of Upper Canada and Lower Canada (the eventual Intercolonial Railway, completed in 1876) and a railway from the Canadas to the Pacific (the eventual Canadian Pacific Railway, completed in 1885) to span the North American continent from ocean to ocean entirely through British territory. Such railways, along with steamships, would provide a communication link with Britain to the east and the British Empire in Asia, part of the 'all-red-route-to-the-Orient.'[42] For Nova Scotians, these rail lines would bring them closer to the mother country and other members of the Empire, thus enabling them to widen their horizons and enhance their culture without even having to leave home. Herein lay the power of railways for Nova Scotians once they were awakened to their potential.

Some British North Americans saw railways as a means by which they could become more independent of Britain and in closer association with the United States through increased trade that would enhance their wealth. Haliburton dismissed such an idea as 'not an agreeable topic': 'I cannot conceive how a separation [of the British North American colonies from Britain] can conduce to the interests of either party.'[43] As he reasoned, 'We owe to steam more than we are aware of. It has made us what we are, and, with the blessing of God, will elevate and advance us still more.'[44] On another occasion, he wrote, 'The communication by steam between Nova Scotia and England will form a new era in colonial history. It will draw closer the bonds of affection between the two countries, afford a new and extended field for English capital, and develope [sic] the resources of that valuable but neglected province.'[45] What steam had achieved for British North Americans, in the form of steamships and steam locomotives, was to make them a great people of British stock, capable of even surpassing Americans. Railways could unite the British North American colonies into a great British nation on the North American continent that could then, through steamships, be tied to the mother country both physically and intellectually, thus enabling Canadians to partake of the finest of Western culture that Britain, the centre of Western civilization, had to offer.

Sandford Fleming has rightly been identified as 'the Father of Communication Technology.'[46] In 1852, he helped build the Ontario, Simcoe, and Huron Union railway (later the Northern Railway). In 1867, he was appointed as engineer-in-chief of the Intercolonial Railway, which linked the colonies of the Canadas with the Maritime colonies, and then in 1871 was appointed the chief engineer of the Pacific Railway, a transcontinental railway that tied Canada together. Ironically, Fleming's latter accomplishment proved to be one of his shortcomings. In 1880, Sir Charles Tupper, Minister of Railways and a friend of Fleming's, dismissed Fleming from his position because he had become a political liability through his independent actions and outspoken stance on the Pacific project. In a rare moment of self-pity, Fleming reflected on this difficult period in his life:

> I indeed felt the weight of the responsibilities that were thrown upon me and I labored night and day in a manner which will never be known, some time after I began to work double times. I had the misfortunate [sic] in two consecutive years, 1872 and 1873, to meet with serious accidents. By the first I came near to terminating my life, by the second I was placed on crutches for 6 or 7 months. During the whole of these periods except when actually confined to bed I never ceased to carry on my work which I need not say was at times very arduous. As a consequence my general health suffered and I was forced to seek for some respite.[47]

It was a temporary setback. In 1884, he was appointed director of the CPR, and a year later, on November 7, 1885, Fleming participated in the hammering of the last spike at Craigellachie. In the famous picture of the event, he is the tall central figure with the top hat and broad beard, a fitting position for the man who was instrumental in this technological feat. As well, Fleming's friend, George Grant, principal of Queen's University, appointed Fleming chancellor in 1879, a position he held until his death. In his inaugural address as chancellor, Fleming made a strong case for putting science at the centre of university education. He was also instrumental in the founding of Queen's School of Mining in 1893 and its faculty of applied science a year later. As well, Fleming was the impetus behind the laying of the underwater cable across the Pacific Ocean that linked Canada to the British colonies in Asia and the Pacific and formed part of a telegraph system that encircled the globe on British territory. Finally, and most importantly, he invented standard time, an important component of technology because it ensured the efficient operation of railways and brought the entire world under a uniform system of keeping time.

Fleming frequently used such noble concepts as the brotherhood of man, world peace, and civilization to describe the world that modern technology was about to usher in. What he often failed to note was his more selfish desire to use the new technologies of communication to unite the British Empire so as to enhance its power in the world. This oversight might be excused, since Fleming saw such noble concepts as world peace and civilization as synonymous with Britain and her Empire. In his mind, the technology of communications was power: the power to create a 'better world,' the power to control that world through a mindset that put the highest premium on technology, and the power to ensure that this world would operate on the 'proper' moral values, values that he believed were the positive by-product of technology: efficiency, orderliness, predictability, uniformity, and consistency. In Fleming's estimation, who better to entrust with such power and responsibility than Britain, the most 'civilized' and 'morally upright' nation in the world? Thus, for Sandford Fleming, technology, power, civilization, and the British Empire were all synonymous.

Sandford Fleming was born in Kirkcaldy, Scotland, on January 7, 1827, one of eight children of Andrew and Elizabeth Fleming.[48] Kirkcaldy was famous for being the home of the cotton spinning machine that revolutionized Britain's textile industry, along with being the birthplace of Adam Smith, the economic theorist who saw a link between the Industrial Revolution and the advancement of capitalism through a policy of laissez-faire. The town also housed the Burgh School, where Thomas Carlyle taught. The Scots in general had the reputation of being particularly adept at technology.[49]

At school, Fleming excelled in mathematics but above all enjoyed drawing. His skills at design caught the attention of John Sang, a local engineer and surveyor, who took Fleming on as an apprentice in 1841 at the age of fourteen. Sang was reputed to be a 'practical and mechanical genius,' who, among other things, invented an elaborate gauge, a converter with a readout, to measure acreage from a map by tracing the perimeter of the area to be measured. Sang was also known for turning out superb engineering students. Upon completion of his apprenticeship in 1845, Sandford and his brother David were sent to Upper Canada, where opportunities for work seemed better than in Scotland. As a gift upon departure, Fleming's father gave him a valuable watch with a built-in sundial; ironically, it would be this traditional way of keeping time that Sandford Fleming would eventually discard. Fleming arrived unaware of his fate or what he might do but with a will to work, a keen mind, and faith in his potential, the same faith he would place in the potential of his technological inventions. He first acquired work in Peterborough, Upper Canada, where he undertook the first survey of the town. Then he moved to Toronto, where he attended evening classes at the Mechanics Institute, a movement begun in Great Britain in the 1820s to provide adult education, particularly in technical subjects, to working people. The Toronto institution opened in 1830 with a library, lecture series, and night classes. Fleming worked for John Stoughton Dennis, a surveyor, in order to be recertified as a surveyor according to Canadian law. He also helped found the Canadian Institute, a professional society for architects, surveyors, and engineers, along with its periodical, the *Canadian Journal*. The Canadian Institute and its journal provided a forum for Canadian intellectuals to discuss and debate ideas, ideas that, Fleming was convinced, could rival those presented anywhere in the Empire, including Britain itself. He saw such institutions as overcoming local jealousies and differences so as to create a uniformity of thought, so essential for technology to function effectively. In 1852, Fleming and Collingwood Schreiber, a fellow engineer, won a competition to design a permanent hall in Toronto for the agricultural fair. They modelled their building after the Crystal Palace of the London Exhibition; theirs became known as 'the Palace of Industry.' In the same year, Fleming embarked on his engineering career building railways, which would lead to his involvement in three major Canadian railways. In the year of Fleming's arrival in Canada, 1845, the country had only fourteen miles of railways. He would, in his lifetime, add thousands of miles more.

Despite his busy work schedule, Fleming took time to reflect on railways in a chapter in his book *Canada and Its Vast Undeveloped Interior* (1878), devoted to railway building in the Canadian west. Fleming focused on one key word in the book title: 'undeveloped.' Much of the Canadian west remained undeveloped still centuries after exploration and settlement. What advances had been made toward conquest and settlement he attributed to the British. But most of the land remained vacant and unused, evidently overlooking the existence of the First Nations and their use of the land. Fleming had that nineteenth-century aversion to idleness and inactivity as evidence of stagnation. Thus, in his mind, the region of the west needed to be conquered, tamed, and ordered before the region could be integrated into the rest of Canada. Railways became the means to achieve this objective; they were for him the *sine qua non* of Canada's existence as a nation, and the country's means to future success, since, as he noted, 'these lines of communication perform their functions independently of climate, connecting all parts of the old settlement, and penetrating wide tracts of land not previously accessible.'[50] Fleming identified two virtues of railways that Keefer had also noted: their ability to overcome the limitations of climate and their ability to open up new areas of settlement and link those settlements to more advanced communities. Fleming also stressed, more than Keefer had done but in keeping with Haliburton's perspective, the importance of railways for linking Canada to Britain and to Britain's vast empire. Here was the special importance of the Pacific railway that Fleming focused on in the latter part of his article. When completed, this transcontinental railway would promote colonization of the western interior, foster imperial trade, and form part of an all-red-route-to-the-Orient, a communication link between Britain and Asia through the use of steamships and steam locomotives, with Canada in the centre. The Pacific Railway would elevate Canada to a position of power and influence in the British Empire, making her the jewel in the British Crown. Fleming's bond with Britain and the Empire came in part from his numerous trips back to the home country, forty-four in total.

In the 1870s, Fleming took up the cause for which he is best known: the 'invention' of standard time. The need for a universal standard for telling time resulted from fast-moving trains that caused train time to be out of sync with astronomical time. Also there was a need for clocks within the same area to register a uniform time. The problem, Fleming noted in a paper entitled 'Terrestrial Time,' was not due to the lack of precision in time-keeping: 'As machines for measuring time and dividing it into minute portions, they [clocks] undoubtedly are unrivalled amongst the productions that come from the hand of man.'[51] Nor did the problem reside with time itself. As Fleming noted in an essay entitled 'Our Old Fogy Methods of Reckoning Time' (1890), 'Time remains uninfluenced by matter, by space, or by distance. It is universal and essentially non-local. It is an absolute unity, the same throughout the entire universe, with the remarkable attribute that it can be measured with the nicest precision.'[52] Rather, the problem of time out of sync stemmed from human error as each town and hamlet set its own local time. To correct this irregularity, Fleming proposed dividing the Earth into twenty-four-hour time segments and then convincing everyone in each time zone to adopt a uniform time. After much deliberation and debate, universal standard time officially came into effect on January 1, 1885, although it took time before all countries agreed to adopt it. Standard time has endured, virtually unchanged, making it one of the longest-lasting inventions of the industrial age.[53]

Standard time was a major triumph for technology because it eliminated one of the limitations to technological conformity – local variations – and instead required that ultimately the whole world conform to a standard of time using Greenwich, in the heart of England, as the uniform base. Fleming was well aware of the importance of his accomplishment. By eliminating local time, he was doing away with all the 'vices' that prevented the implementation of the new efficient communication technologies resulting from the power of steam and electricity: irregularity, confusion, unpredictability, the loss of time, and, on occasion, the loss of life. As Fleming noted in a memorandum that he wrote for the Special Committee of the American Society of Civil Engineers on Uniform Standard Time, 'The great object is to secure accuracy, simplicity and uniformity and no effort should be ommitted [sic] to obtain them without seriously interfering with old customs with which we are so familiar.'[54]

It was more than coincidental that the new technologies that were by-products of steam and electricity – what Fleming called 'the great civilisers of the present century'[55] – required that even time succumb to change. Just as a new age was dawning with the advent of new technology, all had to change to adjust to the new technology and the new age. Change came to be seen as a virtue in and of itself – as evidence of progress and an advancing civilization; not to change was to stagnate and even possibly to fall into barbarism, a perspective Fleming shared with his fellow engineer, Thomas Keefer. Fleming stated, 'It is within the last half century more especially that the bounds of human knowledge have been so wonderfully extended; perhaps in the whole world's annals no fifty years have witnessed such a marvellous revolution.' Fleming concluded that as a result 'it is no longer possible to escape the conviction that we have reached a stage when further reform [of time reckoning] is demanded as a requirement of our condition.'[56] Logic carried the day, and the 'day' – and its time – were the purview of the technocrats.

The civilization that Fleming saw advancing with each new technological invention was, of course, Western civilization and within that civilization the British Empire as the epitome of culture. Within the Empire, Canada stood as the jewel, its finest offspring. It may have been coincidental that a Canadian invented standard time, but Fleming saw it as natural and inevitable. As a Canadian, he saw himself partaking of the virtues and innate superior attributes of the Anglo-Saxon race and British character. As he aptly put it, 'as a member of the great Colonial family,' Canada had an obligation to 'share the cost of establishing the communication of the Empire.'[57] Was standard time not simply another of the 'white man's burdens' to help civilize and anglicize the rest of the world? In Fleming's mind, technology and civilization held a symbiotic relationship.

Having conquered time, Fleming then briefly took up the cause of reforming the almanac. In a paper on the subject given to the Royal Society of Canada, Fleming championed Mr. M.B. Casworth's proposal, presented to the Royal Society four years earlier, of creating a calendar with thirteen months of twenty-eight days each or exactly four weeks. That would give a total of 364 days, leaving one extra day to 'fit in.' He suggested that it be scheduled as 'New Year's Day' between December 31st and January 1st but not associated with either one of those days nor appear as part of those months; it need not even appear on the calendar so as not to ruin its symmetry. The 'new month,' as the additional month should be called, would appear in the mid-summer between June and July, and it would also be the month in which leap year day would appear. Fleming was attracted to the proposal for the same reason that he liked standard time: it eliminated irregularities, asymmetry, ambiguity, and confusion. In urging his fellow members of the Royal Society to take up this cause, he pointed out that if adopted 'any person may for all ordinary purposes carry the Reformed Almanac in his memory.'[58]

Electricity led to another great invention in the realm of communication technology which, although it could not rival standard time in terms of its impact and pervasiveness, has nevertheless become so much a part of modern living that it is now difficult to imagine our society without it. This invention was the telephone. The creation of the telephone seemed almost 'inevitable' in the late nineteenth century, given the determination to speed up communication, the desire to make new technology available to everyone, and the knowledge available from the invention of the electric telegraph. Yet it required an individual who combined scientific knowledge of electricity and the telegraph with an equal scientific understanding of the mechanism of speech along with a passion to provide speech to people deprived of the privilege because of deafness. It also required the skills of an individual who never gave up in his quest for technological success. That person was Alexander Graham Bell.

Bell was born in 1847 in Edinburgh, the second son of Melville and Eliza Bell.[59] He came from a family preoccupied with speech. His grandfather had taught speech and elocution, and his father did the same. His mother was near deaf and required an ear tube to pick up human voices. (Bell would also marry one of his deaf students in later life.) Maybe this explains Melville Bell's interest as a researcher in discovering a 'universally applicable phonetic alphabet: a written system in which symbols represented the positions of the vocal organs and indicated the articulation of all possible sounds.'[60] His 'Visible Speech,' completed in 1864, was immortalized in George Bernard Shaw's *Pygmalion* with Professor Higgins using it to transcribe Eliza's cockney dialect. One of Alexander Graham Bell's early occupations was assisting his father in getting his 'Visible Speech' accepted as the best way to teach the deaf how to speak. Alexander was convinced that this was a superior method to sign language because the latter, a specialized language available only to those willing to learn it, cut the deaf off from communicating with anyone else. It also deprived the deaf of a full understanding of English or other languages with its inability to convey abstract ideas. In preparation for his lifelong work, he took courses in anatomy and physiology at University College in London, although he never completed his degree. He also lacked scientific training and knowledge of mathematics. Thus, unlike many creative geniuses, he was not a man of science; he was an inventor, whose inventions came more through intuition than scientific rigour. For him, scientific experiments were lessons to be applied, not proven. He noted, 'In scientific experiments, there are no unsuccessful experiments. Every experiment contains a lesson. If we stop right here, it is the man that is unsuccessful, not the experiment.'[61]

Bell's penchant for inventing began at an early age. It would last a lifetime, getting him involved in finding technology that would improve telegraphy, enable people to fly, and speed up water transportation through the hydrofoil. He was the first to turn sound into an electrical impulse that could then be converted back to audible speech through a receiver. Alexander and his older brother built a speaking machine at home that was capable of producing human-like cries by 'making gutta-percha replicas of the mouth, throat and nose, a maneuverable tongue, and bellow lungs.'[62]

In 1880, Bell made his last invention in telecommunications. He reflected a beam of sunlight onto a battery cell from selenium, a light-sensitive chemical, from a voice-vibrated mirror that made the current undulate. He then transformed the current into audible sound through a receiver. Bell realized the significance of his invention. He wrote enthusiastically to his father: 'I have heard articulate speech produced by sunlight! I have heard a ray of the sun laugh and cough and sing! ... I have been able to hear a shadow and I have even perceived by ear the passage of a cloud across the sun's disk ... We can talk by light to any visible distance without any conducting wire.' He elaborated further:

> Some of the practical results to be obtained I clearly foresee. When Electric Photophony is practiced in warfare the electric communications of an army could neither be cut nor tapped. On the ocean, communication may be carried on by word of mouth between persons in different vessels when great distances apart – and lighthouses may be identified by the sound of

their lights. In general science discoveries will be made by the Photophone that are undreamed of just now. Every variation of a light will produce a sound. The twinkling stars may yet be recognized by characteristic sounds, and storms and sun-spots be detected in the sun.[63]

His photophone transmitted wireless sound sixteen years ahead of Guglielmo Marconi's radio transmission and presaged modern fibre optics. Bell even predicted the day when handwriting would be replaced by 'electronic writing.' 'The days of handwriting have gone forever,' he wrote in 1899 on the eve of the twentieth century. 'They belong to the 19th century. The 20th century will not tolerate script.'[64] He was a century ahead of his time.

The invention for which he is most famous, however, is the telephone. In 1870, the Bell family immigrated to Canada after the death of Alexander's brother and settled in Brantford, Ontario. Alexander joined them shortly afterward. Within a year, he received a short-term appointment at the Boston School for Deaf Mutes to teach Visible Speech. Two years later, he was appointed Professor of 'Vocal Physiology and Elocution' at the newly opened Boston University. With Harvard University and the Massachusetts Institute of Technology close by, Bell took advantage of being in the hub of American scientific and technological activity. He attended public lectures and read extensively on physics, especially on topics of acoustics and electricity. The knowledge he acquired got him interested in discovering a way to electrically transmit speech.[65]

Bell built on the works of Oersted and Farraday to devise a means by which the mechanical power of the voice could vibrate a diaphragm fixed near an electromagnet, changing the magnetic field and thus inducing the necessary varying current. The idea had come to him while visiting his parents at Brantford in the summer of 1874, but the first successful experiment was performed in Boston one year later. When asked, as he often was, whether the telephone was invented in Canada or the United States, Bell had the perfect answer, one that linked the invention of the telephone to human creativity: 'The telephone was conceived in Brantford in 1874, and born in Boston in 1875.'[66] Bell realized the importance and pervasiveness of his new invention. 'The day is coming,' he wrote in 1875, 'when telegraph wires will be laid on to houses just like water or gas – and friends [will] converse with each other without leaving home.'[67]

Bell first displayed his new invention at the 1876 Centennial Exhibition in Philadelphia. It amazed scientists and dignitaries alike, including the Brazilian Emperor Don Pedro, who was invited to experiment with the new 'talking machine.' His reply was, 'My God, it speaks.'[68] The three judges pronounced Bell's telephone 'perhaps the greatest marvel hitherto achieved by the electric telegraph.'[69] According to Robert Bruce, Bell's telephone became 'the talk of the nation's scientific community, and visiting foreign scientists carried the word abroad.'[70] Very quickly, owning a telephone became the latest craze and a status symbol, declared to be essential for business and a pleasure for the home. Indeed, Queen Victoria had one installed in Buckingham Palace as early as 1879.[71] Bell liked to spread the rumour that he did not have a telephone in his own house. In fact, he had a few telephones, but never one in his study, and so he could honestly say, as he did, that he did not have a telephone 'in my own house within reach of my ears.'[72] Indeed, Bell was never enamoured of his own invention. He despised the intrusion of his infernal invention into his private life and once ripped a phone off his own wall.

Bell had ample opportunities to present his findings on the telephone at scientific conferences and in the leading journals of the day. These addresses were mostly technical, seldom going beyond explaining the mechanics of the telephone to reflect on the significance of this new technology for society.[73] In the tradition of the true inventor, he much preferred talking about the invention he was working on than gloating over past achievements. When the members of the Empire Club of Canada invited him to talk about the telephone, he chose instead to speak on 'The Substance of My Last Research.' He did begin by recalling the steps by which he came to discover the telephone, generously acknowledging others who assisted him in his invention, and then quickly went on to talk about his recent work on aerial locomotion. He recalled his early interest in the aerodynamics of kits at his summer place in Baddeck, Nova Scotia, and how he dreamed of the day 'when we would have flying machines.'[74] He acknowledged the work of S.B. Langley of the Smithsonian Institute, who attempted to fly a steam engine, and that of the Wright Brothers, who had succeeded in flying a short distance with a motor on one of their gliders. His own interest had been to put a motor on a kite. He summoned two engineers from the University of Toronto – J.A.D. McCurdy and S.W. Baldwin – and an expert on motors, Glen H. Curtis of Hammondsford, New York. Together they formed the Aerial Experiment Association, which went on to invent a number of different flying machines. His most recent dream, he concluded in his speech of 1917, was to invent faster flying machines, one that would fly without wings, and even ones without motors. Clearly, Bell was addicted to the wonders and unlimited possibilities of technology. It was, one might say, 'in his blood.'

It was mystifying in light of Bell's invention of the telephone and the photophone that Sandford Fleming should have taken up the laying of a cable under the Pacific Ocean to link the British colonies in Asia with Britain via Canada as his last cause for improving communication. In fairness to him, however, when he first thought about the possibility of a Pacific Cable back in the early 1870s, the telephone had not been invented. When he took up the cause again in earnest at the turn of the century, the telephone had wide continental usage but had not yet been used for intercontinental communications.

Once again, as in the case of railway building and standard time, imperial unity was Fleming's prime motive for pursuing an underwater cable across the Pacific. He noted its importance in an article on the Pacific Cable in the *Queen's Quarterly* of 1898: 'The dominant idea with those who have most strongly advocated the establishment of a Pacific cable has been the unity of the Empire ... A telegraph across the ocean would foster trade and commerce – the life of an Empire such as ours.'[75] Fleming believed that Canada had a vital role to play in this imperial mission, situated as it was between the two great oceans. When completed, he claimed, Canadians would be responsible for 'bringing the mother country into direct electrical connection with every one of the great possessions of the crown in both hemispheres without touching the soil of any foreign power.'[76]

Fleming's dream came true on October 31, 1902. He compared the cable with its electric wires to 'the spinal cord of the human body.' As the 'father' of the project, Fleming had numerous occasions to comment on its significance. In the addenda to an article entitled 'Completion of the Trans-Pacific Cable: Interview with the Father of the Project,' he noted its significance for the advancement of Western civilization in the world:

> The first communication [on the cable] was a greeting to the King from the Fiji Islands and it appeared strange that these islands, so remote, so recently occupied by cannibals, should be the first to transmit, through the newly completed Pacific Cable, a message of respectful homage to the Sovereign of the great British Empire. The fact is significant. Is it not another indication that the civilization of the human family is steadily advancing? In this relation we are reminded that in the British Islands in that part of the earth where the Fiji message was received by the King the inhabitants, now highly civilized, were a few centuries back, a race of painted savages.[77]

Fleming had reason to be proud and, given the effort he put into achieving the Pacific Cable, boastful. He truly believed that the world was on the eve of achieving lasting peace, brotherhood, and a global sense of neighbourhood and community, thanks to the technological sophistication of the British race. Not surprisingly, he became carried away with such eulogies during important imperial celebrations such as Queen Victoria's Diamond Jubilee of 1897. Fleming was in London the year before the event and gave a paper to the Royal Colonial Institute in London, subsequently published in *Canada and Ocean Highways,* in which he noted the achievements – particularly technological achievements – that the Victorian age had provided: 'The Victorian age has witnessed vast strides in the extension and unification of the Empire ... In no other similar period in the history of the world has there been so much advance in material and moral progress ... In the Queen's happy reign we can record countless reforms and applications of science to ameliorate the condition of the human family – postal improvement, telegraphy, photography, cheap printing, telephones, railways, ocean steamships, submarine cables, lighting and locomotion by electricity, and the thousand uses to which science is applied in every day life.'[78]

Fleming was so much a part of these technological inventions and so convinced of their virtues for closer communications, world peace, and human happiness that he was unable to stand back and observe certain ironies in his accomplishments, all the more evident in his latest accomplishment, the Pacific Cable. Less than a year before the completion of the cable, Marconi had performed an experiment in Newfoundland that made the Pacific Cable obsolete before it had even been completed: he had received from England the first transatlantic wireless message on a windy hilltop at St. John's, Newfoundland. In time, no country would benefit more from this new technological invention than Canada. Little wonder, then, that the electric cable companies, including those involved with the construction of the Pacific Cable, attempted to discredit and block Marconi's efforts.[79] On this occasion, Fleming was on the 'losing side' rather than on the 'cutting edge' of new technology. Technology seemed to outpace even the 'Father of Canadian Communications' in his own lifetime.

The completion of the Pacific Cable led to a disagreement between Canada and Australia as the two countries feuded over who should pay what portion of the huge expense for constructing the cable. Fleming also became embroiled in a feud within Canada to keep the Pacific Cable out of private hands, where he feared it would be used for subversive purposes.[80] As well, the laying of the Pacific Cable resulted in greater friction between Britain and the dominions – the exact opposite of what was intended – because Britain, much to the annoyance of her dominions, refused to acknowledge or underwrite the costs of the project, feeling that it did not serve her interests.

Finally, Fleming's promise that this technological wonder would bring about better communication, international peace, universal brotherhood, and a world sense of community rang hollow early on in the twentieth century.[81] Fleming had failed to realize what the noted American Henry David Thoreau had quickly grasped when he heard that a telegraph line was being built between Maine and Texas: that more efficient communication did not mean better personal communication or a deeper understanding between the parties involved. 'We are in great haste to construct a magnetic telegraphy from Maine to Texas,' Thoreau wrote in *Walden,* 'but Maine and Texas, it may be, have nothing important to communicate ... We are eager to tunnel under the Atlantic and bring the old world some weeks nearer to the new; but perchance the first news that will leak through into the broad flapping American ear will be that Princess Adelaide has the whooping cough.'[82]

Neither Thoreau nor Fleming realized the ultimate irony that better communication technology could lead to the *breakdown* of communication, resulting in international warfare rather than world peace and unity. Fleming lived long enough to witness the outbreak of the First World War and the use of the telegraph by Britain to notify her dominions, including Canada, that the Empire was at war. Unfortunately, Fleming never recorded his thoughts on the subject. He died in 1915, before he had to witness the worst of the carnage of a war that had come about when world communication was at its best and the technological wonders of the nineteenth century seemed to hold out the promise of a civilized and peaceful world. Had technology failed those, like Fleming, who had put so much hope and trust in it? Certainly, some Canadian intellectuals came out of the First World War questioning the nature, value, and meaning of technology.

3
Advocates of Technical Education: Technology as Knowledge

Keefer, Haliburton, Fleming, and Bell made an appeal for acceptance of the emerging technological imperative against all manner of resistance on the part of its opponents. The latter saw the new technology as a threat to the moral values that had been part of their upbringing and education, whereas the former saw the technological imperative as the embodiment and fulfillment of that moral imperative. Those who did hear the call of their peers for the acceptance of technology were the advocates of technical education. These educational theorists of technology believed that the emerging new industrial and technological society required a new education system that put a premium on technical education. In their minds, technology was knowledge, and the greatest form of knowledge was technical education. They believed technical education would advance Canada not only materially but also, and more importantly, morally and spiritually, thus ensuring the country's superiority over its competitive and always threatening neighbour, the United States. Technical education would ensure both technological and moral progress. Like Keefer, Haliburton, Fleming, and Bell, advocates of technical education believed that the technological imperative was compatible with the moral imperative. Both were premised on instilling the values of efficiency, orderliness, productivity, autonomy, industriousness, and perfectibility. As 'educational prophets,' the theorists of technical education heralded the new technological age as the fulfillment of the finest of Western civilization as embodied in the moral imperative that had brought that civilization to its present level of greatness. But Canadians had to accept technology as the means to advance civilization to new heights. These educational theorists were 'prophets' as well in their ability to see the values of technical education before others even considered that option. They were able to do so because they, like Keefer, Fleming, and Bell, actually worked in the field they wrote about. They were 'doers' as well as 'theorists.' However, the tension they felt in fighting for the technological imperative as an embodiment of the moral imperative reflected the general need of the time to reconcile these two imperatives. What they failed to realize was the role that technical education would play in weakening rather than fulfilling the moral imperative.

Appeals for technical education took different forms. In the public schools, the demand was for manual training for boys and domestic science for girls. In the universities, it focused on the establishment of Schools of Practical Science or Schools of Technology and later faculties of applied science and faculties of engineering. However, advocates of technical education faced an uphill battle, since the existing education system was based on a belief that the classics, literature, and history were superior subjects for cultivating the moral imperative to science and technology, which were considered to be 'practical' subjects. They put forward a number of arguments in favour of a technical education, including the improvement of manual skills and mental dexterity essential for a machine age, the cultivation of moral values, and as the means by which Canadian civilization could progress to its ultimate level. In an age that believed without question in progress, advocates of technical education argued that technical education was the one and only means for society to progress. Advocates of technical education accepted without question what William Leiss describes as the 'idols of technology': that technology should dictate thought; that modern society had surpassed the 'superstitions' of earlier ages in its thinking; that every breakthrough in technology was a triumph for humanity; and that all other ways of interpreting the human experience were inferior to the mechanical one. These assumptions formed part of the thinking of advocates of technical education and the basis of the technological mindset that took hold in the late nineteenth century. They also guided the thinking of the machine age and offered justification and a rationale for technical education.[1]

Advocates of technical education saw manual training for boys and domestic science, a form of manual training, for girls as important subjects in the elementary and secondary schools to prepare children for the world of work in an industrial age, especially for students who would not go beyond secondary school. However, they deliberately pitched their arguments on a higher level than mere work so as to appeal to those who were concerned about the negative impact of the technological imperative on the moral imperative. They pointed out that manual training and domestic science educated the 'whole child,' cultivated skills necessary for life, instilled proper moral values, and prepared children to be upright citizens able to advance Canadian civilization to new heights.

Domestic science, often identified at the time as 'technical education' or 'industrial training' (and later called home economics), soon outdistanced manual training as a popular program, thanks to the support it received from Adelaide Hoodless, an ardent social reformer. Hoodless was born on a farm in Brant County in 1857. She received little education beyond elementary school. She also had a limited life outside the home until 1890, when she first took up the cause of introducing domestic science into the school curriculum as a result of the death of her eighteen-month-old son in the summer of 1889 due to drinking contaminated milk. She saw domestic science as preparing young girls to be more knowledgeable about such issues. But her larger objective was to better prepare them for the new industrial and technological society that was emerging.[2] She founded the first Women's Institute in Stoney Creek, Ontario, a small village east of Hamilton, and helped Lady Aberdeen, the wife of Canada's governor general in the early 1890s, organize the National Council of Women. Hoodless also founded the Victorian Order of Nurses and was instrumental in organizing the Young Women's Christian Association. But her real passion was promoting domestic science. She would be its strongest advocate right up until her death in 1910, when she collapsed from heart failure while giving an address on women and industrial life before the Women's Club in Toronto.

Although the focus of domestic science was on the home and, for Hoodless, was a means to keep women committed to the home and out of the labour force, the movement to implement domestic science in the schools was not anti-industrial. Indeed, Hoodless wanted to bring the values of the new technological age to bear on the home so as to make it more scientific and technical and therefore more attractive than factories for the modern woman. As well, she believed that a home run on scientific and technical expertise would prepare the leaders and workers for the new industrial age by enhancing technical knowledge, cultivating technical skills, and instilling proper moral values in youth. In an address to the National Council of Women in 1899 called 'Technical Education and Domestic Science,' Hoodless emphasized that they were in an industrial age, one that required 'highly improved quality, as well as ... exceptional skill in industrial branches. There is a demand for advanced methods in all things.'[3] Just as men in factories needed to be re-educated to prepare them for the new age, so too, Hoodless argued, did women have to be retrained for their important roles as wives and mothers, responsible for producing healthy workers who in turn would ensure a healthy and competitive industrial nation. Hoodless thus saw domestic science very much within the context of the new technological age and accepted without question the assumptions that underlay the technological mindset. In arguing for the introduction of domestic science in the school curriculum, Hoodless consistently emphasized the scientific and technical nature of 'domestic science.' In an article entitled 'A New Education for Women,' for example, she outlined the benefits of the 'new education' for the new woman of the industrial age. 'Instead of the primitive industries, she will find scope for all her powers in dealing with scientific questions, such as the science of agriculture, which will enable her to raise poultry, make butter, cultivate flowers, fruit, vegetable, etc., with so much interest that city life will appear vapid by comparison. She will study domestic science, not as cookery, but in its broadest sense, that of home economics, which deals with the vital question of homemaking, not housekeeping, as there is a wide difference.' The end result, Hoodless predicted, would be women who had 'a fuller sense of their responsibility as the caretakers of that greatest of all social institutions – the home.' To date, such skills had been ignored, 'with the result that [a] low standard of offering prevails, and the haphazard methods of acquiring proficiency in any trade ha[ve] only been equalled by the slipshod methods of the homemakers.'[4]

Like other advocates of technical education, Hoodless also emphasized the ability of domestic science as a form of technical education to cultivate the proper moral values for the industrial age. In doing so, she challenged the established belief that studying English, history, and arithmetic was the best means for 'cultured training.' She claimed that such subjects did not interest young children and thus failed to arouse 'their listless minds.' As well, these purely academic subjects developed mental skills but at the expense of physical skills and thus failed to develop the 'whole child.' Then, too, 'intellectual culture' produced 'indolent habits.' By contrast, manual training 'disciplined the will,' thus saving children from 'disorder' and 'despair.' Manual training and domestic science also developed 'cleanliness, promptness, neatness, self respect and self dependence,' Hoodless claimed, qualities that were valued in the new technological age. These technical subjects also 'encouraged energy,' 'lessened truancy,' and 'prevented crime.' Thus, she concluded, 'as a moral influence Manual Training is given a high place.'[5]

When asked how domestic science as a form of manual training was able to have such a 'moral influence,' Hoodless replied, 'The prosperity of a nation depends upon the health and the morals of its citizens; and the health and morals of a people depend mainly upon the food they eat and the homes they live in.'[6] She believed that if the home operated on values instilled through technical education, then its members would acquire the same values. The home and the workplace in the new industrial age were not separate spheres but inextricably linked: strengthen the former and the latter would take care of itself. Therefore, the really vital foundational institution in the modern industrial age was not the workplace but the home. The proper education, a technical education, had to begin there.

Adelaide Hoodless believed that knowledge was power and that technical knowledge was the most powerful form of knowledge in the industrial age. For women, this meant the study of domestic science as the best means of preparing for the new industrial and technological society into which they would enter after schooling. Indeed, Hoodless was of the opinion that the power of a technical education was 'not only as an educational factor, but as a social power.'[7] In other words, domestic science would give women not only a quality education for the new technological age but also the power to transform society into a higher form of civilization. For women reformers like Hoodless, this was truly the greatest value of a technical education. Hoodless had that same moral zeal and passion for social reform that was evident in the writings of Keefer, Haliburton, Fleming, and Bell. She was a firm believer that the technological and moral imperatives were compatible – indeed, they were synonymous.

In the struggle to have science and technology recognized as legitimate subjects of study in Canadian universities, no individual was more committed or determined than Nathaniel Fellowes Dupuis. In a way, this is surprising, given the fact that Dupuis was a professor of mathematics at Queen's College (the forerunner of Queen's University) and mathematics in the late nineteenth century was considered one of the 'pure sciences.' Yet Dupuis had an abiding interest in the practical sciences, which, according to his colleague James Cappon in a memorial to Dupuis, was evident in 'the highly practical and scientific way in which he dealt with the question of daily life and work and the facts of life in general as far as he concerned himself with them.'[8] Cappon attributed this penchant to Dupuis' background as 'a clockmaker's apprentice' and to the fact that he was a self-made man.[9] Whatever the explanation, Dupuis was the major impetus behind the drive to establish the faculty of applied science, and later the Mining School, at Queen's University.

In an address delivered at the opening of the 1872 session of Queen's College, Dupuis noted that throughout history all great inventors and scientists had to fight against established beliefs and entrenched values to have their inventions or discoveries accepted. Initially, the resistance came from the Church and the prevailing religious dogma of the time; today, he wrote in 1872, opposition, at least in Canada, came in the form of a philosophy of conservatism, an attitude 'in opposition to all change, and hence to all progress in scientific thought.'[10] Conservatism seemed endemic to the Canadian mind, Dupuis noted, and was evident in a number of ways. The one way that concerned him, as a university professor, was the refusal to recognize science, and especially practical science, as a legitimate subject of study, because, it was claimed, 'the study of science does not elevate or train the mind'[11] in the way that literature did. Dupuis dismissed the idea as nonsense based solely on an attitude from a bygone age and unsubstantiated in fact. He challenged his audience to prove that 'scientific men of the world are one whit inferior in mental power and acumen to the literary ones.' 'Besides,' he queried, '*where* are the faculties of observation and comparison, those two great faculties of the human mind to which all the active ones may be referred, more peculiarly exercised and strengthened than in the study of any of the natural sciences[?]'[12] None of these spurious arguments against science mattered anyway in the end, Dupuis argued, because the wave of the future favoured a scientific and technical education. The advancements of the modern world and their accompanying material comforts and moral uplifting that practical science provided made the importance of scientific and technical knowledge a foregone conclusion.

In Dupuis' opinion, the only question remaining to be answered was whether practical science should be taught in the universities or in independent schools of technology. He favoured practical science as a legitimate subject in a university curriculum, not simply as a 'bye-subject' or option, as though of secondary importance, as it then was in a number of universities, but as the crowning subject of a student's education. In challenging the traditionalists who claimed that science did not train the mind, Dupuis argued that no subject better prepared students mentally and morally for the world and enabled them to make a greater contribution to society than practical science. Other subjects deemed more worthy of inclusion in a university curriculum because of 'their ability to fit them for becoming eminently useful as citizens of the world' really only 'exalt the individual, to the exclusion of those which not only enoble [sic] the man, but through him, improve and beautify the world.'[13] Dupuis' assumption was that the latter values were better achieved through a technical education. Moral purpose and social reform needed to be the ultimate goals of education.

Dupuis concluded his address by comparing scientists struggling to have practical science recognized as a legitimate intellectual exercise to Galileo struggling three centuries earlier to have his theory that the Earth moved accepted. 'With Galileo we may say that the world moves, and as it moves physically so it also moves intellectually.' For Dupuis, motion meant progress. He ended his speech with an exhortation to progress through technical education: 'It [the Earth] moves, and it is ours to keep pace with its progress, not in the wild and unproven theory which exudes from the brain of every speculating man, but in the staid and noble step of *truth* as she marches from conquest to conquest beaming with a pitying love upon her opponents, and speaking joy and satisfaction to those who follow in her train.'[14]

In the latter years of his life, Dupuis once again took up the cause for greater recognition of science and technology in the university curriculum. In an article in *Queen's Quarterly* in 1896, entitled 'Some of the Factors of Modern Civilization,' he questioned what 'factor' had made the greatest contribution to the advancement of Western civilization. He dismissed religion, claiming that, during the time religion had predominated, society was stagnant and even regressive. Instead, he found the key in education, although not traditional education. The least advancement in education in the present century, he maintained, had been made 'in the subjects of philosophy and classics and a few other allied subjects.' The real advancement had occurred in the sciences, especially the applied sciences. 'It is with the world of physical study, of devised experiment, of invention and the ever-increasing employment of apparatus and machinery,' he argued, that one must look for the great achievements of the modern age. Dupuis listed some of those achievements, from the steam engine to electricity, and even the bicycle. What the bicycle did for society in particular, he maintained, was 'materially decrease the consumption of strong drinks'! (He never elaborated on the connection.) However, the greatest advancements had occurred in the technologies of communication. The telegraph and the telephone had 'made of man a cosmopolitan in both life and ideas' and thus brought all the nations together into 'a common level of civilization,' a view reminiscent of that of Sandford Fleming. Dupuis attributed these inventions to two current developments in education: specialization and application. 'We see that the demands of civilization have created a specialization,' for 'all the great machines so commonly employed nowadays are made under the care of workmen who each have charge of a special part.' While he admitted that such specialization often worked to the detriment of the 'general mechanic,' it appeared to be a necessary evil 'in carrying out the demands of our modern civilization.'[15]

Dupuis maintained that, given the importance of the study of mechanics in the advancement of civilization, technical education should have had a central place in a university curriculum. Alas, it did not. He blamed this failure on those education theorists who believed that scientific, mechanical, and physical training had no place in a course of study because it did not promote culture. Dupuis queried what these educational theorists meant by 'culture.' If they meant merely putting 'a polish upon the individual without increasing his usefulness in the world in which he is compelled to live, or assisting in the onward march of civilization,' then the world could do as well without it. If, however, they meant 'that which enable[s] man to assist in the well-being of his fellow creatures, and to do his part in lifting the world from a lower to a higher plane of civilization,' then technical education had as much a claim to culture as any other subject. For in studying a machine, the student was studying not only physical principles but also 'the thought of the inventor or inventors who gave form to the machine, just as truly as the student of poetry is studying the thoughts of the poet whose work he is reading.' Even in studying a machine as a mere physical entity, Dupuis argued, the student exercised 'his ingenuity, his taste for accuracy, his means of surmounting difficulties by making the most of his appliances and ... his faculty of observation.'[16] Technology, and the study of it through technical education, were, then, as much mental and moral as physical exercise, the training of the body *and* the mind.

John A. Galbraith, a member of the School of Practical Science at the University of Toronto when it began in 1878 and its first principal when it was reorganized in 1889, also fought hard for the recognition of technical education in the university. A prize-winning graduate of the University of Toronto's honours mathematics program, he had spent ten years as an engineer in railway building before joining the School of Practical Sciences as one of its instructors.[17] Prior to making the appointment, the Minister of Education sought the advice of James Loudon, who had a personal interest in the School of Practical Science. Loudon highly recommended Galbraith for the appointment. Later, as president of the University of Toronto, Loudon was reported to have said that Galbraith's appointment 'was the best piece of work he had accomplished either for technical education or for the University.'[18] Galbraith would go on to become the first dean of the faculty of applied science and engineering at the University of Toronto under Loudon's presidency.

Galbraith set out his views on technical education in an address by that title delivered at the opening of the engineering laboratory of the School of Practical Science in February 1892. He pointed out that technical education had taken on two meanings that might be classified as manual training schools and technical science schools. The former derived from the Greek word for 'technical,' meaning an art, handicraft, or trade, and translated into schools for 'the training of apprentices in the arts and handicrafts.' Such schools, he noted, had served their purpose and had proven their worth, since to them 'we owe the greater part of the material progress which has been made since the world began.' The latter, technical science schools, were more recent, a consequence 'of the growing competition for trade among civilized nations, and the recognition of the relations of art and science to production,' and had translated into 'schools for giving ... scientific training to those engaged in industrial pursuits.'[19]

Why the need for the latter? One reason was the need for professional teachers who had both practical experience and scientific knowledge. The former was important to be able to relate theory to practical situations in the interest of students, but the latter became essential, for 'the principal work of a technical school is the teaching of science and not, as many suppose, to turn out fully fledged engineers, architects, manufactures and tradesmen.'[20] (Henry Bovey would make a similar point in 1904, arguing that, as a forum for the study of science, faculties of applied science could be considered among the pure sciences.)

Even the practical work done in a School of Practical Science differed from that of actual life. For one thing, Galbraith pointed out, it was work done in an experimental way, where the student was allowed to make mistakes as a useful way of learning. Second, practical work in the school had to be necessarily general, since it was impossible to relate it to all possible circumstances and thus designed to enable the student to apply it when necessary in practice. Third, even laboratory work was different, designed to experiment with equipment to discover the scientific principles of construction and action underlying the machine rather than to learn how to use it. In general, then, Galbraith noted, 'the work of the [technical science] school is more analytic than synthetic, more destructive than constructive.'[21] In another context, he expressed it thus: 'The life work of the engineer is construction and production. His practical work in the school should be analysis and experiment.'[22] In essence, Galbraith argued, technical education had more to do with science and intellectual exercises than with technology and manual training. This became particularly true as the student progressed through the program. Indeed, one of the unique features of the University of Toronto's School of Practical Science that Galbraith had initiated was to have a student's practical training or shop work completed outside the school, thus saving time in the classroom for the more important theoretical work. In this way, he claimed, the School of Practical Science had been saved from becoming a School of Technology and in fact, he announced in 1901, had just been recognized as a faculty of applied science. 'The result is,' he concluded with pride, 'that the university gains without expense a fully equipped Faculty of Applied Science and in this respect puts itself on an equality with the other great universities of the continent: while on the other hand the School gains public recognition of the fact that its work is of equal rank and dignity with that of the ancient faculties of Arts, Medicine and Law.'[23]

Professor Daniel Wilson stands out as one of the most influential and enthusiastic advocates of technical education. Given his background, this is surprising. Born in Edinburgh in 1816, he began his career by setting up a business as an artists' colour-man and print-seller. Then in 1853, he was appointed to the chair in history and English literature at University College, Toronto. Despite his strong humanistic background, Wilson went on to become a distinguished man of science. He also became president of University College in 1880 and then president of the University of Toronto from 1887 to 1892. Wilson's interest in science lay in the area of ethnology, a field to which he made a major contribution. At the same time, however, he was fascinated with technology. He attributed this interest to his brother George, who held the first chair of technology in Britain – the Regius Chair of Technology at the University of Edinburgh – and who became the director of the Industrial Museum of Scotland.[24] Daniel Wilson became a member of the Canadian Institute in Toronto (founded by Sandford Fleming in 1849 for the promotion of science) in 1853, assumed the editorship of its *Canadian Journal* in 1856, and was its president in 1859 and 1860. Wilson was also instrumental in assisting the governor general, the Marquis of Lorne, in founding the Royal Society of Canada in 1882. Although Wilson was skeptical about including a literary component in the proposed society, he ended up as the first president of its Section 2, 'English Literature, History, and Allied Subjects,' in 1882. He became the society's president three years later.

Wilson had been drawn to science because of its 'secret truths,' and his approach to the study of natural science reflected his sense of awe, wonder, and reverence for nature as God's handiwork. It was this religious sensibility that caused him to take exception to Charles Darwin's unemotional and matter-of-fact study of the natural world. He came reluctantly to accept much of Darwin's theory of evolution in the animal world and even the fact that humans had been on Earth for thousands of years. He was also impressed with 'man's innate capacity'[25] throughout history for creativity, especially in the area of technological inventions. He set down that impressive record in a monumental two-volume study entitled *Prehistoric Man: Researches into the Origin of Civilisation in the Old and the New World* (1862). However, Wilson questioned to the end of his life the idea that man's intellectual capacity had evolved out of material nature. He could not accept the premise of many enthusiasts of technology in the nineteenth century that humans were 'machines' that operated by the same mechanical laws.

Wilson was not one to dwell on the past when it came to appraising technology. He believed that past achievements should be appreciated as evidence of future potential. 'Let us learn by every experience of the past,' he told the convocation class of 1885, 'and make of it a stepping-stone to higher things; for we ourselves "are ancients of the earth; and in the morning of the times."'[26] In his mind, no country held greater future potential than Canada, but only if Canadians had the scientific and technological knowledge and expertise to develop its riches. That required students with a solid technical education. It was time, he proclaimed in an address at the inauguration of evening classes at the School of Practical Science at the University of Toronto in 1881, 'that some adequate recognition be also extended to education of an essentially technical and practical kind.' For one thing, the captains of industry demanded it: 'Economic industry more than ever demands the careful husbanding of all our resources, including the grand industrial army of mechanics, artisans, and all who bring pitch and muscle to the labour-market as an article of hire.'[27]

More importantly, Wilson argued, society itself was at a stage of evolution that required students who had both brain power and practical knowledge 'to solve the vexed problems of social and political life. With advantages rarely ... if ever equalled, you enter on the inheritance of a virgin soil, with all the grand possibilities of a new era. But the willing hand of the industrious toiler will need the help of the keen intellect and the no less busy brain, if we would not be mere gleaners, loitering in the rear of a progressive age.'[28]

He added for good measure 'that in the rivalry among civilized nations for supremacy in the world's marts, the race will be to the swift, and the battle to the strong; and strength in such rivalry means intellectual supremacy.'[29]

Most of the arguments Wilson made in favour of technical education were the familiar ones of the time. But he also put forward a novel reason. Technical education would democratize the university. At a time when most advocates of technical education were assuring colleagues that its inclusion in the curriculum would not lower the standards, respectability, sanctity, and exclusivity of the institution, Wilson was willing to stand up for the right of students from working-class families to benefit from the education a university had to offer. 'The age of an exclusive scholarly caste had passed away,' he proclaimed in his convocation address of 1891, 'and there is an ever growing demand for an educated people.' He welcomed this as a sign of 'a progressive age.' 'We need not only the power of the gifted few; but the wide sympathy of a well-educated community.'[30] In another address, he claimed that the goal of modern education should be to turn each individual into a 'skilled labourer,' since, he noted in words that were reminiscent of Adam Smith's *The Wealth of Nations,* 'it is more and more coming to be seen and felt that the wealth of nations depends on the highest application of knowledge and skill in every department of industry.'[31] This was the reason, Wilson claimed, that the School of Practical Science at the University of Toronto in 1881 had as its mandate not only the teaching of engineers and other professionals but also the training of 'intelligent workmen for all the ordinary applications of skilled labour.'[32]

James Loudon, Daniel Wilson's arch-rival and successor as president of the University of Toronto, also advocated technical education but for reasons different from Wilson's. Born and educated in Toronto, receiving his BA and MA in mathematics and physics from the University of Toronto in 1862 and 1864 respectively, Loudon was appointed in 1875 to the Chair of Mathematics and Natural Philosophy in University College. He was the first Canadian-born professor to hold a chair in Canada. Shortly after his appointment, the Ontario government asked him to report on the School of Practical Science. The views he formulated from his extensive research at that time formed material for a number of later speeches on the subject of technical education.

While Daniel Wilson had favoured an approach to the study of science that was speculative and designed to harmonize science with philosophy and theology, James Loudon insisted that science be divorced from metaphysics and approached from a purely objective and analytical perspective. It was not so much that he opposed technical education being concerned with moral issues but that he believed that a good grounding in the theoretical aspects of technology would better prepare students to address moral issues in later life. The two men also differed in their views of the role of the university. Wilson was a product of the Scottish university system, which believed that a professor was predominantly a teacher and an imparter of wisdom to students who were then expected to serve the public good. In the case of the School of Practical Science, this meant, in Wilson's mind, trained engineers and even educated skilled workers who would use their knowledge and expertise for the material and moral well-being of society. James Loudon, on the other hand, although educated in Canada, greatly admired the German educational system with its emphasis on research. In his mind, university professors should be first and foremost researchers, interested predominantly in discovering and disseminating new knowledge. He believed this should be equally true of faculties of applied science.[33]

Technical education topped Loudon's list of superior aspects of the German education system. In a speech on technical education at the opening of a new science building at the University of Fredericton in 1894, Loudon compared 'this temple of science' to the great walls of Carthage. 'The strength of the ancient city was in its walls. The strength of the modern nation lies in its institutions of learning.'[34] He argued that Germany's victory over France in the Franco-Prussian War could be attributed 'to the German Universities and higher seats of learning.' Today, he noted by comparison, an equally momentous conflict was being waged in the world for industrial and commercial supremacy. Here, too, the winner would be decided by the nation 'which possesses the most abundant resources of pure and applied science, resources which the Universities and technical schools alone supply.' At this time, that meant Germany. Loudon maintained that the marvellous progress of the Germans over the past twenty-five years had gone 'hand and hand with higher scientific, technical, and commercial education, and is directly attributable thereto.'[35]

He admitted that Canada was clearly no rival for Germany, at least not yet. In the evolution of material progress, Canada was only at the stage of realizing its vast potential of natural resources and beginning to establish a capitalist class and technical experts to develop these resources. To date, these technical experts had come wholly from abroad, leaving native-born Canadians in inferior positions. Loudon believed that only a better system of technical education could change that situation.

Too often in Canada, Loudon noted, technical education had been confused either with manual training or with the work of schools for the teaching of trades. It was neither. Rather, technical education could be defined as 'instruction in those principles of Science and Art which are applicable to industrial pursuits.' The emphasis should be on 'theoretical knowledge' as opposed to training or teaching a trade, Loudon maintained, so as to 'render the artisan master of his trade, or the engineer master of his profession.'[36]

Then Loudon proceeded to outline how the Germans accomplished this goal. First of all, they had technical schools, independent of general public schools, which offered specialized technical education from the beginning of schooling, coupled, however, with a complementary broad, liberal education. Those students wanting to go on to further technical studies at the university level had nine technical universities to choose from. Out of such a system came 'an industrial army, perfect in discipline, under skilled leaders, who in turn are amply provided with scientific advisers. It is indeed the presence of these professional experts – chemists, physicists, engineers, etc. – which is the distinguishing feature of German manufacturing enterprise; and it is to this circumstance, coupled with the employment of accomplished agents in the markets of the world, that is due the phenomenal success achieved in many of their leading industries.'[37] Loudon admitted that such compulsion, discipline, and segregation by specialization did not sit well with the more democratic Canadian system of education. Nevertheless, he believed that Canadians ought to rethink their educational system. For the remainder of his talk, Loudon advocated an education system that emphasized the study of English, mathematics, French, and German in the public schools and a good theoretical scientific education at university, leaving practical training for the workplace.

Loudon reiterated the importance of the theoretical aspect of technical education in his convocation address to the University of Toronto in 1899: 'I should like to state emphatically here my conviction that no diffusion of technical training will in itself be effective if we do not take care to maintain the higher and the highest kind of scientific instruction, and if our manufactures do not utilize this expert knowledge.'[38] This required good educational leadership, which, he claimed, only a university-affiliated School of Practical Science could provide. These education leaders in turn had to work with capitalists and politicians to ensure the best system to fulfill the present needs of the country. 'Under the united forces of capital, enterprise and technical knowledge, aided by industry and frugality on the part of the people,' he concluded, 'our country will be enabled to make the best use of its great opportunities, and to win for itself great gain without loss, nay, with added gain, to the world at large.'[39]

In 1904, Henry Bovey, a Canadian engineer and the dean of the faculty of applied science at McGill University, gave an impressive keynote address on technical education at the International Congress of Arts and Science held in conjunction with the Universal Exposition in St. Louis, Missouri. Entitled 'The Fundamental Conceptions which Enter into Technology,' the address drew on the ideas of the earlier generation of Canadian educational enthusiasts of technical education, outlined in this chapter, to offer a persuasive and comprehensive argument for introducing technical education at all levels of schooling. The address was published the following year in the *McGill University Magazine*.[40]

Bovey was an excellent choice for the keynote speaker. He was one of the few members of the early generation of engineers in Canada who had a formal university education. Born in Devonshire, England, in 1852, he was educated in British public schools and at Cambridge University, where he won an open scholarship.[41] Upon graduation, he excelled in the Mathematical Tripos and soon afterward was made a fellow of Queen's College. He decided to become a civil engineer and worked for the Mersey Docks and Harbour Works, where he was promoted to assistant engineer. In 1877, at the age of twenty-five, he became a professor of civil engineering and applied mechanics at McGill University and a year later dean of the university's newly created faculty of applied science. (Prior to the creation of the new faculty, engineering had been a branch of the faculty of arts.) He remained its first dean until 1908. He was also the William Scott Professor of Civil Engineering and Applied Mechanics at McGill from 1890 to 1908. In 1908, he was invited to be the rector of the Imperial College of Science and Technology in London, England, a position he held for only one year due to ill health. He was a founding member of the Canadian Society of Civil Engineers and a member of its council for many years as well as its honorary secretary and treasurer. In 1896 and 1897, he was vice-president and in 1900 the society's president. He was a fellow of the Royal Society of Canada and served as president of Section 3: Mathematics, Chemistry, Astronomy, and Physics in 1896. In his presidential address, he extolled the virtues of engineers as advancers of civilization, concluding that 'there is no department of human life or industry which has not been illumined, and I do not hesitate to say that no one has done more than the engineer to advance the civilization of the world.'[42] He was also a fellow of the British Royal Society and vice-president of its mechanical section.

Bovey began his address to the International Congress of Arts and Science by noting that, as a technologist, he was being asked to be philosophical about a subject – technology – that was considered the most practical of the sciences and therefore required 'no mental foundations of any kind.' He asked: Was this a sign of the times? In an analysis reminiscent of Thomas Carlyle's 'Signs of the Time,' Bovey noted that the eighteenth century had been dominated by 'the philosophic spirit, while that of the present age is admitted to be the scientific spirit; some even call it the age of the application of science.'[43] He pointed out how science in general had moved away from a concern with the external world to explore 'the invisible, the intangible, the inaudible' and therefore was returning to its earlier association with philosophy. Thus, for the technologist – or the 'Technologue' – to do the same was in keeping with the times. Technology, Bovey claimed, had also become as much about how one viewed the world – a mindset – as it had about the working of the world, as much philosophical as practical. So in his talk he aimed 'to discover the controlling ideas' underlying this technological mindset.

Bovey noted three key ideas central to that mindset: 1) 'that Nature works in no arbitrary manner, but by fixed laws'; 2) that 'if we could bring ourselves into the right relation with them [the laws] – into the line of their working – we might hope to gear our small machines [our minds] to the vast wheel of nature, and make it do for us what we could never do for ourselves'; and 3) 'that in the study of the laws of nature and in the attempt to put ourselves into touch with them, there would certainly be revealed more and more of what seems to be the infinite possibilities of our environment.' Bovey accepted the prevailing attitude of his time that the entire world – human, physical, and mechanical – operated by similar laws. Thus, through a study of the laws of mechanics could come the means to understanding the working of the laws of nature and the laws of the mind. As Bovey explained, 'The laws of the mind ... are, equally with heat and electricity, the laws of nature. We must make the laws of the mind work for us instead of against us, just as we are seeking to do with the forces external to us.'[44] In a perspective that came close to one that Marshall McLuhan would put forward a half century later, Bovey pointed out how many technological inventions could be seen as *'extra-organic sense organs'* to show the interconnection of human beings with machines: 'To see a projection of the human eye in the telescope and microscope, which so marvellously extend our vision that it can resolve the misty light of the far-off nebulae into suns, or discern in a clod of clay a world of wonder; to hear in the telegraph and the telephone the tones of the human voice so intensified as to reach round the world, and in the printed page the silent voices of long past generations; to know the express train and the ocean liner as extensions of our locomotor-mechanism; and to discover in a tool or a lever the human arm grown strong enough to perform seeming miracles.'[45]

Bovey argued that, by bringing humans in line with the laws of nature, 'the powers of nature [could be brought] under our control in the interests of humanity.' This was the job of the technologist. His training through observation, experimentation, and analysis was designed to assist in this process. Making a pitch for technical education, Bovey reminded his audience 'that the training of the hand actually stimulates the brain centres. This has given to manual training its true value.'[46] Such a view was in keeping with the psychological theories of the day, particularly William James's theory of pragmatism, which posited the brain to be what it did. The human mind, like the world of nature, was in a state of constant change and motion as it evolved to higher levels of consciousness. An active mind was a creative mind. Stimulation of the mind came from doing, as opposed to mere speculation, which was seen by advocates of technical education as being 'useless knowledge.' Thus, Bovey concluded that the truly creative scientist was the applied scientist, since he was using both his mind and body, applying thoughts to actions, and studying ideas and mechanical devices. The ultimate result of this dual process of action and contemplation was the bringing forth of 'the divine gift of imagination'[47] in students with a technical education. This 'divine gift' may be God-given, but, Bovey argued, it remained only as potential within individuals until brought forth into actuality. This was the power of a technical education: to cultivate the human imagination, a quality up until then considered the reserve of the humanities or the pure sciences.

What then was technology? It was, Bovey concluded at this point in his argument, 'a process of education – a secondary science – a process which has been described ... as an entire system of education by new methods to new uses.'[48] Technical education taught students not only the 'principles of mechanics' but also how to apply those principles 'to the construction of works of utility of every kind.' As well, a technical education trained the mind to think scientifically and, in fact, did much 'to form the scientific mind.' As well, Bovey noted that a new area of technical education had recently come into being that studied 'the chemical and physical properties of matter, into the dynamics of steam, electricity, etc.,' areas that were usually the preserve of the pure scientist, thus, in essence, turning technology into a pure science itself.

So, Bovey queried, if technology was a form of the sciences, what distinguished technology from science? In the popular mind, it was the belief that science aimed only to acquire knowledge without an interest in practical application, whereas technology acquired knowledge to apply it for other purposes – hence the name 'applied science' for technology. As Bovey put it, in a phrase that refigured Bacon's dictum that knowledge was power, 'It is not *knowledge* but *power* which is his [the technologist's] ultimate aim'[49]: the power to use the knowledge of science to bring about practical and utilitarian improvements in the world – in essence, to reform society. So long as this power was used for the benefit of society instead of the individual, for noble ends rather than commercialism, the aim of the technologist, in Bovey's estimation, was a worthy one.

Bovey claimed that such issues were moral, not scientific. While he admitted that the teaching of morals was not the responsibility of a technical school, the cultivation of a '*love* of truth' could be a by-product of a technical education. He reasoned, 'Within its limits, in common with all true scientific teaching, and perhaps in a larger measure proportionate to its appeal to a larger clientele, technology may lay claim to produce moral strength, truth and manliness,'[50] values that underlay the moral imperative.

But a technical education had even more to offer. Bovey went on to argue that it also cultivated qualities of self-restraint through the patient waiting on nature and self-sacrifice through a search for truth. Thus, technology as technical education provided both the material and, more importantly, the spiritual qualities necessary for Western civilization to progress.

Not yet finished in extolling the virtues of a technical education, and returning once again to his earlier argument of the importance of a technical education in cultivating the imagination, Bovey noted (as would Heidegger a half century later) the interconnection of the mechanical arts and the fine arts. Both cultivated the imagination and then used it as its *'instrument'* to discover 'ideal truth or ideal beauty.'[51] In the early stages of a technical education, drawing was the means to cultivate the imagination. It continued to do so at more advanced levels in the fine arts but not in the mechanical arts. Bovey concluded that what the mechanical arts required, then, was a means to continue to foster the imagination in its students throughout their education so as to make them aware of 'the *unity* of [their] own mental being and the mighty unity of Nature.'[52] He left no doubt in the minds of his audience that technical education was capable of achieving this goal.

By the end of his address, Bovey had shown that technology was knowledge, and technical knowledge was power: the power to impart theoretical and practical knowledge; the power to cultivate personal moral values; the power to sustain the noblest of ideals of civilization; and the power to cultivate the imagination. He would surely have agreed with any fellow Canadian engineer who argued that, since technology was the dominant force of the age, and since the engineer was the power behind technology, then the engineer with his technical education was 'the master spirit of the age.'[53]

Part 2
Grappling with the Imperative

4
George Sidney Brett and the Debate on Technology as War: Technology Dethroned

The Great War was a turning point in terms of perspectives on technology. The magnitude and destructiveness of this technologically dominated war led to the realization that technology was more than objects, machines, or knowledge. Technology was a complex process that was associated with war and that raised disturbing questions about human nature and even about the moral imperative that had guided the evolution of Western civilization since its origin. For the Great War appeared to be a civil war within Western civilization. Britain and Germany, considered the most technologically advanced and the most 'morally upright' Western nations, had descended into barbarism at the very time when the Western world was proclaiming to be the most advanced civilization of all times, thanks to technology. Was there a disjuncture between the technological imperative and the moral imperative? Theorists of technology were not yet prepared to admit such a disjuncture, nor were they prepared to denounce the technology that had brought such material advancement to the Western world. But they could not dismiss the Great War as an aberration in the progressive evolution of that civilization. They also had to come to terms with the demonic side of human nature and to what extent that demonic nature had something to do with technology. Some analysts of the Great War saw all four issues – war, human nature, civilization, and technology – as interconnected. During the interwar years, in trying to understand the interconnection, Canadian theorists of technology came to a new level of understanding about the complexity of the technological imperative and its association with the moral imperative.

George Sidney Brett, Canada's eminent historian of psychology in the early twentieth century, reflected at length on the question of the relationship between the Great War, human nature, civilization, and technology. His reflections were part of a debate that had begun during the war years and continued well into the postwar era. The debate drew in a number of other intellectual commentators. What is significant about this debate that focused on technology is that for the first time technology was being discussed as a 'topic' of interest to people outside the field itself, specifically by social scientists and other academics, as well as informed members of the public. At the same time, technology had moved beyond being seen as objects, machines, or knowledge to that of process. For Brett and others who reflected on the Great War, technology was a process of war. For William Lyon Mackenzie King (Chapter 5), technology was a process of industrialism, while for Stephen Leacock (Chapter 6) technology was a process of mechanization.

In the immediate aftermath of the war, Brett first entered the debate on technology as a process of war in his address to the Royal Society of Canada called 'The Revolt against Reason: A Contribution to the History of Thought.'[1] Brett's background prepared him well for the task at hand. He was the son of a Methodist minister from Briton Ferry in South Wales. He was educated at Kingswood, a Methodist preparatory school founded by John Wesley at Bath, where he won numerous honours.[2] Although he had an interest in medicine, he chose instead to pursue classics at Oxford with a focus on the ancient philosophers. Upon graduation, he spent four years in India teaching philosophy and English; these years made him aware of the constructive contribution that Indian philosophy could make to the moral thought of the West. From India he went to the University of Toronto in 1908, where he taught in the Department of Philosophy and Psychology.

Brett had an abiding belief that humanistic moral values needed to guide the sciences, including psychology. Thus, psychology and philosophy should abide in the same department. Historian Michael Gauvreau points out that, as a result of his studies at Oxford, Brett believed that three strands of thought underlay psychology: the older, philosophical psychology based on the introspective 'science of the soul'; the 'new psychology' of experimental science introduced by the German scholar Wilhelm Wundt; and 'applied psychology,' which, Gauvreau notes, 'rested upon a functionalist, activist interpretation of the human mind, and stressed its ultimate capacity to predict and control human behaviour.'[3] If the latter two strands were turning psychology into more of a science than a humanistic study, the first strand was vital, in Brett's estimation, in keeping psychology a human science concerned with ethical issues and moral values. All three strands of thought were evident in Brett's monumental three-volume study, *History of Psychology*, considered his most outstanding contribution to scholarship.

Brett first introduced the term 'technologists' in *History of Psychology*. He applied the term to those philosophers or psychologists who were interested in 'technological problems' and who 'proposed policies for life and worked out their requirements for health and happiness.'[4] They did not propose theories like the theoretical scientists or systematically test the assumptions they used like the experimental scientists. However, Brett argued, these 'technologists' should not be dismissed as unimportant. 'After all,' he wrote, 'most psychology even nowadays is technology rather than theoretical science.'[5] Indeed, he saw Sigmund Freud as 'ostensibly a technologist. By that is meant that his professional concern was to cure his patients and to provide remedies for the discontents of civilization, discontents which seemed to him to be almost inevitable.'[6] Thus, Brett saw 'technologists' as practitioners who applied theories to solve problems and technological psychology as an applied field of science.

Brett had a particular interest in the theories of Descartes, especially his mind-body dualism and the implication of that concept for psychology. He believed Descartes to be 'the father of modern psychology,'[7] because his mind-body dualism was the root of the two dominant schools of psychology, those who studied physiological phenomena as the root to an understanding of the human psyche through mechanistic biology and physiology, and those who studied the science of the mind through introspection. Cartesian philosophy, particularly Descartes' theory of the human body as a machine like that of an animal's body, had a profound influence on Brett's thinking. In his Royal Society address, he presented his belief that the world was becoming increasingly mechanistic and technological in nature.

By way of introduction to the subject of rationalism and irrationalism in his Royal Society of Canada address, Brett noted that, prior to Descartes, philosophers believed that animals as well as humans had souls and therefore felt pain. Descartes challenged this belief, arguing that both animal and human bodies were 'machines.' In terms of studying their physiological nature, therefore, Descartes was 'very much inclined to regard machinery' as the prototype, a perspective in keeping with pre-war views of machines as the 'perfect specimen,' whether natural, animal, or human. Machinery became the ideal against which to judge everything else. 'At this point,' Brett went on to note, Descartes 'could see no distinction between a sound organism and a perfect clock';[8] both moved with machine-like precision. In the case of human organisms, Brett noted, Descartes saw their reflex activities as 'operations of our animal nature, our bodies as machines.'[9] However, Descartes believed that what distinguished humans from animals was the ability of humans to reason. A human was 'an animal organism united with a rational soul,' whereas an animal had only a body, a 'self-contained machine.'[10] The reason for Descartes' distinction, Brett argued, was to reassert the unique power of human beings. Descartes believed that 'the intellect that invented gunpowder was not amenable to the old definitions'[11] of humans as mere animals. (It is interesting to note that when pressed to find an example of human rational power Descartes should choose the invention of gunpowder, often seen as an early technological invention, over visions of God.) By granting humans a soul and an intellect, Descartes elevated them above animals, although he continued to believe that, in terms of their bodies, humans were still like animals. But more importantly, Descartes saw both humans and animals as essentially 'machines.' Thus emerged the mind-body duality in philosophy.

According to Brett, the result of Descartes' dualism was an inclination to challenge reason as a valid human quality and to focus instead on the new perspective of the body as a machine, through experimental science. Brett pointed out that La Mettre went so far as to coin the phrase 'man the machine.' Brett sketched out the negative implications that resulted from viewing humans in this way: 'In a maze of complicated machinery all sense of simplicity vanished; reality could not be seen through the mist of representative ideas; benevolence could not be conceived as even possible in a world of self-seeking tempered only by hypocrisy; law, order and government became inexplicable absurdities in view of the fact that the world, never very intelligible, was now openly regarded as having no intelligence to give its procedures significance.'[12] Indeed, Brett noted, the concept emerged that man was not 'natural' but instead a 'manufactured article.'

Having sketched out Descartes' perspective on humans as machines, a perspective that Brett argued had essentially gone unchallenged over time, Brett then turned to the question of the implication of Descartes' perspective for the post–First World War period. He maintained that Descartes' view of humans as machine-like entities still held sway, reinforced in the late nineteenth century by the 'discovery' of the irrational unconscious, yet another blow to the concept of humans as rational beings. Brett believed that the only way out of this conundrum brought on by science was to realize that science could only provide the means to achieve the ends that humans aspired to; it could not dictate those ends. Also, Brett noted, in the context of the Great War those ends could no longer be selfish ones; they had to be concerned with the good of society as a whole if the Western world were to survive. It was Brett's awareness that the moral and technological imperatives might be at odds rather than in harmony with each other. He ended his talk by linking the philosophical concept of humans as mechanistic beings to the Great War and the ideologies that it spurred:

> Philosophy did not cause the recent war, nor has it caused the subsequent anarchy; but it must of necessity be an active power in war and revolution and peace, because men love to justify their actions and the justification becomes the philosophical creed in which the spirit of action is embodied. Marxism, syndicalism, Bolshevism, and the modern developments are all rooted in a scheme of ideas which is as truly a philosophy as Platonism, Hegelianism, or the new realism. This we must sooner or later face, and first

of all we must be clear about our original question – is life essentially rational or irrational?[13]

Brett left his question unanswered. But the mere fact that he raised the question in the context of the debate in the field of psychology over whether humans were rational beings or irrational and machine-like entities, and in the context of a technological war that had resulted in mass destruction and human carnage of unparalleled nature, revealed how an understanding of technology had moved to a new level of consciousness, complexity, and concern. Clearly, Brett was challenging the perspective on technology put forward by the pre-war generation of Keefer, Haliburton, Fleming, and Bell as marvellous machines that reflected the greatness of human beings as creative individuals who were advancing civilization to a higher level of consciousness.

In 1925, Brett presented a second address to the Royal Society of Canada in which the subject of technology was introduced, this time speaking on the topic 'The History of Science as a Factor in Modern Education.' By way of introduction, he noted the impact of specialization on modern education, the result, he claimed, of the 'industrial expansion and solid commercial prosperity'[14] of the mid-Victorian age, when education was used to make industry more expansive and commerce more prosperous. The sciences rose to prominence among the subjects of study, so much so that the modernists classified historical ages in technological rather than political terms. Anthropologists, for example, talked about 'the copper, the bronze or the iron age,' and historians were known to refer to 'the horse age, the steam age – or in the case of warfare, the bow age, the gun age, and so on.'[15] However, Brett was concerned with the increasing tendency in the modern age of science 'to ignore the organic relation between science and the totality of human life.'[16] Here, he believed, was where the humanities, especially a study of the history of science, could have a salutary effect. A proper history of science should focus less on great scientific discoveries and instead make students realize 'that the spirit of discovery is only a species of curiosity' that is innate in all minds and not unique to only a few. The scientific mind was also not the only type of mind to cultivate, Brett cautioned. Science and technology were important subjects to study but not the only subjects – contrary to what advocates of technical education argued – or necessarily the most important. As well, a critical history of science should dethrone some of the myths of modern science (what William Leiss would later identify as the 'idols of technology'), such as the beliefs that science develops on its own trajectory 'independent of social forces'; 'that science has always been the benefactor of society rescuing it from political strife or religious mania'; and that science has been 'free from superstition, bigotry, and the kind of narrow-mindedness which thinks to build without proper foundations.'[17] Brett appealed for a perspective on scientific and technical education that realized their limitations as much as their strengths in an ideal education system that needed to address major social problems and moral dilemmas that technology and science had created.

In his Conclusion to a review of two books on science and technology, the one focusing on Bacon and the other on Spinoza, Brett also re-emphasized the need for balance, in particular to offset the obsession with science and technology in the modern age. He pointed out that Bacon and Spinoza offered opposing perspectives for the modern world, one advocating the ideal of power and the other the ideal of universal harmony:

> The contrast between Bacon and Spinoza illuminates the complexity of modern life. The ideal of power, achieved by the conquest of nature and the use of limitless resources, has been partly realized as time goes on. The ideal of universal harmony and peace, described by Spinoza as the logical product of intellectual development, does not seem to make similar progress. The material products of scientific activity ... are worthy of the admiration they excite; but even more admirable is the spirit of the men who achieved this work, and hope for the future welfare of mankind seems now to depend less on the multiplication of discoveries than on the universal acceptance of rational principles in conduct.[18]

Brett was challenging the belief set forth by Keefer, Haliburton, and especially Fleming in the nineteenth century that the way to achieve the ideal of 'universal harmony and peace' was through technology. It was precisely this faith in the power of technology through 'the multiplication of discoveries' to create a better world that ironically stood in the way of achieving universal harmony and peace through 'the universal acceptance of rational principles in conduct.' Brett believed that technology and spiritualism, the former expressed as the technological imperative and the latter as the moral imperative, were in tension or opposition, if not diametrically opposed. While in his review he did not point out that this tension was made evident in the Great War, the association was there in the realization that the Great War had not ushered in the 'ideal of universal harmony and peace' that Spinoza had envisioned and that postwar reformers had anticipated. As well, the tension harkened back to Brett's earlier address to the Royal Society of Canada in the aftermath of the war in which he presented the two sides of human nature: rationalism and irrationalism.

Brett had earmarked one of the major concerns of the war and postwar generation in regard to technology, namely, that technology brought out the demonic or irrational side of human nature, resulting in technology being used for destructive ends. For Brett, this was the lesson to be learned from the Great War. Contrary to the belief of the pre-war era, including the ideas of Keefer, Haliburton, Fleming, and Bell, that technology and rationality went hand in hand in the march of civilization to the 'promised land,' Brett and his generation saw technology and human irrationality as more likely to lead the world to Armageddon unless technology could be brought under human control. In the war and postwar era, technology was dethroned.

Brett's dualism of power and universal harmony and peace, materialism and spiritualism, irrationalism and rationalism, technology and thought, reflected the debate that occurred shortly after the outbreak of the Great War. The debate centred on the question of the role that technology played in enabling Germany to be so militarily successful. The two sides differed as to what aspect of technology contributed to Germany's success. One side saw Germany's superiority in technical knowledge as the reason, while the other side saw technology in the form of industrialism and the resulting advancement of the machinery of war as the explanation. Both sides agreed, however, that it was technology that made Germany such a powerful country. It was the German technological imperative that Canadian analysts of the war ironically both admired, as proof of the power of technology, and despised, since it was this very technology that brought Western civilization to the brink of destruction.

Lindsay Crawford initiated the debate in his 'Current Events' column in *Canadian Magazine*. He argued that it was Germany's superior technical knowledge, which came from its scientific mind, that accounted for the success of 'the German military machine.' It was most evident in the German propensity for organization. Indeed, he wrote, this was 'the one great lesson which Germany has taught Britain – the value of organization.'[19] J. Squair agreed with Crawford that it was Germany's superiority in the sciences, especially the applied sciences, that accounted for its quick rise to power in Europe. 'She was the only nation taking full industrial advantage of the discoveries of science. France, England and America were mere rule-of-thumb workers in comparison.'[20] Robert Falconer, president of the University of Toronto, attributed Germany's success to 'Prussian militarism,'[21] which combined technological might with a strong military spirit. The two went together, Falconer argued, in that technological might cultivated a technological mindset that put a premium on militarism. Falconer admitted, if only surreptitiously, to a certain admiration for 'the organisation and efficiency'[22] that went along with the technological mindset of the German military. What he could not accept and thus condemned was the use of that organization and efficiency for evil and destructive ends. T. Brailsford Robertson, professor of biochemistry at the University of Toronto, argued that the Germans had cultivated only 'the materialistic aspects of science to the almost total exclusion of its idealistic and spiritual values.' What he hoped the war would show was the importance of science 'as the pre-eminently creative factor in civilization ... For that order of society which attains the greatest harmony of its social consciousness with scientific thought must inevitably attain the domination of the world.'[23]

C. Lintern Sibley was one of the few Canadian analysts who refused to admit to German superiority in science and technology. In 'Britain's Intellectual Empire,' he argued that all the great advancements and discoveries in the physical and applied sciences from Newton's discovery of the laws of gravitation, to James Watt's invention of the steam engine, to Faraday's discovery of electric induction were 'British.' The same was true in communication technology. He mentioned William Watson, who sent the first electric shock; Francis Ronalds, who discovered the electric telegraph; and a British newspaper, the *Morning Chronicle* in London, that sent the first news report by telegraph. While he conceded that the telephone and the wireless telegraph were not invented by Brits, he did nevertheless ingeniously include them too. It was this intellectual superiority in the sciences within the empire that justified Britain's physical empire too. 'History shows,' he noted in terms reminiscent of the ideas of Sandford Fleming, 'that to the Anglo-Saxon race has been allotted a position of profound responsibility in regard to the evolution of civilization, and history reveals that, despite all her faults and shortcomings, the Anglo-Saxon race has on the whole performed its task both nobly and for the general good of the universe. In other words, the British have won for themselves an intellectual empire commensurate with the vast breadth of their territorial domination,'[24] both the by-products of technology.

Others argued that it was not German superior technical knowledge that enabled them to have the upper hand in the early stages of the war and to hold out for so long but rather their superior war machinery, the by-product of the country's industrial might. Francis Miller Turner argued that the two words 'Germany' and 'science' had been used so frequently together they had become synonymous, and rightly so.[25] But what made the Germans so outstanding in the sciences was not their interest in the pure sciences, something in which the British were superior, but rather their concern with the applied sciences linked to industry. Turner pointed out that this ability to apply science to industrialism was equally true of the United States and accounted for the mercurial growth of both of these countries in the late nineteenth and early twentieth centuries. However, he noted with concern, 'the very civilization which we have been centuries in making, and which is our pride and boast, by its incessant and insatiable demands, is to be the instrument of our destruction.'[26]

A.B. MacCallum, president of the Royal Society of Canada in 1917, presented an equally ambivalent perspective to that of Turner on the role of technology in the war in his presidential address entitled 'The Old Knowledge and the New.' On the one hand, he admitted that German technological might accounted for the country's success and wished that the Allies had been equally successful in building up their military might so as to keep Germany in check. On the other hand, he realized that it was this utilization of technology for destructive ends – the result, he argued, of a technological mindset that measured progress in mechanical terms – that led to the barbaric conditions that brought into serious question the idea that the world was progressing. As he noted:

> For many generations to come there will obtrude in the minds of all who will look back on this great war the memory of a nation which, nursed and participating in a civilization at least a thousand years old and boasting of a culture higher than that of its environment, developed after a few weeks of the stress of war a condition of mind and ethical standards that must have characterized the human race in the long night that preceded the dawn of our civilization. This will chasten all high hopes and beliefs as to the permanence of the forces that make for human progress which we so fondly held in the past.[27]

The only hope he could see in the whole sordid situation was that both sides would realize the limited nature of the world's natural resources and thus the need to use them constructively instead of destructively.

Stanley Mackenzie also contrasted the ideals that enthusiasts of technology had put forward as justification for faith in technology in the nineteenth century with the reality of the role that technology played in the war. In his address to the Royal Society of Canada a year after MacCallum's, he noted, 'Fostered by us as the great servant of civilization and the promoter of prosperity and comfort to men, it [technology] is equally potent for the horrors of war, and the same power that has banished diseases and ameliorated the terror of wounds has poured out the deadly gas and high explosive [sic] on innocent women and children.'[28]

It was this ambivalent attitude toward technology, both admiring the advances that technology had brought to the world and yet realizing that those very advances were what had generated such a protracted and destructive war, that characterizes the analysis of technology during the war and interwar years. Theorists were not yet prepared to question whether technology was inherently flawed, that it had an imperative of its own that could be used for destructive ends; they still believed that it depended on how it was used. But after the Great War, it was no longer possible to have an implicit faith that technology would always be used to create a better, more peaceful world that characterized the thinking of Keefer, Haliburton, Bell, Fleming, and Bovey in the pre-war era.

The debate over the role that technology played in Germany's military success also raised questions during the war and interwar years about the role that technology, in the form of industrialism and as technical education, played in Canada. Clarence Warner, president of the Ontario Historical Society, argued that it was Canada's phenomenal industrial growth at the turn of the century (which he attributed to the unification of the country at the time of Confederation) that in turn enabled Canada to be in a position to help Britain and her Allies in time of war. In highlighting Canada's achievements, he wrote, 'We were making great progress along material lines. Our great industries were turning every wheel. Our railroads were adding new mileage and new equipment to handle an ever-increasing business. Then the word came that Britain was at war with the greatest fighting machine that the world had ever known. We immediately gave our best thought and work in an endeavour to take our share in the burden.'[29] C. Lintern Sibley predicted in 1915 (besides the fact that the war was about to end) that the war would benefit Canada materially because it had shown the world the tremendous contribution that Canadian manufacturers had made to the Allied cause. In essence, the war elevated Canada to the status of a technologically advanced country.[30]

Others, like Harold Garnet Black, worried that it was precisely this materialism that made North Americans, including Canadians, 'Philistines,' the term used by Matthew Arnold to condemn those lacking in culture. Black pointed out that, at a time when North Americans were more materially advanced than Europeans and therefore should have more time to be cultured, they were less so because technology demanded more and more time to keep apace. 'The chief trouble with all of us,' he wrote, 'is that the daily duties and activities of our modern world, with its rush and whirl and ceaseless clatter, are usually so pressing and exacting that they leave little time for the cultivation of the mind. Your professional man, your city-dweller, throws down his napkin before his wife and children are half through breakfast, dashes off to his office, spends a busy day, and hurries home for dinner at half-past six.'[31] Black's perspective contrasts with that of Keefer, who found the increased activity that the locomotive brought to the Sleepy Hollows of British North America exhilarating and intellectually invigorating.

The issue of technical education also came under scrutiny in light of Germany's success in this area of education which, as already noted, many Canadian analysts believed accounted for Germany's military prowess. Invariably, the issue of technical education raised the question of values to which Canadian society should adhere and to what extent those were or should be technologically driven. On the one side of the debate were those analysts who worried about the negative role that technical education was already playing in Canada and pointed to the German situation for their evidence. On the other side were those who pointed to German success as justification for more emphasis on technical education in order for Canada to remain competitive. In 'The Teacher and the New Age,' H.T.J. Coleman noted, 'In these years of social unrest and upheaval, of the revaluation of all values, men and women in increasing numbers are asking what our schools have accomplished and are judging these accomplishments by certain rather definite standards of achievement.'[32] What was behind this period of questioning in 1918, Coleman claimed, was the belief, true or not, that the German schoolmaster – 'the village teacher as well as the University professor' – was responsible 'for German industrial and military efficiency.'[33] Coleman appealed for better organization in Canadian public schools, too, but only to a limited extent. The danger of over-organization was evident in Germany, in his mind, where there was no opportunity for debate and dissent; the child became sterile. Interestingly, in setting out the role that the teacher should play in Canadian schools, Coleman used the metaphor of the teacher and the student as 'mechanical entities,' despite the fact that he was arguing against such a perspective:

> The teacher is not, let us repeat, merely a mechanic though he must make use, at times, in common with the rest of mankind, of mechanical appliances and mechanical routine. He is not merely a scientific practitioner, though there are many important laws of human nature and human society which he should know and should apply but which, with all our boasted educational progress, have had, as yet, no adequate recognition in the ordinary schoolroom. The teacher is primarily an artist and, so far as the potentialities of his work are concerned, the greatest of all artists for he deals with the most plastic and most wonderful material in the universe – the child soul.[34]

J.K. Robertson, founder of the Manual Training program, claimed that the war had accelerated the move toward science: 'The stoppage of certain supplies, the production of which depended on the work of applied scientists in enemy countries; the nature of modern warfare itself, with its ghastly scientific means of destruction, its aeroplanes, its submarines, its long-range guns, and on the brighter side its highly equipped hospitals and laboratories – such things have brought about such a recognition of science that it is certain the next generation will be highly scientific.'[35] Robertson saw two dangers in such a trend: the study of science at the expense of classical and literary studies, and the study of applied science at the neglect of pure science (points that Brett raised in the postwar era, as noted earlier). However, Robertson went on to state that, if science were properly taught to foster a spirit of investigation, a search for the unknown, an open-mindedness, and a love of nature, then such concerns were unnecessary. 'The world is in the melting-pot these days and much dross will be purged away,' he stated. 'Is it too much to hope that, in the purified life which emerges, pure science and the Humanities will be found arm in arm, closely united in their common aim to advance the cause of Truth and Humanity?'[36]

Ira MacKay, professor of law at the University of Saskatchewan, argued that the current system of education in Canada already owed its nature to the military model of technical efficiency influenced by the Germans: 'We teach our children in squads and companies, as if their minds could be made to move like their arms and legs. The reason why we adopt this plan is purely because the military formation is the simplest, cheapest and most primitive type of formation in which large numbers of minds may be easily handled and made to present a superficial appearance of efficiency and thoroughness.'[37] What the system fostered, he argued, was subservience to the state, a German model, rather than respect for the individual, a British ideal. What he proposed was not to use education to serve the state but to use the state to serve education. 'The motto of our imperialism,' MacKay claimed, 'should not be world-supremacy but world service.'[38] 'Our next move,' he concluded, 'must clearly be made in the direction of Imperial educational reconstruction, looking towards future Imperial solidarity.'[39]

Discussion of the importance of technology in winning the war and the lessons to be learned in terms of educational reforms continued into the postwar era. In his inaugural address to the class of 1919, C.H. Mitchell, dean of the faculty of applied science and engineering at the University of Toronto and himself a brigadier general in the war, attributed the winning of the war to technology and the expertise of the engineer: 'The long, tedious and laborious efforts of these hosts of scientific workers all mobilized in the Empire's struggle and concentrated in the national "will to win" the war, could only have resulted, as it did, in evolving those numberless inventions by which, in a very large measure, the war was finally won.'[40] Dr. Jewelt, chief engineer of the Western Electric Company, pointed out the negative influence of the Great War on postwar research. It had drained the universities and industries of trained scientists and technocrats, thus creating a critical shortage: 'We cut off completely the possibility of further advances into the realms of the unknown and likewise destroyed our chance of developing new men to carry on the investigational work of the old when the latter were worn out.'[41] Jewelt predicted that there would be the need in the postwar era for the establishment of research centres and institutes to pool resources, knowledge, and human power, along with the creation of a parent organization, the National Research Institute, to regulate and oversee research.

In his presidential address to the Royal Society of Canada in 1920, R.F. Ruttan used the German success in the war years as justification for the need in Canada for greater co-operation among scientists, especially practical scientists and technocrats, in postwar Canada. Ruttan attributed Germany's success, particularly at the outset of the war, to its ability to apply the single-mindedness and efficiency of the scientific methods and organization to all aspects of the state. 'Germany did not owe her great strength in 1914 so much to her scientific knowledge,' he argued, 'as to the power she had attained by the organized combination of national effort to the one end ... The political, military, financial and scientific resources of the nation were fitted into a gigantic, perfectly working machine, characterized, to quote McAndrew's hymn, by "Interdependence absolute, foreseen, ordained, decreed."'[42] Ruttan recommended the same pooling of expertise through research institutes in Canada. He admitted that this may appear to curtail the creativity of the individual inventor and genius but argued that such individualism in the past was really more myth than reality. Great scientists and inventors had always co-operated. The difference now, in the postwar world, was that co-operation was imperative. The world demanded it.

Thus, the Great War led to antithetical perspectives among Canadian intellectuals on the issue of the role technology should play, in the form of industrialism and as technical education, in dictating Canadian values during the war and in the postwar era. Some saw Canada's industrial growth as enabling the country to play an important role in the Allied cause and appealed to Canadians to continue in this vein in order to continue to play an important role in the world. Others saw the emphasis on industrialism as detracting from the more important task of culturally advancing. For them, technological development was the antithesis of cultural development, the technological imperative in tension with the moral imperative. In terms of technical education, dissent also prevailed. On the one side were those who advocated more technical education to keep Canada competitive, emphasizing in particular the need for collaborative effort through research institutes, like the Germans. On the other side were the detractors who saw technical education as already leading to a regimented and narrowly focused educational system that turned out robotic students rather than critical thinkers. What the educational system needed, the detractors argued, was more emphasis on the humanities and social sciences, following the British rather than the German system.

Canadians, like people in the rest of the Western world, emerged from the Great War confused as to its meaning and significance for the postwar era. In their reflections, Canadian intellectuals revealed their ambivalent view of technology, seeing it at times as a hope for a better future and at other times as a portent of worse to come in its destructiveness. What appeared to unite both perspectives was the realization that the new age emerging from this technologically driven Great War was one of chaos, instability, and ceaseless motion, which Canadian intellectuals in general found disquieting and troublesome – a striking contrast to the enthusiasm for change and progress that marked the thinking of the pre-war era. In an article entitled 'The Effects of War on Literature and Learning,' E.F. Scott, a war chaplain and the father of the eminent Canadian poet Frank Scott noted, 'The war has let loose a number of incalculable forces, and how they will operate we cannot tell. Thus far it is impossible to guess whether the main result will be good or evil, whether we are entering on an age of wider freedom and brotherhood, or on a period of chaos, which will finally throw us back on barbarism.'[43] Scott himself was pessimistic. He pointed out that the war had destroyed a generation of youth who might have brought about a renaissance of learning. The war had also become an obsession, hindering writers from getting beyond the immediate present to look at larger issues, as all great literature did. 'In the region of abstract thought,' Scott wrote, 'no work, apparently, is being done at all.'[44] Worse yet, the war had speeded up the world, thus depriving creative genius and great inventors of the leisurely life they needed to be productive.

The fear that Western civilization might be entering 'a period of chaos, which [would] finally throw us back on barbarism,' as Scott put it, was a pervasive theme in postwar analysis by Canadian intellectuals. James H. Lindsay, for example, contrasted the world of Newton and of Darwin – a world of stability, orderliness, and certainty based on the laws of nature – with the present world of movement, chaos, and uncertainty in an article entitled 'On Thinking Biologically' in the inaugural issue of the *Dalhousie Review* of 1921. After outlining the world of Darwin and Newton, he noted that 'the modern outlook was the complete opposite. 'Becoming, not Being, is the note of the modern world ... Nothing is fixed, nothing is final. All is movement, change, variation, development – not always progress.'[45]

University of Toronto's president Robert Falconer used the metaphor of the automobile, the technological craze of the twenties, to describe the point Western civilization had reached in the postwar era: '[Society might] forget that there is danger of a slow decline setting in and the car of civilization may fall into such disrepair and its power be so reduced that it will be unable to take the hills, many of which lie before this world in its long forward road.'[46] E.H. Blake was convinced that Western civilization was already 'in the midst of one of [those] periodic disintegrations, and that the downfall ... ha[d] actually begun.'[47]

Harry Elmer Barnes saw a ray of hope for the West if the 'New American History' of Charles Beard, J.H. Robinson, and the economic historian Thorstein Veblen was accepted. Barnes argued that, in their use of psychology and sociology as means to study the ideas underlying the development of Western civilization, these historians revealed just how wide a gulf existed between 'our social vision' and 'our unparalleled technology and material culture.' The best of these historians have recognized 'that the great need of the present is to bring our attitudes, reactions, and interpretations in the field of the social sciences up to something like the same level of objectivity and scientific candour which now pervades natural science and technology.' The role the historian could play would be to emphasize 'how contemporary waste, inefficiency, exploitation, and war are ruining western civilization'[48] and to expose the inability of social scientists to cope with these all-important problems.

In a review of the book *The Conquest of War,* Richard de Brisay, one of the founding editors of *Canadian Forum,* was reminded of Edward Carpenter's metaphor of modern industrialized civilization as 'a disease through all the stages of which mankind must pass before it emerges to a healthy life.'[49] He feared, however, that 'the disease' had taken such a hold that it might prove fatal. He quoted the author of *The Conquest of War* as saying, '"The world is sick with a terrible intermittent fever"' and noted, '[the author] believes that the next bout may finish us off.'[50]

Colonel R.W. Leonard, a prominent civil engineer and mining promoter, observed that the Great War had forced Canadians to doubt and question the values upon which Western civilization had rested for centuries. In his presidential address to the Engineering Institute of Canada in January 1920, he reminded his fellow engineers, 'A few years ago it would have been unnecessary to deal with the fundamentals of our civilization; they were so imbued in our beings that it was almost silly to formulate them into words. We rested securely in the belief that "Truth is might and will prevail."' No longer was that the case. The Great War had brought truth into question to the extent that 'downtrodden Truth would appear to need the assistance of some Minister of Propaganda if she [was] to hold her place and prevail before the structure of our civilization [was] badly shaken.'[51] Leonard was expressing a perspective that would become commonplace in the interwar years – that is, that the technological imperative appeared to be taking over the moral imperative to the point at which society seemed to be existing in a void, storm-tossed in a sea without a compass to guide it.

A.R. MacDougall argued that the materialism of the technological age even brought into question the values that should guide Canadians. In a series of questions, he got to the heart of the dilemma facing Canadians in the postwar era in which technology had become the new imperative:

> What, after all, should be a proper estimate of value? What is the summum bonum towards which we should direct our civilization, the ideal which is as yet visionary in all lands, but to which some approximate rather better than others? Is it industrial efficiency and the fulfillment of the sociologist's dream? Or are these but means to the end to be achieved? It must be apparent to all thoughtful persons that, despite our ingenuity in Canada in constructing railways over thousands of miles of territory, and bringing to

> but nine million people – spread over such a vast area – a high degree of material civilization, we have not, as a people, as a nation, arrived at a satisfactory solution to the problem which these questions should provoke. It is a homely truth that the standards which we ascribe to a civilization must not be destitute of either religious or ethical considerations, unless we are to descend into rank materialism.[52]

The devastation associated with technology and the Great War raised an even deeper and agonizing question: what within the human psyche allowed humanity to go mad? Other questions followed. What role did technology play in the destructive human process? Were humans the rational, orderly, peaceful, and moral beings that society believed them to be or irrational, mechanistic, destructive, and amoral entities? These were questions that George Brett attempted to answer in his addresses to the Royal Society of Canada. But Brett was not alone among philosophers and psychologists in discussing these questions.

Within the Canadian philosophical tradition, there emerged two 'schools of thought' – British idealism and German realism, the latter carried over into North America in William James's philosophy of pragmatism – that presented opposing perspectives on these critical questions. The idealists clung to the idea of mind as a spiritual entity, distinct from the physical nature of man, and maintained that humans were rational beings, created in God's image. Realists and pragmatists argued against the distinction of mind and body, seeing both as 'machines' that could be, and needed to be, studied scientifically. They claimed that no evidence of human rationality existed; indeed, Sigmund Freud's discovery of the 'unconscious' and 'irrational' side of human beings appeared to prove otherwise. Some questioned whether irrationality was restricted to certain 'neurotic' beings or, as the Great War seemed to suggest, whether certain societies were 'sick' or indeed humanity as a whole.[53] The role of technology in shaping and dictating man's irrational nature was also questioned.

Some, like Newton MacTavish, argued that the Great War had no impact on human nature. 'Values will change, as they have changed, and the currents of trade will shift. Forms of government will change, as they always have changed, and the voice of the people will be heard, as it always has been heard, with increasing force and determination. But human nature will not change. It will be just as grasping, just as selfish, just as keen to get on at the cost of others as it ever has been.' Human beings would only redirect that selfishness in the postwar era from fighting in battle to fighting in society as a result of 'social unrest, class differences, even here in Canada, and racial and religious grieving.'[54]

J. Clark Murray noted that for some theorists (not necessarily including himself) psychology had a great deal to do with the war. In a review article entitled 'Pragmatism,' he pointed out that some philosophers saw the war and this philosophy's rise to popularity as interrelated. The war could be seen as 'a terrific expression of the pragmatist tendency of thought, which finds the real significance of life in energetic action rather than in speculative occupations. Militarism then becomes a phase of Pragmatism.'[55] Murray noted that the aim of pragmatism in general was 'to dethrone reason from its position of supreme authority as final court of appeal in all questions with regard to reality or truth.'[56] D. Fraser Harris, professor of physiology at Dalhousie University, noted that the war made respectable 'the doctrine of mind or consciousness *as a cause*,' a source of 'nerve-energy.' He went on to explain, 'The hospitals during the war soon became crowded with men whose troubles were evidently largely mental and whose cures were entirely so.'[57] Harris admitted that at that time scientists could not explain the connection, but few could deny that it existed.

George Brett re-entered the debate on the meaning of technology in the context of the Great War in an article called 'The Modern Mind and Modernism,' written in 1928 and published in the *Canadian Journal of Religious Thought*.[58] The article addressed the need for a new religion for the new age, but its real value lay in setting out the essence of the modern age and its mindset. In essence, he argued that the new age – what he called 'modernism' and that he characterized as an age that science and technology both shaped and dominated – created a 'new type of man,' one who wanted 'activity and rest, aggression and peace, sin and repentance,'[59] to satisfy both sides of his complex nature. Religion offered only rest, peace, and repentance and therefore only satisfied half of his nature, leaving his active and aggressive side unfulfilled. Yet fundamental to human beings was the desire and need for change.

Brett emphasized that change was the operative word for the modern age. Francis Bacon, whom Brett described as 'the wisest, brightest, meanest of mankind,' was the first to identify the value of change. Brett noted that Bacon had been the first philosopher to be aware of the difference between philosophical knowledge and technical knowledge, or what Bacon identified as 'the mechanical arts.' The former was handed down from master to pupil to ensure the continuity of thought, whereas the latter – technical knowledge – was 'transmitted from inventor to improver.' 'In that sentence,' Brett pointed out emphatically, 'Bacon formulated the spirit of the new world. The inventor destroys while he creates. By his example he inspires others, not to copy nor to preserve, but to improve.'[60] Brett maintained that this view became transformed and sanctified in the concept of progress. Change meant progress, and progress meant ipso facto that the world was getting better.

Brett noted that the value of progress had become a sacred truth in the modern age, an idol to worship and not to question. And progress in the modern world was invariably judged in terms of technical change. 'Progress,' he wrote, 'is most easily seen in small things and those which are produced mechanically, that is to say, by the arrangement of interchangeable parts.' He elaborated further:

> Man has not invented steam or electricity: he has invented the means by which these things are controlled, limited and adapted. Some parts of the human race have elected to move in new ways: the steam engine and the motor car have transformed the methods of travel and also the ideas of space and time. For the present, at least, man has a new sense of power and feels confident that nature can be conquered ... The general name for this knowledge of the means to fulfil purposes is science, and this is essentially an age of science because it is more and more occupied in acquiring or extending the vast mass of detailed knowledge by which we make the resources of the universe available for practical purposes.[61]

Brett observed how this new scientific spirit, initially mechanical, spread to other areas of the sciences and even to the social sciences and religion: 'The idea of God was transformed by the idea of mechanism and became essentially an idea of law. The form of thought called Deism was a temporary acceptance of the view that nothing is absolutely real except mechanical necessity and nothing is truly operative except laws of nature.'[62] Thus, technology had become the means to define God, to define humans, and to judge good from bad. And the essence of technology was change. 'Whether we like it or not,' Brett noted with a sense of resignation, 'we cannot give up a belief in progress even if we would: it has become a destiny rather than a choice,' or a choice made for us by the very technology that has caused us to believe in progress in the first place. Brett pointed out that 'we do not control this progress'[63] – anymore, he might have added, than we control the technology that demands progress. Both control us.

Having sketched out the essence of the modern world as he saw it, Brett then returned in the latter part of his article to address the implications of this new world order for religion. Here he reflected once again on the meaning of the Great War. That war, he argued, left the human race in a state of suspension, 'like souls in purgatory hang between the truths that are abandoned and the promises that cannot be fulfilled.'[64] It created a state of turmoil and anomie from which the world had not recuperated. He concluded, 'The problems of the present day are still the problems that come from trying to reconcile the history of Christianity with the history of the Great War.'[65]

Brett identified the dilemma, and agony, of the interwar era, namely, that analysts had to confront the negative side of the optimistic and naive faith in technology that prevailed in the pre–First World War era. Indeed, they had to come to terms with the realization that this very naivety could be seen as the cause of the Great War, because blind faith led to an unquestioning acceptance of the truism of the day that technology would be the means to create the perfect human species, the ideal society, and the highest form of civilization. Faith in technology allowed technology to secure a supreme position of power – the new god – that could not be questioned, challenged, or really understood any more than the God of religion: indeed, technology became the new religion, identified and described in language previously reserved only for a higher deity. For the pre-war generation, as Brett so aptly pointed out, technology became the means to define what it meant to be human, what it meant to be 'natural.' Technology also dictated the values by which to judge the ideals the world should progress by. And progress necessitated change. Keefer, Haliburton, Bell, Fleming, and Bovey had realized this a generation or two earlier and were thrilled by the changed world that technology was ushering in. In their minds, change ipso facto was good, because technology judged it to be so.

For Brett, change was indeed the operative and defining term of the modern world of technology. Technology and change were interchangeable terms, symbiotic in their relationship. Yet for the interwar generation, it was change – constant motion, speed, activity, chaos, instability, uncertainty, doubt, all synonymous words in their mind with technology – that was the most inevitable, and agonizing, aspect of the modern world of technology in which they lived. The upheaval of the Great War and the chaos and uncertainty the war caused in its aftermath created *not* a sense of excitement and anticipation, as it had for the pre-war generation, but rather a sense of fear and resignation that the modern world of technology engendered.

5
William Lyon Mackenzie King and Frederick Philip Grove: Technology as Industrialism

In 1918, as the Great War drew to a close, William Lyon Mackenzie King, one of Canada's emerging prominent politicians and social reformers, published *Industry and Humanity*. It is a study of industrial relations in the early twentieth century against the backdrop of the Great War and the industrial unrest in the world, most evident in the upheaval of the Russian Revolution. King saw all three events – the Great War, the Russian Revolution, and industrial unrest – as interconnected. They were a result, he argued, of the wrong values governing the modern industrial and technological age. Materialism had taken precedence over spiritualism, economic concerns over human values, and profit over social needs, causing the moral imperative of the modern world to be out of balance. In the book, he explored the roots of the industrial age, examined where and how it went wrong, and offered his solution as to how the technological imperative might be reconciled with the moral imperative.

Over two decades later, Canadian novelist Frederick Philip Grove published *Master of the Mill*. The novel presented three conflicting visions of the mill based on the implementation of new technology and explored the impact of the new technology on the lives of its owners, the workers, and the surrounding society. Each vision failed because the mill, representing the technological imperative, ended up bringing about changes contrary to those of its 'master,' leading the reader to conclude that Grove saw technology as eliciting values that were out of sync with the moral imperative.

Both King and Grove grappled with the negative qualities of the technological imperative and worried about the possibility of reconciling the emerging technological imperative with the traditional moral imperative. The dilemma created a tension in their minds, in keeping with the tension in society as a deeper and more disturbing view of technology as a complex process emerged in the interwar years. For both thinkers, the problem and the solution resided with the individual. Individuals needed to regain or retain their moral values against the onslaught of a technological imperative that challenged those values. Both thinkers realized that this Herculean struggle was a constant; the modern technological world offered no solace or reprieve from the endless change and activity that technology as a process of industrialism inaugurated. More than ever, it meant the need to return to the spiritual roots of the moral imperative, while doubting the possibility of doing so. King and Grove were modern versions of the biblical Adam, awakening out of their state of innocence to find themselves banished to a technological wilderness in which they felt alienated and their moral values challenged. What they did not realize was the role they played in weakening the moral imperative. By the very act of sounding the alarm bell about the dangers of technology, they reinforced the belief that the technological imperative was dominant. Their fear became the reality. While both men wanted to believe that human beings controlled technology, King's interpretation of Mary Shelley's *Frankenstein* and Grove's depiction of the mill, representing technology, as a 'character' unto itself in *The Master of the Mill* hinted at darker foreboding.

In *Industry and Humanity,* King is imprecise as to what he means by 'industry' in the title of his book. It would appear that he does not mean industry as a by-product of machines. He makes only passing references to machines and notes only a few technological inventions that started the Industrial Revolution. It is evident from reading his work that King had a limited understanding and appreciation of technology as objects or machines. (Rumours had it that King opposed having a telephone in his home, seeing it as an 'alien' object.) Instead, he saw industrialism as a process of change that began with the Industrial Revolution and culminated in the Great War. As a process, industrialism affected all aspects of society. He identified four broad areas of change brought on by industrialism: the mechanical arts, industrial organization, the division of industrial processes, and industrial areas. In the mechanical arts, he noted three types of changes occurred: 'the use of new tools and implements, the adoption of new processes, and the application of new powers.'[1] He argued that these changes were the most revolutionary because they in turn initiated changes in the other three areas. Under 'industrial organization,' King noted the shift from a domestic to a factory system of production, from hand tools to machines, and the rise of capitalism. Under 'the division of industrial processes,' he listed the shift to unskilled labour in the highly mechanized factories and the introduction of female and child labour. 'Division of processes and division of labor ignore personality, and are "dehumanizing" to the extent to which they make Labor's part in Industry mechanical, and tend to destroy initiative and resource.'[2] This in turn led to a change in attitude: now capital and labour saw their interests as being opposed instead of supportive. Finally, under 'industrial areas,' King noted the changes brought about by the application of new sources of power, most notably electricity. What King saw as the common denominator in all four areas affected by industrialism was change, the same distinguishing feature of the modern age noted by Thomas Keefer, Sandford Fleming, and George Brett. 'Ancient and mediaeval civilizations differ from modern in that their many activities were markedly circumscribed,' King wrote. 'Contrasted with the present, theirs was a society essentially stationary ... It is this transition from an unchanging social order to one permeated with constant change, a transition in industrial relations from "certainty" to "uncertainty," from "stability" to "instability," that makes the Labor Problem of to-day wholly different.'[3]

For King, the most interesting and important of the changes industrialism brought about was in 'the division of industrial processes' between capital and labour, because this was the 'human aspect'[4] of industrialism. Since he believed that a positive change in the industrial process could only come about through a change in attitude on the part of those involved in that process, he maintained that the place to begin understanding industrialism was in examining relations between capital and labour. However, he argued that to look at industrialism as 'a labour problem' only was misleading, while to see it as an issue between capital and labour alone was to overlook two other groups affected by and therefore concerned with industrialism, namely, managers and the community. These four groups – labour, capital, managers, and community – together make up 'humanity' in the equation of 'Industry and Humanity' in his book title. King saw a tension between technology as one component of industrialism and human beings as the other component. Given that he believed changes in the industrial system must start with 'the human aspect,' he implied that technology, the industrial process, was still under human control and could be improved by changing the attitudes and values of the individuals involved in the process.

What were the values on which the system of industrialism functioned? This was the uppermost concern in King's mind. In general, he fundamentally believed that industrialism was creating a modern technological mindset that was continuing to shape and govern the new age. This mindset and the values it perpetuated needed to be questioned in the post–First World War era to see whether they were the right values to govern what he hoped would be a new age emerging out of the carnage of war. King believed that industrialism could be directed to either beneficial or destructive ends, depending on the values applied to it: 'There is nothing inherently beneficial or baneful in any factor, force, or form of organization to be found in the whole phenomena of Industry or the State. Everything depends upon whether its use is, or is not, made to accord with right ideas of social progress.'[5] Unfortunately, King argued in 1918, negative values of wealth, competition, and power appeared to predominate, accounting for the outbreak of the Great War and the industrial unrest following the war. It was necessary to replace these negative values, which he defined as 'materialism,' with positive spiritual values. What those spiritual values should be and how they might be applied to industrialism were his reasons for writing *Industry and Humanity*. As King noted by way of introduction to the book: 'It is the main purpose of this study to point the way to a change of attitude in industrial relations, and to suggest means whereby a new spirit may be made to permeate Industry through the application of principles, tried by time, and tested by experience.'[6]

In *Industry and Humanity,* King dwelt at great length on the presence of the dual forces of materialism and spiritualism – or what he often referred to as good and evil – that affect everything that happens. He was thus attracted to Louis Pasteur's contrary and opposing 'laws' that Pasteur said operated in the social world similar to the laws of nature in the physical world: the 'Law of Peace, Work, and Health,' and the 'Law of Blood and Death.' For King, these opposing laws were the equivalent of the opposing values that vied for supremacy within the industrial system. 'As respects the phenomena of Industry,' he wrote, 'the perception of ... an order demands, above all else, fine discernment between *economic* and *human* values; between the ends which *Wealth* and the ends which *Life* were meant to serve. The unplumbed depths of contrasts so profound are to be estimated only by the unfathomable difference between *matter* and *spirit.* It is impossible to express relationships born of such distinctions in terms of either class or nationality. A material *versus* a spiritual interpretation of Life alone defines the issue.'[7]

This dualism runs through many commentaries on technology. As the noted cultural critic Marshall Berman points out in *All that Is Solid Melts into Air,* his landmark study of modernity, Karl Marx presented a dualistic perspective similar to King's, a material *and* a spiritual component to his thinking on technology, often overlooked by analysts of Marx.[8] Berman argues that it is impossible to separate one from the other because together they expressed the totality of modern life for Marx. King's perspective on technology also had a spiritual and a material component to it, as evident in *Industry and Humanity,* coming out of his dualistic view of life. Living a 'very double life,'[9] King was cognizant of the opposing forces of the spiritual and material operating within the world and human nature. Too often, analysts of King make the same mistake as critics of Marx and stress only one side of King's dualism. Those who overemphasize his spiritual side dismiss him as a woolly headed, impractical man of the world and see in *Industry and Humanity* only vague platitudes regarding reform, without any practical side to his thinking. Those who focus on his practical and political side see him as an opportunist, devoid of ideas, and thus dismiss *Industry and Humanity* as a political tract written only to gain political points. Both perspectives fail to see the integration of spiritual and material concerns in King's thinking that lay at the heart of his understanding of technology as a process of industrialism and that makes the book an important study of technology and modernity.

In *Industry and Humanity,* King stressed the spiritual side of the duality because he believed that the material side had already been emphasized too often at the expense of the spiritual, which was exactly the reason industrialism had gone awry. Even in his emphasis on the spiritual, however, King dwelt at length on its interrelationship with the material as essential for understanding industrialism. He realized the importance of showing this interconnectedness as the theme of *Industry and Humanity* while working on the book in 1915. He explained, '[I am] working up the spiritual interpretation of the universe as against the material. I am sure that in a failure to recognize this lies the fundamental cause of error in much of the effort at social reform and that it is also at the bottom of conflict. If I can make this clear and broadly reveal the practical application of Christian principles to industry I will have helped to make a real contribution.'[10]

King began *Industry and Humanity* with what might have appeared to many readers of the day as a strange, almost disturbing subject: a discussion of Mary Shelley's *Frankenstein: The Modern Prometheus,* published exactly a century earlier in 1818. What did this novel about a monster created by mechanical means in a laboratory – a monster that King described as 'a depraved wretch, a man-machine, whose delight is in carnage and misery' – have to do with industrial relations in the twentieth century? To King, absolutely everything. King saw the tale of Frankenstein as 'a parable, all too realistic, of the War that has destroyed so large a portion of mankind.' In turn, the war resulted from industry being 'directed to the transformation of the world's resources into instruments of human destruction.'[11] Thus, there were lessons to be learned in drawing parallels between the novel and the present.

The first lesson from the novel was the realization that technology has a negative side that could as easily be used for destructive ends as for constructive ones, just as Frankenstein's monster ended up causing havoc in the world. The novel spoke to the modern generation of what could go wrong if technology were used for the wrong purpose.

The second lesson was that technology had a way of taking on a life of its own beyond the control of those who created it. King noted that the theme of the novel dealt with the innocence of Dr. Frankenstein, a 'youth of fine sensibilities' who 'created a living monster, endowed with powers which prove greater than his own.' King asked rhetorically: Was this not equally true today with regard to the war and to industrial unrest? Did these forces not appear to take on a life of their own? Why had this occurred? The reason, he maintained, was that 'the instruments which man has created appear to have become more powerful than human genius to control its own inventions.'[12] Technology had the power to go beyond the will of the creator, to have its own imperative. It had the potential to be out of control: Prometheus unbound, as the subtitle of the novel, 'The Modern Prometheus,' denoted. King believed the message was even more important in 1918 than it was in 1818.

There is a strong element of fear in *Industry and Humanity* – fear that technology was out of control; fear of industrial unrest, fear of revolution, fear that the world was about to descend into chaos. It is a fear of disorder, of things being consequently out of control. Robert Wiebe, an American cultural historian, sees 'the search for order' as the guiding principle of the rising middle class in the United States during the period of industrialization from 1877 to 1920. Seeing change all around them as disruptive and unsettling, members of the middle class developed a new system of values based on 'continuity and regularity, functionality and rationality, administration and management.'[13] Certainly King, a member of Canada's rising middle class and aware of the changes occurring all around him as Canada was undergoing its industrial revolution, was concerned – indeed obsessed – with the need for order. In the first entry in his diary, he quoted the following passage from Goethe's 'Faust' as having special meaning for him:

> Make good use of your time, for fast
> Time flies, and is forever past;
> To make time for yourself begin
> By order – method – discipline.[14]

King feared revolution more than anything else because of the upheaval and disorder that it would cause. 'I fear revol'n in the ctry yet,' he wrote in his diary in 1897, 'another 1793 – as in Fr[ance], growing Democracy vs growing Wealth and tyranny of rings and combines.'[15] He believed, as did many other social critics of the immediate postwar era such as Stephen Leacock and Salem Bland, that world revolution was imminent following the Russian Revolution and that Canada was not immune to such revolutionary tendencies.

King's perspective on industrial unrest in the postwar world and his solution to that unrest were based on the premise that the natural state of the world was one of order. To be out of order was to go against the laws of nature, laws that King argued applied not only to physical nature but also to human nature. As he wrote in *Industry and Humanity:* 'If such an order exists in Nature; ... if all material things of the heavens and of the earth are thus related in a perfect harmony which the human intelligence is able to grasp; is it conceivable, is it rational to believe that underlying the social relations of men and of nations, an order is not discoverable somewhere, obedience to which will bring as perfect a harmony?'[16] Without such a belief, it seemed futile to King to hope for reconstruction. 'But for some chart and compass,' he wrote, 'presupposing an order somewhere beneath all the apparent confusion, effort at reconstruction of human relations in their international and industrial aspects might well be abandoned.'[17]

The example King cited to 'prove' his point about order in the social world was the apparent collaboration between James Watt, the inventor, and Adam Smith, the theorist. Each assisted the other in bringing about success for the well-being of society as a whole. This example (which King had borrowed from Arnold Toynbee's *Lectures on the Industrial Revolution* without acknowledging Toynbee as the source) is a further reminder that for King industrialism was a dual process, an interconnection of ideas and mechanical devices, a combination of both spiritual and material aspects. It also reinforced King's belief that, when both the material and spiritual aspects of industrialism were in harmony, in order, then industry would serve humanity, and technology would be used for beneficial ends. King believed the opposite equally held true: in a chaotic world, industry would end up causing havoc. So the second lesson of *Frankenstein* was to realize the need for order, to keep technology under control, and to remember that the potential was always there for technology to get out of control – Prometheus unbound.

It was the third lesson that caused King to describe the novel as 'hideous.' He realized the story was an allegory on human nature, particularly conveying the 'dark side' of man's dual nature. Frankenstein's 'man-machine' represented man's alter-ego, the irrational nature Sigmund Freud had unearthed in his psychological experiments. King believed that the Great War, the most technologically advanced and destructive war to that point in time, was the most recent example of man's destructive nature unleashed through the power of technology. The industrial unrest, coming in the aftermath of the war and most marked in the chaos and bloodshed of the Russian Revolution, was another portent of things to come unless humankind could bring technology and the dark side of human nature (which King saw as interconnected) under control. King noted that 'man [is] "at once so powerful, so virtuous and magnificent, yet so vicious and base"; appearing "at one time a mere scion of the evil principle, and at another as all that can be conceived of as noble and god-like."'[18] It was this duality of human nature that distinguished humans 'from brutes and from God,' he pointed out. In King's estimation, this duality was not a defect, since it lay at the heart of human genius and creativity. This was especially true in terms of technological inventions. To illustrate his point, King chose one of the more recent discoveries revolutionizing the industrial world: electricity. 'Electricity uncontrolled or miscontrolled may destroy a community,' he wrote; 'properly controlled and directed, it may transform cities and towns with heat and light, rapid means of transit and communication, and a thousand and one useful and ornamental devices.'[19] Whether or not technology would be destructive depended on the intentions of those who used it.

Yet here King ran into a contradiction. He believed that those in control of industry needed to have the right spiritual values and intent in order that technology might serve humanity. Intentions were everything. Yet the disturbing aspect of the Frankenstein story was that Dr. Frankenstein did *not intend* to create a monster when he created his 'man-machine.' Equally, King believed, or at least wanted to believe, that the intentions of the world leaders in 1914 were not to begin a global war any more than the leaders of industry wanted to create industrial unrest throughout the Western world. 'Surely, Industry is something *other than was intended* by those who contributed to its creation,' King wrote, 'when it can be transformed into a monster so demoniacal as to breed a terror unparalleled in human thought, and bring desolation to the very heart of the human race!'[20] So if good intentions were uppermost, then why did technology often get used for evil purposes? Why did Frankenstein's monster end up being destructive if the scientist had intended it to be otherwise? As well, why did the Great War end up causing such carnage if that was not the intention of the world leaders in 1914? And why was there industrial unrest in the world if that was not the intention of the world leaders in 1918? In all cases, why did technology seem to end up being used for destructive ends rather than for the constructive and beneficial ends that were intended?

Here King presented the final lesson of the *Frankenstein* novel, the most disturbing of all: that, in the modern world of technology, the propensity to use technology for destructive ends was greater than the propensity to use it for beneficial ends. King realized that technology was power: the power to control both the physical world and human beings. It had its own imperative. It was technology that unleashed man's demonic nature that resulted in the carnage of the Great War. It was technology that caused the owners and workers of industry to distrust each other to the point of threatening world annihilation, just as it was technology that caused Frankenstein's alter-ego to take control in the form of a monster to roam the world destroying all that was good. And just as the scientist Frankenstein was destined to spend a lifetime trying to undo the wrong he had inadvertently produced in the world through the creation of his 'man-machine,' so too, King argued, were human beings in the twentieth century destined, or doomed, to a life of endless struggle to undo the wrongs that the Great War and the subsequent industrial unrest had unleashed in the world through technology.

In presenting his image of the industrial world as one in which technology was constantly used for purposes of power and dominance, King drew on Louis Pasteur's germ theory. He noted that according to Pasteur's research disease in the human body resulted from 'invading organisms.' Once inside the body, these germ cells vied with healthy corpuscles for supremacy. King noted that Pasteur found the same was true in the social body. As mentioned earlier, here the struggle was between the forces of good, represented by the Law of Peace, Work, and Health, and the invading forces of evil, represented by the Law of Blood and Death. King went on to explain the significance of Pasteur's discovery for his own understanding of what was happening to the world in 1918:

> Within nations and organizations, as within the lives of individuals, Good and Evil are forever contending. In some nations, and in some organizations, as in the lives of some individuals, evil influences are permitted to gain control. In other nations and organizations, they are held in check. *The horror of the situation is that, in individuals, organizations, and nations alike, the Good, itself, once contaminated, may turn to Evil.* The fallen angel may become a malignant devil. Germs and vice increase in virulence in proportion to numerical support and organization. *When Evil is the master, whether in the human body or in the body politic, the infection is certain to spread. It cannot be controlled till the forces that contend against it are able to hold their own.*[21]

King had come to understand that the relationship between technology and values was not simply a one-dimensional relationship: the values that humans applied to technology dictated the way in which technology was used. He realized that technology as the industrial process had its own imperative – like Frankenstein's monster – that equally shaped the human mindset and therefore the values of society. It was a symbiotic relationship.

Once unleashed, or out of control or order, technology was well nigh impossible to bring back into order. In the modern world of technology, the propensity and power to destroy the world were greater than the possibility of benefiting the world. Why? Because, King realized, technology brought out the worst in human nature – its evil side – by creating what was in essence a 'technological mindset': an attitude of indifference to the plight of others, a coldness of heart, and a lust for power instead of freedom.

This was the link between modern technology and Shelley's *Frankenstein* that King found most disturbing and that makes his view of technology so novel and frightening compared to the thinking of the theorists of technology of the pre-war era in Canada. King saw himself as the Mary Shelley of his generation of Canadian thinkers, warning the world of the dangers of technology once it got out of control, a technology that failed to operate on the basis of rationality because it represented the demonic and irrational side of human nature and because it had an imperative that overrode that of the moral imperative.

King maintained that constant vigilance was required to keep technology under control. As he noted in *Industry and Humanity,* 'Resistance to evil is strengthened by struggle. Without persistent endeavor, atrophy is certain to follow. Not to act heroically in whatever pertains to political and industrial well-being, is, by so much, to forsake the forces which seek to relieve mankind from the scourges that beset it, and to aid the forces which seek their satisfaction in Blood and Death. It is not to a life of repose, but to one of vigorous action, that the call comes to the men and women who love Peace, Work, and Health, and who would conserve these blessings for mankind.'[22]

The alternative to a healthy industrial state, then, was 'atrophy.' This reminds one of Thomas Keefer, who used the term 'apathy' to describe his fictitious hamlet of Sleepy Hollow on the eve of the coming of the railroad. Both terms denote a state of inertia. But the image these terms project with regard to technology shows the changed perception of technology in the fifty-year interval between Keefer's *The Philosophy of Railroads* and King's *Industry and Humanity.* For Keefer, apathy denoted a state of idleness, stagnation, and lethargy, an unprogressive state of being. Keefer believed the technology of railways would overcome apathy by generating action and advancing progress through the introduction of new ideas, an air of sophistication, a refined culture, and a feeling of freedom. It would turn the little hamlet of Sleepy Hollow into a lively, dynamic, and progressive town. For King, the ceaseless activity that was a by-product of the modern industrial age and that was required to avoid atrophy was less a positive and progressive state of being, less a feeling of exhilaration, promise, and progress, and more an inevitable state of constant motion from which there was no escape. It was like Dante's state of purgatory, where people were constantly in a state of motion, for to stop was to perish. For King, then, the modern world of technology was fated to be one of endless change, motion, uncertainty, and instability – in a state of chaos or disorder. He realized that technology had both created this state of ceaseless activity and now was the means by which it continued as the 'normal' state of being. There is a sense of resignation toward technology – a sense of being trapped by the very technology that promised freedom, liberation, and repose – that is absent in Keefer's perspective. The promise of the morrow that Keefer saw in technology had become a tarnished dream for King and his generation. The Great War during which technological warfare had resulted in massive killings and untold destruction, and the looming threat of industrial warfare that King feared, were certainly factors explaining King's more jaundiced view of technology in the early twentieth century. Equally important, given King's dualistic perspective on life, was Freud's 'discovery' of the irrational side of human nature. King saw technology as a process of industrialism, with its impersonalization of the individual, its large-scale and desensitized organizations, and its system of values that put economic and material goods above human and spiritual concerns, as the curse of the modern age of technology, a modern hell. It was a world in which humans were alienated from nature, from each other, and even from themselves. It was an inhumane world.

When King sought an explanation for this modern predicament, he looked neither to the Bible nor to the writings of theologians but to the modern theories of psychology, particularly William James's theory of pragmatism, a psychological theory that was especially appropriate for the modern industrial age. King quotes James at length in *Industry and Humanity,* especially James's explanation that both the Great War and the industrial unrest the war unleashed were due to 'human blindness.' King found the following passage from James's *Talks to Teachers on Psychology: And to Students on Some of Life's Ideals* particularly insightful: 'One half of our fellow-countrymen remain entirely blind to the internal significance of the lives of the other half. They miss the joys and sorrows, they fail to feel the moral virtue, and they do not guess the presence of the intellectual ideals. They are at cross-purposes all along the line, regarding each other as they might regard a set of dangerously gesticulating automata, or, if they seek to get at the inner motivation, making the most horrible mistakes.'[23] James's theory that unrest in the world was the result of human indifference and insensitivity of human beings toward each other – treating one another as mere machines or 'gesticulating automata,' as he described it – fit well with King's own perspective that industrial unrest resulted from the dehumanizing nature of technology that alienated and desensitized individuals to each other. It was King's realization that the technological imperative and the moral imperative were in a state of tension, that the values that were operative in the modern world of technology were not the moral values most needed for that world to function humanely.

While King garnered the insights into the theme of technology as industrialism from reading Mary Shelley's *Frankenstein,* Frederick Philip Grove acquired them through the writing of his own novel *Master of the Mill* (1944), in which technology as a process of industrialism is the central theme. Grove begins his novel with the mill well established and the current owner of the family business, Samuel Clark, a senator, reviewing his life as 'the master of the mill' as he prepares to die. At the outset, Grove presented opposing views of the mill – the liberator or the enslaver of humankind – reflective of the contradictory views of technology of the interwar generation, evident in George Sidney Brett and William Lyon Mackenzie King. Grove wrote:

> To many people, as the old man was aware, that mill stood as a symbol and monument of the world-order which, by and large, was still dominant; of

> a ruthless capitalism which had once been an exploiter of human labour but had gradually learned, no less ruthlessly, to dispense with that labour, making itself independent, ruling the country by its sheer power of producing wealth.
>
> To others, fewer these, it stood as a monument of a first endeavour to liberate mankind from the curse of toil; for it produced the thing man needed most, bread, by harnessing the forces of nature ...
>
> To still others, fewer again, the old man among them, it was the abode of gnomes and hobgoblins, malevolent like Alberich, the dwarf of the Rhinegold, but forced, by a curse more potent than their own, to do man's work. The uncanny thing about it was that these gnomes and hobgoblins – or were they jinn? – had the power of binding man to their service in turn, or to the service of the machines, as he, the old man, had been bound.[24]

Here was Grove's version of the theme of Shelley's novel of technological monsters taking on human form, promising freedom from want by 'harnessing the forces of nature.' In *Master of the Mill,* the servants of humankind end up as masters, binding humans 'to the service of the machines, as he [Samuel Clark], the old man, had been bound.' Very early on in the novel, Grove makes the reader aware that the title of the novel is to be taken ironically, that the mill *is* the master. All who come into contact with it, including the three generations of the Clark family who own it, become enslaved to it. Such a theme was popular in many late-nineteenth- and twentieth-century novels relating to technology because the theme of master/servant could easily be understood as a reversal of roles, in which the servant became the master.[25] Also, nineteenth-century enthusiasts of technology, as noted in the views of Keefer, Haliburton, and Fleming, heralded machines as perfect servants, delivering humankind from necessity and want. Thus, for twentieth-century critics of this perspective, it was useful to use the master/servant metaphor in dealing with the theme of technology but to reverse it, making humankind the servant, serving the needs of the machines in their relentless drive to dominate as the technological imperative. Grove used the metaphor to show in the novel that giving technology the power to produce more and cheaper goods for the material comfort of humankind ultimately meant the abandonment of traditional values and the suppression of the spiritual side of human nature – the moral imperative – to satisfy the relentless drive of machines to control humankind. Equally, in 'liberating' humankind from material want, technology 'enslaves' humankind to a worse form of bondage, to mechanical values that are the antithesis of the spiritual component of human nature. Here was Grove's rendition of the antithesis of materialism and spiritualism within industrialism that concerned King in *Industry and Humanity*. It was also his attempt to come to terms with the tension between the technological imperative and the moral imperative.

Grove presented the theme of liberator/enslaver in a dialogue between Samuel Clark, the mill's owner, and Bruce Rogers, his foreman. When asked by Clark to account for the discontent among the mill workers when he introduced new machinery into the mill to ease their burden of work, Rogers replies, 'They want to feel human. They're slaves ... Slaves to the machines. They must keep pace with the machines; or something happens. When they quit, at the end of their shift, they are deadbeat ... It's nervous exhaustion. Physical rest doesn't remedy it.'[26] When pressed to explain further, the foreman replies:

> Suppose a new hand starts work with us. He's an ordinary human being; he laughs and jokes as he goes to work. But within less than a year something comes over him. Whatever he does, he seems to do automatically; in reality, the pace forces him to be constantly on the watch; it isn't that he becomes a machine; that would be tolerable if undesirable. What he becomes is the slave of a machine which punishes him when he is at fault; the machine seems to watch for the chance. All the time. The men are tempted to yell and to curse at it. And then he is spoiled for anything else.[27]

The attributing of human qualities to machines reinforced the analogy of the relationship of technology to humankind to that of master and servant.

When Samuel Clark asks Bruce Rogers if there is any way out of the conundrum, his 'servant' foreman replies, 'Before the machine we're all equal, as we're supposed to be before the law. My father thinks we're still working under the eyes of a boss who appreciates good work. But the machine is the boss now; it doesn't promote; it just exacts motion.'[28]

Like Brett and King, Grove saw the power of technology in its ability to elicit – indeed to exact – 'motion.' There is no free will, no inherent desire, on the part of the individual to accept motion; it exists as a given, a necessary condition and by-product of technology. Keefer could write on the eve of Canada's technological revolution in the mid-nineteenth century of the freedom of Canadians to overcome their inertia, their apathy, by adopting railways and in so doing begin to progress to a higher state of civilization, but for Grove and his generation motion was not a choice; it was a binding requirement of technology. And once accepted, humans had no power to control it or reject it. *Motion,* he realized, became the source of technology's power, the means to enslave all who got caught up in the frenzy of industrialism. By moving to the rhythm of the machines, humans took on the qualities of the machines as their servants. Rather than autonomous beings, humans become automata, mimicking machines, and taking on the qualities of the machine; men and women become its subordinate parts, whose 'functions' were strictly 'mechanical.'

Grove has the three-generation owners of the mill represent a different view of technology. Rudyard Clark, the man who turned the one-man grist mill of his father Douglas Clark into a mammoth operation through dishonest financial means, represents those who see the power of technology to bring wealth and riches to those who use it for their own selfish benefit. Samuel, Rudyard's son, represents those who see technology as an agent of social reform. He automates the mill in hopes of freeing its employees from physical drudgery and in so doing 'direct the fortunes of the mill for the good of mankind.'[29] His son, Edmund, represents those who want to use technology to create a 'Brave New World' in which machines run everything and people live in 'a toilless heaven,' a world in which, Edmund proclaims, there would be 'only one master, one god: the machine.'[30] Grove questioned the validity of all three perspectives by having the representative of each perspective fail to achieve his vision. Rudyard is slave to the mill rather than its captain and dies prematurely of a heart attack while being blackmailed by his accountant, who knows of his illicit acts that had enabled him to turn his mill into a profitable enterprise. Samuel succeeds in turning his mill into a self-regulating operation that eases the physical drudgery of the workers only to be attacked and abused by the workers, who are convinced that he wants to put them out of work. Edmund is killed at the young age of twenty-five by a striker's bullet when the workers rebel against his attempt – upon wrestling power from his father – to implement his fully automated mill where workers were no longer needed. By the end of the novel, only the mill remains, a testimony to the power of technology. Like Mary Shelley's monster, the mill is a human invention out of human control that ends up destroying its creator, user, and even defender. The mill as symbol of technology reigns supreme; all who come into contact with it become its victims and slaves.

Yet Grove was not prepared to present only this negative perspective on technology. He refused to simply dismiss technology as a dangerous and harmful thing. At the end of the novel, in those final moments of reconciliation and lucidity before succumbing to death, Samuel Clark presents his final testimony on the mill:

> Five hundred years from now it [the mill] would remain; and probably it would stand as the monument of a time which, in retrospect, appeared to have been a great time: a time of the building of a novel sort of civilization, raised on the idea of a magic liberation of man from the curse of labour, setting him free for greater and higher tasks. There would still be men working in that mill; but they would not be slaves; they would be masters, benevolent masters, bound to their task by some twist of heredity which made their work the supreme fulfilment of their being.

> Yes, that mill would live on even though it might crumble; it would live on in the thoughts of man, for it had demonstrated the possibility of a new way of life.
>
> The mill ... That mill was the justification of his life: the mill he had loved and hated.[31]

Like King, Grove had an ambivalent perspective on technology: he held on to the pre-war optimism that technology would create a better world, a higher form of civilization, but was all too aware of its dark side, which brought out the demonic side of human nature. For King and Grove and the generation of the interwar years, technology threatened to become master and to make human beings its servants.

William Lyon Mackenzie King's perspective on technology and his compulsion to warn the world of the dangers of the technological imperative to the moral imperative that had guided Western civilization for centuries came from his background and the ideas that he had acquired from his education. The fact that King was the grandson of William Lyon Mackenzie, the leader of the Rebellion of 1837 in Upper Canada, reinforced his faith in the moral imperative. King maintained that his upbringing instilled in him a moral responsibility from an early age to carry on in his grandfather's footsteps. He saw himself as a rebel with a mission to right the wrongs of the world as he believed his grandfather had done. In his diary, he reflected on the importance of his grandfather to his own sense of purpose: 'Reading the life of my dear grandfather I have become a greater admirer of his than ever prouder of my own mother and the race from which I am sprung ... I have greater desire to carry on the work he endeavoured to perform, to better the conditions of the poor, denounce corruption, the tyranny of abused power and uphold right and honourable principles.'[32] King believed that, in his grandfather's days, the 'wrongs' were the injustices that the farmers experienced at the mercy of the large landowners and the commercial interests; in his days, it was the industrial workers who suffered at the hands of the factory owners and the capitalists.

King spent a great deal of time as a youth reflecting on his future career. He was pulled between a more spiritually oriented career, such as the ministry or as a settlement-house worker, and a more practically oriented one such as an economics professor or a politician. This once again reflects King's dualistic perspective on life. What he never doubted was that the career he chose had to enable him to fulfill his desire to help the poor and the labouring class. In a number of diary entries, he made comments like the following: 'I am going to make a careful study of the poorer classes and the worst social evil with a view to remedying to some degree the latter, and bettering the condition of the former';[33] 'I feel more anxious than ever to work at Economics most thoroughly and seek to learn all I can of the masses, the labouring classes and the poor, to understand their needs and desires and how to alleviate them, and better their conditions.'[34] In the end, he chose a career first in the civil service when the position of deputy minister of the newly created Department of Labour was offered to him by William Mulock, a family friend and a minister in the Laurier government, and then a political career when he ran for Parliament for his home constituency of Berlin (later Kitchener), Ontario. However, he continued to cultivate his spiritual side, as is evident in his involvement in séances in later life. To the end, he attempted to marry these two worlds.

It is likely that King's middle-class background played a decisive role in his choice of career. The rise of the middle class, a by-product of the industrial age, saw its purpose and sense of power in shaping society by instilling the proper moral values, *their* middle-class values, on the other classes within society. Their advantage as a class was their 'middle' position, identifying with neither the upper class of capitalists and entrepreneurs nor the lower class of workers but trying to find a means of reconciliation. Here too the middle class was at an advantage because what they had to offer were 'arguments and exposition,' as Christopher Lasch put it,[35] that their university education, with its mixture of theoretical and practical perspectives, provided. As a well-educated youth from a middle-class professional family, King certainly believed that he and other reform-minded members of his social class were the promise of the future for solving the social problems, particularly the 'labour problem,' as it was called, of the day. In a eulogy King wrote about his friend and confidant Henry Albert Harper, who drowned attempting to rescue a young woman who had fallen through the ice on the Rideau Canal, he quoted with approval Harper's perspective on what political economy courses should teach so as to prepare the next generation of youth to deal with social problems in an industrial age:

> I feel more and more the necessity of emphasizing the importance of the scientific study of economic and political problems in a country in which every man has the franchise, and is supposed to be in a position to express an intelligent opinion upon public questions, and particularly at a time when labour and kindred problems are prominent in the public mind.
>
> The poor downtrodden have more to hope for from men who, having a specialized training in the operation of social forces, apply themselves to the proper remedy, than from all the windy, ultra-radical demagogues.[36]

King's educational background prepared him well for his self-imposed mission. He obtained an undergraduate degree in political economy at the

University of Toronto at a time when the leading professors in the department were shaping the discipline to better enable it to deal with the new industrial age.

In the late nineteenth century, the discipline of political economy was undergoing a shift from a moral science to a social science, from a study concerned with the moral imperative to one that dealt with the technological imperative. In 1892, in his book *Political Economy and Ethics,* J.G. Hume, one of the early enthusiasts of political economy at the University of Toronto and a member of the philosophy department, appealed for 'a closer alliance ... between the study of Political Economy and Ethics'[37] by situating the discipline in his department. However, G.W. Ross, the Ontario minister of education, decided to create a separate department of political science. Ross maintained that the discipline had to deal with 'social and constitutional problems which require particular attention in a rapidly expanding country like Canada'[38] and believed that this could best be done if the discipline was free of moral persuasion. Thus, the belief that political economy should be a social science rather than a moral science was in keeping with the further belief that political economy should be a discipline that addressed the social problems the modern world of technology imposed.

William James Ashley, the first holder of the Chair in Political Economy and Constitutional History (the chair title was suggested by Ashley himself) agreed with Ross's perspective. Ashley was King's professor for one year at Toronto before he took a position at Harvard University (and the two men would reconnect when King enrolled as a PhD student at Harvard in the fall of 1897). Ashley believed that the discipline had to offer the working class something better than the platitudes an earlier generation of economists had provided and help integrate workers into the new industrial order. In his inaugural address on assuming the chair at Toronto, Ashley expressed quite frankly that the subject of political economy 'stank in the nostrils of intelligent working men. Mechanics' Institutes had been fed up on it for half a century to show artisans that everything in the industrial world was for the best; or, at any rate, that it could not be improved by combination, or by the interference of the State.'[39] Political economy had to become a realistic subject that addressed the problems of the industrial and technological age head on, to offer practical solutions if it was to be a meaningful subject for the modern age of technology. Ashley explained what he meant in more specific terms in a letter to his fiancée: 'Therefore, it seems to be that the work of the *Economist* should be, (i) the investigation of economic history – no facts are too remote to be without significance for the present, and both Lassalle and Marx have given a great impulse to investigation in this direction, and (ii) the examination of modern industrial life *in the piece.*'[40] Ashley was an advocate of trade unions. While he admitted their weaknesses in 'remedying social inequalities,' he believed them to be 'the only means towards that end.'[41]

James Mavor, Ashley's successor as chair and the other professor who greatly influenced King at Toronto, was of the same opinion as Ashley concerning the direction of political economy. It needed to be practical in approach and to address current social problems on an empirical and investigative level. He saw the role of political economy as producing a generation of social scientists who had both empirical knowledge of the industrial society along with an understanding of the major perspectives on social questions so as to go out into society as agents of social change and social reform.[42] To this end, he introduced original research even at the undergraduate level. Mavor's assistant, S.M. Wickett, noted, 'At Toronto the honor economic students are obliged to write as many as four 'term essays' on specified subjects. The first of these essays regularly treats of the "Social Conditions and Resources" of the district whence the student comes, and at times furnishes some valuable sociological material, such as typical household budgets, etc.'[43] Such an approach appealed to young Willie King and reinforced his social reform zeal while at the same time offered a practical side to social change.

Mavor's views on how best to solve the social problems of the industrial age changed at about the time King had him as a professor. In his early years, Mavor actively participated in a number of campaigns for social reform and advocated Fabian socialism as a solution to social ills. However, such activities only caused him to question the validity of this approach. By 1900, he had become one of the fiercest critics of socialism; he had also abandoned his social reform zeal and activities to devote himself instead to empirical studies of industrial society as an impartial observer, what he believed to be the true role of the social scientist. As well, he had come to the conclusion that the best means to achieve social reform would be through the moral reform of the individual. This meant that initiative for change had to come from the individual, not from the state. Furthermore, the most effective leaders of social change would have to come from the educated middle class rather than from the ranks of the working class.

Mavor had also altered his perspective on the industrial age from a negative to a positive one. He admitted that the modern industrial society had lost the old medieval sense of constraint, social hierarchy, and stability, but he argued that it gained material well-being and individual freedom instead. Mavor noted that 'the standard of comfort of the mass of people of the western races has risen during the past fifty years,'[44] thanks to technology. Thus, Mavor became an apologist for industrialism and no doubt conveyed that enthusiasm to his Toronto students, including King. Where Mavor and King disagreed was not on their perspectives of society but in their approaches to change. As a student, King was still the young idealist who believed in social reform and advocated a moderate form of socialism; Mavor had nothing but contempt for this viewpoint by the time King had him as his professor.

The differing opinions of King and his professor reflected the changed perspective occurring with regard to social change within the social sciences. The new emerging social scientist believed in a dispassionate approach to social problems and the application of scientific theory, devoid of moral undertones, in resolving them. Problems arising as a result of the technological imperative required more sophisticated and rational answers than moralists and social reformers could provide. As an idealist and moralist, King still clung to the belief that social reform and the moral imperative were not only reconcilable but also necessarily linked. He believed that, without the undergirding of the moral imperative, the ethical values of the good society would be lost in an antiseptic application of scientific theory to social problems. King never lost this moral compulsion as he left his student days behind and entered the 'real world of technology,' as is evident in *Industry and Humanity*. But by raising the consciousness of his fellow social reformers to the dangers of severing the technological imperative from the moral imperative, King inadvertently contributed to that perspective.

However, the greatest influence on King's perspective on the industrial age and its impact on society while he was at the University of Toronto came not from the lecture halls but from the writings of Arnold Toynbee Sr., the founder of Toynbee Hall, a settlement house for the poor in London's east end. King had first read his *Lectures on the Industrial Revolution of the Eighteenth Century in England* in the summer of 1894 while vacationing in the Muskokas. He recorded his enthusiasm in his diary: 'I was simply enraptured by his writings and believe I have at last found a model for my future work in life.'[45] What he 'found' in the *Lectures* was a lucid account of the origins of the Industrial Revolution in England in the 1760s that emphasized the interconnection of ideas and practical inventions. One of Toynbee's favourite stories was about the close and beneficial friendship of James Watt and Adam Smith, an account that King used in *Industry and Humanity,* as already noted. Toynbee maintained that Watt's steam engine made possible 'the realisation of that freedom which Adam Smith looked upon as a dream, a utopia,'[46] but equally Smith's vision of the new industrial and capitalist society created the right intellectual climate for Watt's invention to be accepted. King cited the story as an example of how ideas and machines were both requisites of industrialism, and when both factors came into play in a positive way 'a world-wide development of human intercourse' resulted.[47]

Toynbee's approach to the study of industrialism impressed King in two other ways: he had an ability to relate the past to the present and to show the connection between material and spiritual concerns. Concerning the former, Toynbee emphasized that one goes to the past 'to seek large views of what is of lasting importance to the human race ... You must pursue facts for their own sake, but penetrated with a vivid sense of the problems of your own time ... You must have some principle of selection, and you could not have a better one than to pay special attention to the history of the social problems which are agitating the world now, for you may be sure that they are problems not of temporary but of lasting importance.'[48] King used a similar approach in *Industry and Humanity,* relying heavily on Toynbee's work for the historical perspective to the present age. Concerning Toynbee's other approach – the interrelationship of material and spiritual concerns – King noted in his diary what he liked about Toynbee's study of the relationship of church and state: 'One little passage impressed me greatly, it was in speaking of this world seemingly black with sin and degredation [sic] yet thro' it all we could see an undercurrent of the love of God. To my mind this has always been & is one of the most perplexing problems. What does all this evil mean, what is to be the result and what is to become of the larger masses of ignorant and thriftless persons! The idea of a hell too seems hard for me to grasp as I read of the great love the Father hath for all his children.'[49]

King kept going back to Toynbee in later life. While at the University of Chicago, he wrote in his diary on August 30, 1897: 'The rest of the day ... was spent in reading again "The Industrial Revolution" by Toynbee. I covered about 130 pages in all. The first chapters on the Old Political Economy I regard as splendid, they are good examples of his work. I was inspired by them. The Revol'n itself is full of facts but needed his pen to weave them together. The book I find I know very well, yet it has lost little in interest to me.'[50] Twenty years later, when preparing to write *Industry and Humanity,* he reacquainted himself with Toynbee, once again describing him as 'the finest influence in my life.'[51]

Toynbee's and King's views of industrialism show a number of similarities. Toynbee acknowledged some of the positive attributes of industrialism but mainly emphasized the negative repercussions, especially 'an enormous increase of pauperism' and 'a rapid alienation of classes' in hopes that such an awareness would lead to an improvement on the attributes of industrialism. King would do the same, setting out in the early part of *Industry and Humanity* all that was wrong with the present industrial age as justification for offering a solution in the remainder of the study. Toynbee also maintained that 'the most vital' consequences of the Industrial Revolution lay in the impact of 'external forms of industry upon its inner life,'[52] the negative impact that industrialization had on human values. 'These efforts were terrible,' Toynbee argued, breaking down the old personal relationship between master and journeyman based on the transmission of skills and replacing it with an impersonal one of capitalist and labourer based on the cash nexus. King would deal with this issue in the context of the tensions between material concerns and spiritual ones, the technological imperative and the moral imperative. As a 'Christian economist,' Toynbee also emphasized that industrial growth had to be tempered by a sense of moral and social responsibility. King could not have agreed more. Both Toynbee and King accepted a greater role for the state as the guardian of the industrial order, although neither believed that the state should overpower the role of the individual. Seeing the two as compatible institutions, Toynbee preferred an alliance of church and state. Interestingly, King seldom mentioned the church as a social institution, but he did concur with Toynbee that no distinction should exist between sacred and secular concerns when addressing social problems brought on by industrialism.

However, there were also differences between Toynbee's and King's perspectives on industrialism. Toynbee denied the existence of immutable laws of social science. He saw such 'laws' as only 'facts of human nature' based on empirical observation that were 'capable of modification by self-conscious human endeavour ... They are [therefore] relative for the most part to a particular stage of civilisation.'[53] Thus, Toynbee had an optimistic view of the future of industrialism, believing it to be 'a transitory phenomenon,' only 'the foundations for a new and higher stage of social existence.'[54] King was less optimistic. The laws of society, such as Pasteur's Law of Peace, Work, and Health and Law of Blood and Death, may be vague, but for him they were immutable and dominant. As noted earlier, King was also far less optimistic about human potential for good, and about social progress, probably as a result of writing in the shadow of the Great War, which for him belied any optimism in human nature. Indeed, Toynbee's writings disappointed King – and revealed themselves to be relative to an earlier and different age – in failing to provide some understanding of how this new industrial age, which held such promise for the morrow as the great age of technology in Toynbee's time, could wreak such havoc on society, both at the time of its advent in the Industrial Revolution and thereafter, a question at the centre of King's analysis of industrialism. Here King had to turn to modern psychologists like William James for the answers.

King graduated from the University of Toronto believing that his political economy degree had provided him with an understanding of the current theorists of the day on the new industrial age along with a practical knowledge by dealing with social issues, thus preparing him well to understand and guide the new industrial and technological society that was emerging in Canada. From Toronto, King went to the University of Chicago for further study, although he never acquired a degree from that institution. What he did acquire there, through his work as a volunteer in Jane Addams's settlement house, Hull House, was an awareness of the negative social repercussions of modern industrialism. In fact, shortly after arriving in Chicago, he moved from residence into Hull House to experience first-hand on a daily basis the plight of people from the slums of industrial Chicago who frequented the place.

King first met Jane Addams while a student at Toronto. She had come to give a series of talks in the city. He was taken by her from the start. He recorded in his diary on July 20, 1895, that he had gone 'to the Pavillion to hear Miss Adams [sic] of Chicago deliver an address on "the Settlement idea": I have never listened to an address which I more thoroughly enjoyed.' The next day he attended another of her talks in which she 'spoke on her work in Hull house.' What impressed him was the link she made with Toynbee and how rewarding such work was. 'The history of the movement, taking it back to the influence & practical work of Arnold Toynbee was more than delightful to me. I love Toynbee and I love Miss Adams [sic]. I love the work in which the one was and the other is & which I hope soon to be, engaged in.'[55] From the talks King concluded that 'there is [sic] certainly many burdens for the social reformer to remove.'[56]

Interestingly, King's initial enthusiasm for Hull House did not last. He moved out of Hull House and back into residence shortly after the initial switch, feeling the commuting distance to the university and his volunteer time were adversely affecting his studies. There is also a notable absence of references to his Hull House experience in *Industry and Humanity*. It was as though his intellectual studies and his practical experiences were not of equal importance and that the latter were only to reinforce his more intellectual approach to social questions and social reform.[57]

While King's desire to work at Hull House appeared to be his chief motivation for going to Chicago, he did meet one professor who greatly impressed him: Thorstein Veblen, the *enfant terrible* of American economists. Veblen was the first American economist to come to terms with the nature, impact, and significance of technology. He argued that technology and capitalism were intertwined to the detriment of the former. By technology, Veblen meant not only industrial arts proper, or production, but also the habits, skills, and values that underlay the technological age, what he referred to as 'the technological habits of thought.'[58] What he meant by the phrase was the complex interaction of factors, habits, and conventions that together created a powerful economic and cultural entity. In this respect, Veblen noted, technology was an 'institution' like capitalism. Veblen was critical of capitalism as being controlled by an elite group of financiers who were only interested in pecuniary profit as opposed to social improvement and who as the idle rich lived off the industrious workers. Technology, on the other hand, brought about material progress through productivity for the betterment of society. Indeed, Veblen saw the conflict between these two 'institutions' – technology and capitalism – as the cause of economic crises, depressions, and even war. Veblen clearly saw the problems in the industrial age as a product of capitalism, not technology. He attacked the capitalist system for providing an abundance of superfluous goods for a class of 'drones' – the idle rich – who with their predatory instincts purchased these superfluous goods as a reflection of social status and for causing a working class that suffered the repercussions of an economy of waste. Veblen put his faith in technology. He believed, as John Diggins notes in his study of Veblen's ideas, that 'men must work with the machine and abide by its processes, for the "discipline" of the machine purges all the anthropomorphic residue in man's nature. The "matter-of-fact" frame of mind is the "cultural incidence" of machine technology.'[59] Thus, Veblen advocated a technocracy, a government in which engineers dominated with their technological expertise and with their penchant for making 'things' instead of money.

King took Veblen's courses on economic theory and socialism, and they stood out as high points in an otherwise dreary intellectual landscape at the University of Chicago. He pronounced Veblen's lectures on socialism the best he had ever listened to.[60] What appealed to King about Veblen was his approach rather than his ideas: his bombastic and irreverent nature, his denunciation of the industrial elite of the day for its lack of social conscience, and his rationalist and trenchant analysis of economic and social questions. These appealed to King's 'rebellious' nature. What he did not like was Veblen's dismissal of religion as an irrelevant factor in understanding or dealing with social issues. Thus, when twenty years later the more 'conservative' King came to write *Industry and Humanity,* he made no reference to the writings of his popular Chicago professor.

Veblen did assist King in terms of ideas by requiring his student to be exposed to the ideas of all types of socialists, including Marxian socialists. King was obliged to read Karl Marx's *Capital* and Marx and Engels's *The Communist Manifesto* for Veblen's class. King found Marx hard going at first; it took him two hours to read twenty-six pages, but he found his writings 'very logical after getting into it.'[61] After reading Engels's *Socialism, Utopian and Scientific,* King admitted that 'there is much in it. There is something about Socialism which interests me deeply. There is truth in it – it is full of truth; yet much that is strange and obscure.'[62]

King would never return to the writings of Marx and Engels, but he did retain many aspects of Marx's view of the world while rejecting his solution to the problems that beset the industrial working class.[63] Marx divided the industrial world into the two dominant classes of the bourgeoisie and the proletariat, or the capitalists and the working class, or the oppressor and the oppressed. 'Society as a whole,' he and Engels wrote in *The Communist Manifesto,* 'is more and more splitting up into two great hostile camps, into two great classes directly facing each other: bourgeoisie and proletariat.'[64] King also saw the world divided into 'classes' or interest groups that were often in opposition to one another, although he believed they need not be. He would also have agreed with Marx that industrialism was a radical break from the past, destroying in its revolutionary change all remnants of the old feudal order, including the close association of master and journeyman, and replacing it with an order based on 'no other nexus between man and man than naked self-interest, than callous "cash payment." It has drowned the most heavenly ecstasies of religious fervor, of chivalrous enthusiasm, of philistine sentimentalism, in the icy water of egotistical calculation.'[65] King could identify with Marx's concern for the instability resulting from such dramatic change, captured in Marx's phrase, 'All that is solid melts into air.' While King did not coin a similar memorable phrase, *Industry and Humanity* is replete with references to the instability of the modern industrial world. Both men saw industrialism as ending the old local and rational approach to issues and replacing it with an international perspective and a disparate approach. King could also accept Marx's theory that ideas – what Marx called 'man's consciousness' – played as decisive a role in social change as material factors and that the former were the by-products of the latter and changed 'in proportion as material production is changed.'[66] Marx saw the world as having meaning from a materialist perspective only, whereas King saw materialism as only one component, the least important, of society. He believed that materialism had to be reconciled with the spiritual aspect of industrialism. Where King most definitely parted company with Marx was in the latter's denunciation of religion as 'the opiate of the masses' and his fervent belief that only bloody revolution could change the industrial system and end the injustices committed against the working class. King would have shuddered at such a thought, even in his more radical student days let alone in his more 'mature' years.

From Chicago, King went to Harvard on a fellowship, thanks in part to the support of his former Toronto professor Ashley, who had preceded King to Harvard and assisted King in getting there. Once there, however, King questioned Ashley's support of him. More appealing and memorable was King's association with three other Harvard professors: Frank W. Taussig, William Cunningham, and Edward Cummings. Taussig was King's PhD supervisor. According to economic historian Joseph Dorfman, he was 'first and last a follower of the classical tradition of Ricardo and John Stuart Mill.'[67] Thus, he exposed King to the traditional economic perspective on social issues that had not been emphasized by King's Toronto and Chicago economics professors. Taussig specialized in the wage fund theory, which argued, in essence, that the wages labourers received were controlled by the capitalists who dispensed them from the 'fund' when it proved profitable. The theory showed workers to be at the mercy of capitalists who, in turn, operated according to traditional capitalist principles. King offered his own reflections on the subject after reading Taussig's writings on it: 'As to the fact that wages must be repaid out of product I see no escape. Whether or not product is the measure of wages, or capital the measure, wh[ich] Taussig states to be the real point I am not sure. I believe that, as complexity of industry increases, it becomes increasingly true, that the laborer is dependent on the product of previous labor, viz. capital for his subsistence.'[68] King questioned whether capitalists needed such large accumulation of capital: 'I can't see why smaller accumulations in the hands of a number should not have as good an effect, if not better on Industry.'[69] Nevertheless, Taussig's argument caused him to rethink Toynbee's position on the subject and to admit, '[There is] much that Toynbee has missed, I remember when I thought the chapter [by Toynbee] a good one, now I would think it very faulty in parts.'[70] What King liked about Taussig, besides his ideas, was his support of his graduate students in an otherwise impersonal and intimidating academic environment. Taussig also reinforced the belief of King's Toronto professors, Ashley and Mavor, that political economy should address current social and economic issues. In proposing an approach to economics in the schools, Taussig emphasized that the discipline should combine 'a history of industry, especially during the last two centuries, ... with an explanation of the principles,' while the instructor provides 'abundant illustrations ... in the familiar facts of everyday life.'[71]

William Cunningham, a visiting professor at Harvard from Cambridge University, replaced Ashley, who was on leave. He was Ashley's mentor, a close acquaintance of Toynbee, and a specialist on the Industrial Revolution. Like Toynbee, Cunningham approached the topic from a historical perspective and within the context of Christian morals. His most famous study was the *Growth of English Industry and Commerce* (1882), but his most revealing study of the Industrial Revolution was his *Outlines of English Industrial History* (1895). Here he emphasized the importance of material prosperity for individual well-being and national grandeur. But he warned against making material prosperity the end rather than simply the means to a greater spiritual goal. Still, it is evident that, in his evaluation of the social and cultural impacts of the Industrial Revolution, Cunningham saw it as essentially beneficial.[72]

King was immediately attracted to Cunningham, a warm and caring individual, and liked his approach to economics, being that of a Christian idealist, since it offered a counter-perspective to the more utilitarian viewpoints of his other Harvard professors. Upon first attending Cunningham's course, King wrote, 'He speaks of the need of higher ideals and self-discipline as the great factor in individual progress. He is right. This man is having an influence on me. He is the sort of personality I have sought most, & I meet him now at the close of my course. A Christian Economist. The Harvard men present the Utilitarian point of view most strongly. Taussig is a strong utilitarian. It is well that Cunningham asserts so strongly the Christian point of view[;] it goes to show the completeness in life which w'd be wanting but for religion.'[73]

King took two sociology courses from Edward Cummings, one on the 'Principles of Sociology,' which he found to be a waste of time, and the other on 'The Labor Question in Europe and the United States,' which he felt had 'lots of good material.' King disliked Cummings's derisive comments about religion, but he found his analysis of the 'labour problem' interesting. He was taken by Cummings's definition of the labour problem as encompassing three aspects: the material, intellectual, and spiritual. As to the last of these, Cummings defined the spiritual as 'an existence which, ceasing to regard the individual as a means, looks to him, however deformed, and however defaced, as an end in himself.'[74] Such a perspective fit King's own social reform outlook and his empathetic approach to the working class.

Besides his course work, King was clarifying and enriching his understanding of the nature and impact of the industrial age through general reading. He was attracted to the writings of Charles Kingsley, a British social reformer, particularly his novel *Alton Locke: Tailor and Poet, an Autobiography* (1889), a story about a tailor who is involved in the 'sweating system' and experiments with different forms of socialism for a solution. King described it thus: '[It is] a wonderful book. I find my own feelings so admirably expressed in many of its pages I could not get but great inspiration from reading it. I feel I must get to work and *do* something.'[75] Kingsley certainly painted the sweating system in bleak and draconian terms. In a prefatory section entitled 'Cheap Clothes and Nasty,' he wrote, 'What is flogging, or hanging, King Ryence's paletot or the tanneries of Meudon to the slavery, starvation, waste of life, year-long imprisonment in dungeons narrower and fouler than those of the Inquisition, which goes on among thousands of free English clothes-makers at this day? ... In short, the condition of these men is far worse than that of the wretched labourers of Wilts or Dorset ... We have, thank God, emancipated the black slaves; it would seem a not inconsistent sequel to that act to set about emancipating these white ones.' What influence Kingsley had on King's own reporting on the sweating system in Canada as a reporter for the *Mail and Empire* in the summer of 1897 is difficult to say, but no doubt after reading the novel he would have been predisposed to see the sweating system as one of the evils of industrialism. And King would have been attracted to Kingsley's spirited defence of the church as the best agent for social change of the system.

King also read two French social theorists, one a psychologist, Gustave Le Bon, the other a sociologist, Edmond Demolins. Le Bon argued in *The Psychology of People* that ideas shaped civilizations and that the popular idea underlying modern European civilization was the equality of all beings. Le Bon questioned such a belief and argued instead in his study that individuals, like nations and races, each had their own position in a social hierarchy and that society only functioned properly when this hierarchy was recognized and allowed to function naturally. With regard to the United States, Le Bon described the Americans as a great people because they were democratic and industrious, creating 'great towns, schools, harbours, and railways.'[76] He placed a special emphasis on the 'genius of invention' of the Americans as indicative of the greatness of their nation and race who, like the inventors of 'printing, gunpowder, steam, or the electric telegraph,' have transformed the world and 'hastened the march of civilization.'[77] King concurred with Le Bon that each person should 'do that for which he is best fitted with a consciousness of individual responsibility to God for the use of his time & talents.'[78]

Edmond Demolins, the other French social theorist to influence King, looked to the individual for effective social action. 'Social salvation ... is like eternal salvation,' he wrote, 'an essentially personal affair; everyone must shift for himself.' Demolins was a student of Frédéric Le Play, the French engineer turned sociologist in later life, who was concerned about the impact of new technologies on civilizations.[79] Le Play's biographer describes him as 'the technologist with a social conscience.'[80] Le Play emphasized that technical decisions had moral consequences and that, in making those decisions, the delicate ties of personal relations must be considered. The best means to do so was through social surveys, and Le Play was 'the father of the social survey in sociology.'[81] Le Play admired English unions for their organized ability and statesmanship and encouraged the co-operative movement in France. He believed workers' efforts at self-help were their only means to protect themselves from the worst aspects of industrialism. At all times, he was concerned about the negative impact of technology on traditional values and social mores; his aim as a social scientist was to find ways of mitigating the effect of technology on 'the suffering classes' (Le Play's own phrase).

Demolins did not always agree with his mentor. He admired Le Play for his concern for the poor in the growing industrialization of society but criticized him for drawing premature conclusions. Their disagreement over the relationship of analysis and prescription caused Demolins to leave La Société d'économie sociale, founded by Le Play, to create his own society, La Société internationale de science sociale. Demolins' most famous study was *Anglo-Saxon Superiority: To What Is It Due?* in which he compared and contrasted English and French modes of education and traced their influence on the national character. In particular, he emphasized the spirit of the self-reliance of the English while denouncing the French for their tendency 'to increase the army of functionaries and to swell the intellectual proletariat.'[82] King liked Demolins' emphasis 'upon indiv'l enterprise & initiative, self-reliance etc., as against state and etc.' He saw this as 'the underlying thesis of the book.' He wrote, 'Here is an element of character and conduct & is rightly, I believe, at the bottom of the great social questions. Individ'ism vs Socialism considered from this point of view seems to give everything to the individualistic point of view. This sort of reasoning has won me over to the individualistic from the socialistic camp.'[83]

His formal studies completed, King set off on a research trip to England, France, and Germany to observe first-hand the workings of the most industrialized countries of Europe. Already predisposed against socialism, he became disillusioned with the ideology during his research trip, believing it to be too extreme. Instead, he embraced the co-operative movement in England as a viable alternative to state socialism because of the emphasis of the former on workers helping themselves and on individual initiative. He noted the virtues of co-operation for the working class in his diary: 'Co-operation has in it all the virtues, claimed for Socialism, without its defects; it is individualistic, all self help, self initiative, and self dependence, no government protection. I am greatly taken with the movement as the best thing seen yet to put the working class on a higher level, to make them good citizens and men, and to raise them above the plane of industrial strife which destroys and enslaves.'[84]

King's formal education at Toronto, Chicago, and Harvard had introduced him to the ideas of some of the major thinkers on the roots, nature, and impact of the modern industrial society. His readings and professors tended to emphasize an emotive as opposed to a rational approach to the subject. Thus, he came to see industrialism as being as much a value system that needed to be spiritual in nature as a material complex based on mechanical inventions. Certainly, his education reinforced and strengthened his natural inclination to see the world as consisting of the forces of good and evil, the spiritual and material, the idealistic and realistic. His own 'very double life' had already inclined him in this direction, and his formal education provided the intellectual justification and rationale for such a perspective. As well, his involvement in Hull House, his stint as a journalist for the *Mail and Empire,* and his study abroad on a Harvard fellowship provided a practical perspective to balance the more bookish approach to social issues arising out of industrialism that his academic program provided. Clearly, however, the more abstract and academic approach prevailed in the end, as evident in his discussion of 'the labour question' in *Industry and Humanity* and in the highly moralistic tone in the tome in general. Still, the combination of his academic and practical approaches to industrialism meant that he would emphasize both aspects, seeing industrialism as a mindset based on certain values and principles that it perpetuated as well as a material conglomerate driven by the machines that undergirded the system.

King's years in the Department of Labour as the deputy minister and as editor of the *Labour Gazette* journal from 1900 to 1908 reinforced his theoretical and practical approaches to industrialism. Much of his time was spent reading and writing government reports or writing articles of a factual nature for the *Labour Gazette*. This mundane work did not bother King, because, like his friend and confidant Henry Albert Harper, he saw it as part of a greater cause: to study social problems in the industrial age with a view to finding a solution that would ensure that industry served humanity. King quoted with approval Harper's perspective on the role of the civil servant in the Department of Labour: 'The work on the *Labour Gazette* allows opportunity for a careful and searching analysis of the industrial and social life of the Dominion. Already I can see the practical usefulness of the work ... With the added responsibility there has come to me an increasing sense of the usefulness of the work which we are doing. I believe we can do much towards determining the direction of social progress.'[85]

Still, King continued to read widely about industrialism while he was deputy minister, particularly those authors who dwelt on the importance of spiritual concerns in a society becoming increasingly more materialistic and secular. He gravitated in particular to the Victorian critics of materialism, most notably Matthew Arnold and Thomas Carlyle. He delighted in Arnold's attacks on middle-class philistinism in *Culture and Anarchy* and completely identified with Arnold's concern for faith in a society dominated by material values. Arnold used his writings to attack not only a class but also the values the members of that class espoused, which he believed were a by-product of the industrial age. *Culture and Anarchy* was in a way a warning about the chaos that industrialism unleashed in society, as the title of the book indicated and which Arnold immortalized in the concluding lines of his famous poem 'Dover Beach':

> And we are here on a darkling plain
> Swept with confused alarms of struggle and flight,
> Where ignorant armies clash by night.[86]

Arnold looked to the 'respectable' section of the working class, trade unionists, and others to provide both a counter-pull to the anarchistic element and guidance for the masses. Such analysis, both of impending anarchy in an industrial society and looking to an educated working class for the solution, reinforced King's own moralistic reflections on the industrial age.

Thomas Carlyle's more stinging and virulent attacks on the new industrial order compared to Arnold's also appealed to King. King could identify with Carlyle's fear in his day that the recent outbursts of violence among workers were the handwriting on the wall of impending revolution, for he too saw such a threatening possibility in labour unrest in his day. King would also have been attracted to Carlyle's emphasis on the importance of the spiritual aspect of life, particularly the author's appeal for a return to 'the finer moral and ethical values' of an earlier age. Carlyle's *Past and Present* was a study in contrasts: the present industrial age juxtaposed against the past feudal age.

Such contrasts complemented King's own perspective on the world. What Carlyle found to be the most identifying of 'the signs of the time' (in his essay by that title published in the *Edinburgh Review* in 1829) was a philosophy of 'cash-payment nexus' coming out of the emerging industrial age that denigrated the value of the human agent in society. Given King's similar perspective on the new industrial age, it is not surprising that he should describe Carlyle as 'the greatest soul the British Isles has yet produced.'[87] What is surprising in a way is King's failure to quote either Arnold or Carlyle in *Industry and Humanity;* it is as though Arnold's and Carlyle's reflections on industrial Britain in the nineteenth century had no bearing on industrial Canada in the twentieth century, despite King's attempt to see developments in his day as having a continuity with the past and in spite of the fact that he was concerned about the same issues raised by these earlier social critics, namely, the impact of industrialism on the working class and especially the impact on traditional moral and ethical values. It may have been King's desire to present a more modern 'social science' perspective that discouraged him from referring to moralists of the past, fearing that such moralizing would detract from the necessary objective, impartial, and authoritative approach of the modern social scientist.

In the end, when offering a solution to the problems of the modern industrial and technological age in *Industry and Humanity,* King was a traditionalist. He believed that the solution lay in regenerating the individual rather than in reforming society. Improvement had to come through a change in attitude on the part of each individual, from one of 'mistrust born of Fear' to one of 'trust inspired by Faith.'[88] Although there are hints in the book that the external world of technology shaped the internal world of the human mind to the point at which attitudes almost appeared to be 'fated,' he recoiled from such a perspective. King needed to believe in the indomitable human spirit and in the power of the individual to shape his or her life for the better and thus shape society for the good of all. For King the moralist, the story of Frankenstein had to have a happy ending. He believed that out of the chaos of the present age, out of the ashes of the Great War, out of the industrial unrest, would come a new beginning. The antithesis to chaos – that is, order – would re-emerge as the pendulum moved back from its destructive extreme to the constructive alternative. But the change would have to come not within a changed world order or from any change within the industrial system itself but from a changed perspective on the part of human beings: 'The epoch of scientific research will be followed by an epoch of reflection, and of search, perhaps, after the invisible realities of life.' Humankind would come to realize that it could not survive on bread alone, on the basis of material comfort; it needed the spiritual component. 'The materialistic interpretation of life has failed to give us progress according to any true meaning of the word,' King concluded. 'It has brought only death and desolation in colossal measure. We must begin anew with a spiritual interpretation of life, and out of the human service which it inspires seek to reconstruct our dismantled world.'[89] In this process of social reconstruction, society could utilize its powerful agencies of 'Discovery and Invention, Government, Education and Opinion'; but most of all the change had to come from within the individual. It would have to come from a change in attitude, from seeing technology (industrialism) only as a source of destruction and as an agent of materialism to seeing it as a potential for good and as an agent of spiritualism. This was as optimistic a conclusion as King could make about the modern age of technology. But given his negative sketch of the new industrial age that he put forward at the outset of *Industry and Humanity,* even this tempered conclusion was upbeat. In the end, the book showed King as a divided soul in his judgment of the modern world of technology: he wanted to have faith in its promise for the morrow, but he had the same fear he blamed others for having and that he saw as standing in the way of that faith. King clearly showed himself as a student of the modern age of technology, pulled between wanting to see technology as a positive force in the future but aware of its negative influence in the past and present.

Frederick Philip Grove was also a divided soul on technology. While he held out hope that technology might in the end liberate humankind, he was equally skeptical because, as an agent of change, technology invariably went beyond the immediate and obvious impact on society. In a reflective article, Grove noted his aversion to technology:

> Though I am a lover of music, I abhor the gramophone and the radio; though I love to talk to my neighbours, I will not tolerate the telephone in my house; though I am keenly interested in the events of the world and the thought of my contemporaries, I read neither the daily press nor the average magazine and very few books (though, truth to tell, I read about as many good ones as I can lay my hand on – a thing which is more fully explained in a little unpublished work entitled In Praise of Poverty); for exercise I resort to walking without carrying steel-tipped sticks with which to drive a hard little ball before me; for entertainment, to almost anything but games.[90]

He explained why he shunned modern technological conveniences by recounting in his putative autobiography *In Search of Myself* (1946) the adverse effect purchasing his first car in 1922 at the age of forty-three had on him:

> Just before leaving Eden, we bought a Ford car ... If I were to give a separate title to that part of this record which deals with it, I could only call it

> *Externalization* – meaning the externalization of a life which, so far, had been concentrated on the realization of purely internal aims. It meant the temporary degeneration of all my powers; and I am only just recovering from it, when, quite, possibly, it may be too late.[91]

Grove went on to recount how materialism could adversely affect not only an individual's but also a nation's moral values. He explained (in a perspective reminiscent of that of Harold Innis, writing about the same time) how he came to realize the danger of technology because he lived 'on the frontier' (what Innis would call 'the margin') of American civilization, the most materialistic society because it was the most technologically advanced:

> We are apt to forget the cost of material progress, the cost to the nation and to mankind, the cost in human happiness, in human life ... Every sky-scraper erected in the United States, every canal dug through every isthmus, every air-line opened up exacts its toll of human life – and who will evaluate the worth of a life that is lost? It exacts its toll in human happiness as well; for the sheer physical labour required to bring such things about can be supplied only by some sort of slavery. I am profoundly distrustful of what is called civilization. Perhaps one has to have lived – as I have done; as I am doing – on the frontier, or beyond the frontier, of a life, that is reasonably secure in order to understand why I call the present civilization the consolidation of barbarism; at least if security of life is acknowledged as the first postulate of what can legitimately be called civilization.[92]

Grove also warned that technology was destroying the very moral and spiritual values on which Western civilization was built. What technology ultimately came to represent for Grove was materialism, and materialism was the valuing of objects over ideas, machines over people. Grove came to associate that materialism with American civilization – that 'glaringly new and purely material civilization of our neighbours to the south,'[93] as he put it – as opposed to the spiritual characteristics of European civilization. Grove made this contrast explicit in his collection of essays, *It Needs to Be Said*. In the first essay, 'A Neglected Function of a Certain Literary Association,' Grove argued that the Canadian Literary Association was neglecting its need to resist the debasement of American literature that seemed on the whole to be aimed at the lowest common denominator – that is, mass society – and to continue to emulate European literature that valued quality over quantity, universal truths over immediate truisms, and everlasting spiritual values over current trends based on mere materialism. Grove went on to explain the roots and innate values underlying both of these opposing forces on Canadian literature and in so doing appealed to his fellow writers to choose European spiritual values over American materialism, the latter a product of a 'mechanistic, almost brutal spirit.'[94] He had come to realize the necessity of having to live *in* the modern world of technology but not having to be *of* that world, to accept technology without accepting the values that it imposed. Like King, Grove still clung to the moral imperative.

In their dualistic perspective on technology, *Industry and Humanity* and *The Master of the Mill* were tracts of their time. In the Introduction to a collection of essays on the perspective of technology in the interwar years entitled *The Intellectual Appropriation of Technology: Discourses on Modernity,* editors Mikael Hård and Andrew Jamison argue that 'the twentieth century has been marked by a schizophrenic relation to technological development. The technological optimism that served as the main story line through most of the nineteenth century in both Europe and North America was transformed, in the decades surrounding World War I, into a widely felt spirit of disillusionment.'[95] Technology was called into question and critically appraised for its ability to create the utopian society that technophiles of the nineteenth century in general envisioned. Certainly, William Lyon Mackenzie King and Frederick Philip Grove reflected this emerging critical view of technology in the post–First World War era. They saw technology as a process of industrialism, as promising but also foreboding. Technology was a double-edged sword that could cut a swath in either direction: toward beneficial or destructive ends. In its contradictory nature and the magnitude of power it unleashed, technology became for King, Grove, and their generation a sign of modernity, a portent of things to come. The Great War and the industrial unrest that followed it were both by-products of the modern world of technology. Such forces and the technology they represented could be seen either as the promise of the morrow, the beginning of regenerated humankind that would lead ultimately to a regenerated society, a utopia, or as the beginning of the end, the destruction of the world through further wars and social upheaval, an Armageddon. The modern world of technology was now seen as a world of contradictions, paradoxes, and tensions, an uncertain age. In their uncertainty of the age and their questioning of technology, King and Grove provided a perspective on the modern world of technology that shows a richness and depth absent in the writings of earlier exponents of technology and that prepares the way for a more sophisticated understanding of technology among the theorists of technology in the post–Second World War era.

6

Stephen Leacock and Archibald Lampman: Technology as Mechanization

In 1920, Stephen Leacock, the Canadian political economist-cum-humorist who taught political economy at McGill University in Montreal from 1901 to 1935 and produced some thirty-five books of humour in his lifetime, published a little book entitled *The Unsolved Riddle of Social Justice*. It first came out as a series of articles for the *New York Times* in the fall of 1919. All but forgotten today, it was an important study in its day on the impact of technology as a process of mechanization on society. In it, Leacock explored the nature of 'the age of machinery and power,'[1] an age that he believed began with the Industrial Revolution and culminated in the Great War. Leacock used his expertise as an economist and political scientist to show how mechanization shaped all aspects of modern society, including the emergence of 'technological values' that he believed were in conflict with traditional moral values. Leacock identified motion as the essence of the age of mechanization, a perspective in keeping with his contemporaries George Sidney Brett, William Lyon Mackenzie King, and Frederick Philip Grove and one that the international scholar Siegfried Giedion would develop more fully a quarter of a century later in *Mechanization Takes Command*.

The noted Canadian poet Archibald Lampman, one of the 'Confederation poets,' grappled with the same issue of technology as a process of mechanization in his poetry. Writing at the turn of the twentieth century when Canada was undergoing its industrial revolution, he used his poetry to raise concerns and alarm bells as to the possible negative repercussions of the emerging technological imperative on the quality of society and on the moral values that had been the mainstay of Western civilization. He believed that the industrial age was cultivating an insatiable appetite for wealth and an acceptance of the noisy industrial apparatus that would supply that wealth. Lampman identified many of the issues that Stephen Leacock would develop more fully in his serious writing and in some of his humorous writings a generation later when these issues had become more pronounced with the advancement of industrialization and mechanization. Lampman was also the first Canadian poet to address the issue of technology in poetry and to offer a social commentary on the impact technology was having on the individual and society. By the twentieth century, the ideas of poets and novelists on social issues, including technology, became commonplace. Thus, thanks to Lampman, the subject of technology moved into the realm of the arts and letters and from beyond the confines of the ivory tower into the public domain.

King, Grove, Leacock, and Lampman were all wrestling with a new phenomenon at least within Canada: the industrialization of society through the introduction of complex machines that formed part of a larger social and economic system – industrialism – that was affecting all individuals in some way and necessitating the restructuring of society. King and Grove were interested in industrialism as a social and economic system that was having an impact on individuals by undermining their moral values, values that had guided the motives and aspirations of individuals in Western society for centuries. Leacock and Lampman focused on the process of mechanization itself: how the mode of operation of machines as a process of mechanization had an impact on society, particularly on the poor and the working class. Both Leacock and Lampman feared that technology as a process of mechanization was satisfying the demands of only the wealthy class and not addressing the needs of the poor. Like King and Grove, they were concerned about the impact of social changes that technology was bringing about on moral values. They noted that technology desensitized individuals to the plight of the less fortunate in society and in so doing undermined the concern for others that both men believed to have been at the heart of the moral imperative of Western civilization.

Lampman identified motion, a process of change that was transforming the world in a multiplicity of ways, as the defining characteristic of technology. He dwelt in particular on the negative impact of the motion of machines on nature. He pointed out that industrialization as machines had invaded the tranquility of nature and was depriving humans of a wellspring of spiritual rejuvenation. In his poem 'At the Ferry,' for example, Lampman introduced the mechanical sounds that invaded an otherwise peaceful scene, 'noises upon listless ears': 'The rumble of the trams, the stir / Of barges at the clacking piers; / The champ of wheels, the crash of steam' cause a 'troubled dream.' In the distance the 'murmur of a railway steals, / Round yonder jutting point the air / Is beaten with the puff of wheels.' Near at hand is a mill that 'with changing chant, now hoarse, now shrill / Keeps dinning like a mighty hive.'[2] Railways and mills symbolized for Lampman the new industrial age.

In 'The Railway Station,' Lampman explored the inner world of these 'temples of industrialism.' Frantic with motion, the railway station allowed for no rest or repose, no spiritual rejuvenation, no religious experience, unlike traditional 'temples' of faith. 'The flare of lights, the rush, the cry, and strain. / The engines' scream, the hiss and thunder smite' – these were the sounds and sights that bombarded those who entered. As 'the great train' moved, 'labouring out into the bourneless night,' it carried within 'its dim recesses' the souls whose 'histories,' 'thoughts,' and 'dreams' became lost in the whirl of 'hurrying crowds.' Lampman juxtaposed the railways as conduits of communication, which Keefer and Haliburton had praised in their writings, with the loneliness and silence of the passengers, drawn into themselves, their souls distressful and anguished, without peace, just as their bodies were without rest.[3]

Lampman also focused on the emerging industrial cities. In 'The City,' written in 1892, he described the city as a pleasant place on the surface with its 'gay and wide' streets, its walls 'high in heaven,' and 'fair as the hills at morning,' bathed in sunshine. But the sunlight was 'a gloom of warning / On a soul no longer free.'[4] And beneath the glitter lurked the cursed money lenders, as mechanical in heart and mind as the machinery that kept the city moving, indifferent to the needs of the less fortunate of its citizens. The repetition of the opening stanza as his concluding stanza reinforced Lampman's critical view of the industrial city.

Canst thou not rest, O city,
  That liest so wide and so fair;
Shall never an hour bring pity,
  Nor end be found for care?[5]

As fervently as Leacock would do a generation later, Lampman attacked the wealthy class, the new plutocracy, that industrial civilization spurred on. In his poem 'To a Millionaire,' Lampman depicted the millionaire as 'a creature of that old distorted dream / That makes the sound of life an evil cry.' He then contrasted the millionaire's 'dream' of wealth with the reality for the 'vain multitudes' whose wretched lives were the by-products of the millionaire's dream.

But I
Think only of the unnumbered broken hearts,
The hunger and the mortal strife for bread,
Old age and youth alike mistaught, misfed,
By want and rags and homelessness made vile,
The griefs and hates, and all the meaner parts
That balance thy one grim misgotten pile.[6]

In 'The Land of Pallas,' Lampman envisioned an idyllic and utopian world, a time 'Of peaceful days; a land of equal gifts and deeds, / Of limitless fair fields and plenty had with honour.'

> And men lived out long lives in proud content
> unbroken,
> For there no man was rich, none poor, but all were
> well.
> And all the earth was common, and no base
> contriving
> Of money of coined gold was needed there or
> known,
> But all men wrought together without greed or
> striving,
> And all the store of all to each man was his own.[7]

In his idyllic world, he imagined all children as being equal: 'But all the children of that peaceful land, like brothers, / Lofty of spirit, wise and ever set to learn / The chart of neighbouring souls, the bent and need of others.'[8] Lampman pointed out that to make this dream a reality would require a social conscience for the emerging industrial age, particularly sympathy for the children of the poor.

At the end of the poem, Lampman leaves this dreamlike world to return to 'reality':

> A land of baser men, whose coming and whose going
> Were urged by fear, and hunger, and the curse of
> greed.
> I saw the proud and fortunate go by me, faring
> In fatness and fine robes, the poor oppressed and
> slow,
> The faces of bowed men, and piteous women bearing
> The burden of perpetual sorrow and the stamp of
> woe.[9]

The City of the End of Things, considered Lampman's finest poem, depicted his dystopian view of the industrial city. People toil robotically in factories without purpose until they become mechanical in body and mind, devoid of 'flesh' or 'bone' but with 'iron lips.' Yet it 'was not always so.' Once, it had been a city of 'pride,' he wrote. But then it achieved its industrial 'might.' A will to power through technology took hold, dispensing the men who had built it up to such heights. Only four of the city fathers remained, three 'in an iron tower,' the fourth, the 'idiot' at the city gate. Eventually the 'hand of Time' would leave only the 'grim Idiot at the gate,'[10] a portal to a city that had gone mad. Literary critic W.E. Collins points out that for Lampman the city was 'an epitome of a mechanical universe which rolls round without purpose, heart or mind, grinding out life and death mercilessly and forever.'[11]

Lampman looked to socialism as the means to bring humane qualities to the modern industrial and technological society. In his essay on socialism, he described it as 'the cause of love and hope and humanity,' in contrast to 'competition,' which was 'the cause of anarchy, pessimism and disbelief in a possible manhood for human nature just emerging from its barbarous infancy.' He posed a riddle that underlay the modern industrial and technological society, not unlike the 'riddle' that Leacock would identify in *The Unsolved Riddle of Social Justice*. On the one hand stood 'a million human beings – the noblest product of the forces of life – most of whom would work or would once have worked for a living, if work were possible – either unemployed and starving or toiling for a wage utterly inadequate to the maintenance of a fair and human existence – many of them living in conditions of unspeakable horror and degradation.' On the other hand stood 'miles upon miles and tract upon tract of [land] unoccupied and untitled, ready to produce food and every material of life for multitudes vastly times greater than those now subsisting upon it.' Why could the two not be brought together, he queried, 'this starving people, a land full of fruit and material waste and unutilized'?[12] The answer came from the cold and calculating economists: the 'law of supply and demand.' This law – what Lampman described as 'a wrong and inhuman principle' that lay 'at the bottom of our whole industrial system' – took precedence over human needs. He went on to explicate what beliefs lay behind the economic theory of supply and demand: 'the principle that the private individual may take possession of the common earth and use it in any way he will for his own advantage; the principle that one man may buy other men into slavery, for this in the process of competition is what it amounts to in the end.'[13] Thus, for Lampman (and also Leacock), the fundamental problem of the industrial age rested on the fact that the wrong values were being applied to society – that is, material values of competition, wealth, and power as opposed to spiritual values of love, hope, and brotherhood.

Lampman identified the problem that Leacock would pursue in greater depth: To what extent did the mechanical age produce cold-hearted robotic human beings with their 'technological values'? Lampman saw his mission in his poetry to signal the potential dangers to the stability of society if such attitudes prevailed. He reminded his readers that society was at a crossroad with two alternatives: 'the competitive plan and the collective.' The first had already been tried and found wanting; it was time to try the latter. He warned, 'If this should fail then we shall have to agree with the pessimists and acknowledge that it is all a mistake; that life is a failure and not worth living.'[14]

Lampman argued that the great leaders of Western civilization had not been the wealthy entrepreneurs – 'the worldlings who filled their coffers with gold and sat in the proudest seats' – but the intellectual, spiritual, and inventive individuals, such as Socrates, Erasmus, Galileo, Columbus, and Watt and Arkwright. Regarding the latter two, Lampman wrote, 'the fruit of whose discoveries in [word illegible] was reaped by other and meaner men.'[15] Lampman ended his essay on an optimistic note, believing in the inevitability of socialism essentially prevailing in the modern world of technology: 'The change will work itself out gradually and intelligently from possibility to possibility.'[16]

Stephen Leacock's foray into the world of mechanization began with *The Unsolved Riddle of Social Justice,* a study about motion. The book examines the radical changes that came over the world with the advent of machines and a process of mechanization. It is also a book 'in motion,' because it is written in a way that enables the reader to become a part of the motion, caught up in the radical change that Leacock describes. He evokes and enacts the rhythm and frantic pace that machines impart to every aspect of modern life. The reader is drawn into the maelstrom of the modern technological world, into its stream, hurtled along, out of control, both excited and frightened by the onward rush. Leacock used active 'mechanical' verbs and short staccato sentences to capture the sounds and rhythms of the machines that lay at the heart of the new mechanical age. 'The machine penetrated everywhere,' he wrote, 'thrusting aside with its gigantic arm the feeble efforts of handicraft ... The quickening of one part of the process necessitated the "speeding up" of all the others ... Mechanical spinning called forth the power loom. The increase in production called for new means of transport. The improvement of transport still further swelled the volume of production.'[17] Everything was forced to change to accommodate the Machine. Nor was there any end to the changes and activity. Leacock predicted that, in a world of full mechanization, 'the wheels would never stop. The activity would never tire. Mankind, mad with the energy of activity, would be seen to pursue the fleeing phantom of insatiable desire.'[18] For Leacock, technology was like a stimulant: once a person used it, the urge to use it all the time – its 'insatiable desire' – brought the individual, or all of humankind, under its spell. The reader senses the power and magnetism yet potency and danger that machines engender; he or she is caught up in the action, a participant in the frenzied activity of the age Leacock analyzes. It is a 'moving' book by a passionate social critic of his time.

However, Leacock was unsure of where he stood on this new age, on which side of the evaluative ledger of the modern mechanical world he would place himself. Not only was this modern age paradoxical – a 'riddle,' as the title of the book conveys – but also Leacock's thoughts are full of contradictions about the age. He vacillated between praise and condemnation. In this respect, his views reflected the period in which he wrote: the indecisive, agonizing, paradoxical, and contradictory verdict of the critics of technology in the interwar years. In the aftermath of the most technologically sophisticated yet most destructive war in history, neither Leacock nor his contemporaries could return to the optimism, the faith in technology, the unquestioning belief in progress, that prevailed in the thinking about technology in the late nineteenth and early (pre–First World War) twentieth century.

To the generation of the interwar years, blind faith in technology was not possible. They had not yet reached the point of seeing technology as all pervasive, as beyond human control, as some theorists of technology in the post–Second World War era believed. What makes the critics of technology in the interwar years – George Brett, William Lyon Mackenzie King, Frederick Philip Grove, and Stephen Leacock – so fascinating is the ambiguity, the uncertainty, the tensions and contradictions their writings convey on the subject of technology; theirs is an uncertain world in which technology still held out a sense of hope but equally a sense of dread and fear. Theirs was the first generation to have to come to terms in a fundamental way with the contradictory nature of technology: its constructive but equally destructive nature. George Brett linked the contradictory nature of technology to the rational and irrational spheres of the human mind. William Lyon Mackenzie King associated it with the duality of human nature, its creative and demonic sides. Stephen Leacock attributed it to the conflicting values underlying modern society, values of materialism versus spiritualism. But they were all united in their awareness of the paradoxical and contradictory nature of technology as a process that was coming to dominate society – an imperative – and forcing individuals to compromise their morals to fit the new technological age.

Leacock revealed his uncertainty and concern about the modern technological age in his opening sentence. 'These are troubled times,' he wrote. The upheaval of the Great War had left the world 'desolated'; following on its heels was the threat of worldwide industrial unrest that could be more destructive than the war itself. Looming over the horizon was the spectre of the Russian Revolution, which had thrown the Russian nation into 'chaos' and threatened to spread, like an 'infection,' to the rest of the civilized world. Who was to blame for this industrial unrest? Leacock was uncertain. In one breath, he blamed the workers for demanding excessive wages; in the next breath, he criticized the capitalists for wanting unreasonable profits. The battle between them appeared to be a by-product of the Great War itself, based on the same unreasonable demands, expectations, and distrust on both sides, a point in keeping with King's analysis of the war. The war, Leacock noted, had cost the world billions of dollars of debt. Reason would dictate that such destructiveness, waste, and debt would bring the industrial world to a grinding halt, but it did not. In fact, the world economy seemed to be moving along as usual, another contradiction. How could one make sense of such contradictions? What role did technology play in these contradictions? The world seemed to be on a precipice – 'on the brink of the abyss' – ready either to leap to destruction or to draw back to more sober judgment and a re-evaluation of aims and aspirations. Leacock was unable to predict which way the world would go. Within a single sentence, he conveyed his contradictory views of the outcome of the 'vast social transformation' that he saw the world undergoing. It was one 'in which there is at stake, and may be lost, all that has been gained in the slow centuries of material progress and in which there may be achieved some part of all that has been dreamed in the age-long passion for social justice.'[19]

This was the nature of the modern world of technology. 'This is a time such as there never was before,'[20] Leacock wrote. Like King, Leacock saw it as an age that, more than the mere discovery of America, truly deserved the term 'New World.' And it had begun a century and a half earlier in the 'wonder-years of the eighteenth and early nineteenth centuries,' in the Industrial Revolution. 'Here began indeed ... the magic of the new age.'[21] However, its culmination was the social upheaval, the wonderful potential yet deadly fears of destruction, of the present day.

Before the reader can recover from the contradictions and mayhem of the times, Leacock introduces the greatest contradiction of all: the 'riddle' of the modern age embodied in the title of his study, *The Unsolved Riddle of Social Justice*. It is encapsulated in Leacock's simple but poignant phrase: 'With all our wealth, we are still poor.'[22] The phrase captures not only the enigma of the modern age of technology but also hints at Leacock's answer to its 'riddle.' For the contrast or contradiction, the juxtaposition, in the phrase is not only between great material wealth and material poverty but also and more importantly between the great *material* potential of the age – its 'wealth' – and the great *spiritual* bankruptcy – its 'poverty' – an observation in keeping with King's analysis. For Leacock, the essence of the modern world of technology was a tension and contradiction between its social values, based on material wealth, and its moral and spiritual values, based on quality of life. In Leacock's estimation, the two were out of sync. Why was this the case? And how could the two be reconciled? Those were the questions Leacock attempted to answer in *The Unsolved Riddle of Social Justice*.

Leacock began his analysis of the riddle of the modern age of technology by setting out the parameters of the paradox of the age. On the one hand, the modern 'age of machinery and power' had the ability to produce sufficient material goods to satisfy the basic needs of the entire human race. It could liberate humankind from want of the basic necessities of life: food, clothing, and shelter, while at the same time providing for 'more comforts.' Rather than all being equal in poverty, as in past ages, now all could be equal in prosperity. What made this possible was the wonderful power of machines to increase 'man's power over his environment.' 'The first aspect of the age of machinery,' Leacock noted, 'was one of triumph. Man had vanquished nature,'[23] a hope as old as the dream of Francis Bacon.

On the other hand, this technological age appeared to lack the ability or, more accurately, was unwilling to bring to fruition the promise of abundance to all. The mechanical process appeared designed to stop producing basic goods before there was a sufficient quantity to satisfy human needs and instead produced superfluous goods. It was at this point in his analysis, having set out so vividly the contradictory nature of the modern technological age, that Leacock introduced his paradoxical phrase: 'With all our wealth, we are still poor.' He elaborated on the nature of this 'industrial paradox':

> After a century and a half of labour-saving machinery, we work about as hard as ever. With a power over nature multiplied a hundred fold, nature still conquers us. And more than this. There are many senses in which the machine age seems to leave the great bulk of civilized humanity, the working part of it, worse off instead of better. The nature of our work has changed. No man now makes anything. He makes only a part of something, feeding and tending a machine that moves with relentless monotony in the routine of which both the machine and its tender are only a fractional part.
>
> For the great majority of the workers, the interest of work as such is gone. It is a task done consciously for a wage, one eye upon the clock. The brave independence of the keeper of the little shop contrasts favourably with the mock dignity of a floor walker in an 'establishment' ... The life of a pioneer settler in America two hundred years ago, penurious and dangerous as it was, stands out brightly beside the dull and meaningless toil of his descendant.[24]

The problem of the modern world of technology, Leacock observed, was that it did not deliver on the promises its exponents had claimed for it. Rather than freeing humans from the drudgery of work, technology had forced people to work 'as hard as ever' and at jobs that were monotonous and unrewarding. Rather than liberating humankind from the vagaries and oppressive force of nature, 'nature still conquers us.' Rather than a higher standard of living, a better quality of civilization, technology had left 'the great bulk of civilized nature, the working part of it, worse off instead of better.' In essence, Leacock was questioning the assumption of the nineteenth century that technology was the *sine qua non* of progress.

Who or what was to blame for the failure of technology to deliver on its promises, its moral bankruptcy? Again, Leacock was uncertain. At times, he blamed technology itself. It seemed to operate autonomously, to have its own imperative, independent of human action or will. At one point, discussing the rationale behind machine production, Leacock wrote, 'If the world's machinery threatens to produce a too great plenty of any particular thing, then it turns itself towards producing something else of which there is not yet enough. This is done quite unconsciously without any philanthropic intent on the part of the individual producer and without any general direction in the way of a social command. The machine does it of itself.'[25]

In general, however, Leacock still believed in 1920 that technology was within human control and that its failure to achieve its potential of satisfying human needs was the fault of the society that controlled it. Ultimately, he believed individuals made choices and humans were in control.

If humans were in control of technology, then why would they choose to have the productive process designed to produce only a limited amount of essential goods and then turn to producing luxury and superfluous items? Yet again, Leacock was not entirely certain. At times he claimed that the process of mechanization created a mindset that was 'mechanical' in nature: calculating, indifferent, cold-hearted, and acquisitive. While Leacock was not one to condemn the present age by glorifying the past, he did ultimately come down on the negative side of the present age for its 'dulling' effect on the minds of the workers. 'Even when we have made every allowance for the all too human tendency to soften down the past,' he wrote, 'it remains true that in many senses the processes of industry for the worker have lost in attractiveness and power of absorption of the mind during the very period when they have gained so enormously in effectiveness and in power of production.'[26] Equally, during the war the industrial system worked to direct human energy to destructive rather than constructive ends; during peace, the same 'wasteful' process was at work, producing superfluous goods instead of essential needs. The process itself – mechanization – seemed to create the conditions that placed the machine and its productive capacity ahead of humans and the spiritual needs of the populace, especially of the working class. Machines turned individuals into robotic beings – 'shifting atoms' – who were desensitized to the needs of others. 'An acquired indifference to the ills of others is the price at which we live,' he wrote. 'A certain dole of sympathy, a casual mite of personal relief is the mere drop that any one of us alone can cast into the vast ocean of human misery ... We feed well while others starve. We make fast the doors of our lighted houses against the indigent and the hungry.'[27]

Ultimately, however, Leacock came down on the side of human values and action rather than machines as shaping the mindset of the mechanical age. As a corrective to the strong emphasis on the mechanical side of the era of machine production in the first chapter of *The Unsolved Riddle of Social Justice,* Leacock went on in his second chapter to examine the 'intellectual development of the modern age of machinery and the way in which it has moulded the thoughts and the outlook of mankind.' He analyzed the tenets of capitalism along with the economic theory of laissez-faire that he noted 'grew up, as all the world knows, along with the era of machinery itself.' Its counterpart on the political side, he pointed out, was political democracy. All three were interconnected and 'represent[ed] the basis of the progress of the nineteenth century.'[28] Those forces and beliefs shaped the mindset of the industrial age. Leacock praised the philosophy of laissez-faire: 'It opened a road never before trodden from social slavery towards social freedom, from the mediaeval autocratic regime of fixed caste and hereditary status towards a regime of equal social justice.'[29] Nevertheless, he realized that the philosophy had had its day and needed to be replaced by a more humane outlook.

Having set out the problem – the riddle – of the modern world of machines and power, Leacock was ready to offer his solution. But before doing so, he reviewed and criticized the other current perspective of the age, the counter-perspective to laissez-faire, namely, socialism. In discussing the roots and the nature of the machine age, in setting out opposing perspectives on the age, and in offering his own solution to the paradox or 'riddle' of the age, Leacock drew upon the ideas he had acquired as a student of economics and political science and upon his own innate knowledge as a keen observer of the social problems of the day.[30]

Leacock's educational background prepared him well for his role as social critic. Born in England in 1869, he had at the age of six immigrated with his family to Sutton, Ontario, on Lake Simcoe, where his father, Peter, had already settled on a hundred-acre farm after unsuccessful attempts at farming in Maritzburgh, South Africa, and in Kansas.[31] Leacock witnessed the transition in the technology of communication and trade in the region from a reliance on water transportation through the use of lake steamers as the means to ship goods to land transportation by means of the railroad, the new mode of transportation and communication in the late nineteenth century. While Leacock grew up in an affluent family and had opportunities beyond those of most children of his age, including being able to attend the prestigious Upper Canada College, he believed that his family was poor. He was also very conscious of money and felt the need to have a sufficient amount to be secure. This may account for his consistent concern for the plight of the poor. It may also explain his fascination for understanding why in a land of plenty of natural resources and unlimited opportunities for work and success and incredible advancements in technology there was such widespread poverty. Certainly, failures abounded in his extended family. In addition to his father's dismal failure as a farmer, largely due to a drinking problem, Leacock recounted at different times the failure of two of his uncles and a brother to make a go of it, all of them ending up in a state of poverty. Thus, the paradox of the industrial age had a personal meaning for him that might have spurred his interest in 'solving it.'

As a student at Upper Canada College, Leacock outdistanced the rest of his classmates to become 'head boy' in 1887, the year he graduated. He won a fellowship that he used to attend the University of Toronto the following year. By the end of his first year, he was granted third-year status. In spite of such success, young Leacock felt that he had to withdraw from university due to financial exigencies, as he claimed in later life, and instead attend teachers' college for three months in preparation for teaching. He taught high school at Uxbridge for a brief time and then became junior master of modern and ancient languages at Upper Canada College. In 1891, he re-enrolled at the University of Toronto to complete his BA in modern languages. For the next ten years, he taught at Upper Canada College but also studied economics. He was assisted by his teaching colleague Edward Peacock, who had studied economics at Queen's University. Peacock had written two important studies: *Trust, Combines, and Monopolies* (1898) and *Canada: A Descriptive Text-Book* (1900). Peacock assisted Leacock in working through Alfred Marshall's *Principles of Economics* (1890), and in return Leacock taught him French. In later life, Peacock recalled that he was obviously the better teacher, since Leacock 'is now the head of the Economic Department of McGill and I still speak no French.'[32]

In 1899, Leacock borrowed $1,500 from his mother to attend the University of Chicago to study political economy. In his letter of introduction to the university, Leacock claimed that he had not taken political economy at Toronto because of the lack of a program on the subject, although W.J. Ashley had been appointed to the Chair of Political Economy and Constitutional History in 1888 and taught courses thereafter until his departure to Harvard in 1893. Leacock studied at Chicago from 1899 to 1901 and received his PhD magna cum laude in 1903. In 1900, he married Beatrix Hamilton, the granddaughter of the Toronto millionaire Sir Henry Pellatt, onetime owner of the palatial Casa Loma. The following year, Leacock was appointed a sessional lecturer at McGill University. He would stay at McGill for the remainder of his teaching career, only leaving in 1935 due to forced retirement.

Chicago was 'a raw place' during his student years, Leacock recalled some years later. Located on Lake Michigan, it was the gateway to the American west at the turn of the century. A series of railways entered and exited the city, making it a hub of commerce and industrialism. Six years before he arrived, Chicago had hosted the great Columbian Exposition, and the 'White City' had gone all out to impress the rest of the world with its industrial might and cultural sophistication. But beyond the exposition grounds were the shanties where the working poor lived. This had not changed by the time Leacock arrived, and he would have undoubtedly been aware of the extremes of wealth and poverty in this modern industrial city, with both groups segregated into their own enclaves and ghettos. It was also a city plagued by strikes, due to the large labouring class and the volatile economy.

The University of Chicago never impressed Leacock. He found it too large and impersonal. By the time he arrived, the university, although only seven years old, already had the largest number of graduate students of any American university, thanks to a generous endowment from John D. Rockefeller. In hindsight, Leacock would criticize universities, 'such as Cornell and Chicago,' that took such endowments and then prostituted their academic standards by being pressured to offer practical subjects in the curriculum and to run the institution as a business. Leacock believed that this was one of the negative repercussions of the 'age of machinery' and a by-product of the emphasis on a technical and commercial education.[33]

Still, by the turn of the century, the University of Chicago had emerged as the leading school of the new political economy that would slowly take hold in most major North American universities. What distinguished the new economics was a belief in the need for an interventionist state to offset the weaknesses of the old laissez-faire economy, which had emphasized the right of the individual to pursue his or her own economic interests uninhibited.

In Leacock's mind, Thorstein Veblen, the University of Chicago's most renowned professor, did not rank high as a teacher. The one time Leacock mentioned Veblen, in his book *My Discovery of the West,* during a discussion of Social Credit theory, he recalled Veblen's poor teaching methods.[34] However, Leacock did find Veblen's ideas on the nature and impact of the industrial age 'priceless.'[35] What he remembered of importance was Veblen's central argument in his major work, *The Theory of the Leisure Class,* namely, 'that human industry is not carried on to satisfy human wants but in order to make money.' He recalled as well that Veblen had emphasized the need for this 'leisure class' to 'find ways of spending it [money] in "conspicuous consumption."' (Leacock used Veblenesque phrases, such as 'the conspicuous consumption of the leisure class,' 'higher learning as an expression of the pecuniary culture,' and 'pecuniary emulation,' in his own writings.) He especially remembered that Veblen had been damning in his indictment of the financiers and their allies, the captains of industry, the 'nouveau riche' of the new industrial age, calling them the 'drones' of society.

Leacock was as trenchant in his criticism of the idle rich as Veblen, at one time describing them condescendingly as 'the crude, uncultivated and boorish mob of vulgar men and over-dressed women that masqueraded as high society – the substitution, shall we say, of the saloon for the salon.'[36] On another occasion, he described the 'leisure classes' as a 'sort of flower on a manure heap.'[37] In discussing the impact of the values of the boorish but powerful class of people in the context of the lonely professor, who had been a member of the intellectual elite before the rise of the nouveau riche, Leacock wrote, 'I had set out to make the apology of the professor speak for itself from the very circumstances of his work. But in these days, when money is everything, when pecuniary success is the only goal to be achieved, when the voice of the plutocrat is as the voice of God, the aspect of the professor, side tracked in the real race of life, riding his mule of Padua in competition with an automobile, may at least help to soothe the others who have failed in the struggle.'[38] Leacock clearly admired Veblen as a critic and as a theorist of technology. However, he parted company with his Chicago professor when the latter 'turned from destructive criticism to the attempt to reconstruct society.'[39] Veblen favoured a technocracy, rule by engineers. Leacock had as little faith in the engineer as he did in the capitalist class, believing that engineers were as motivated by selfish interests and as determined to shape the new technological society to their own ends as any other special interest group.[40] He believed that only a democratic government, representing the interests of all groups in society, would genuinely use technology for the betterment of society as a whole.

Leacock did his PhD dissertation at the University of Chicago on 'The Doctrine of Laissez-Faire.' The topic was, as Carl Spadoni notes, 'intellectually *au courant* in the economic climate of the 1890s,' as the agrarian economy was giving way to industrialism in America and as a faith in free enterprise was giving way to a belief in greater governmental involvement among economists. Both changes necessitated a re-examination of the philosophy of laissez-faire, which Leacock's brilliant dissertation provided. Leacock showed himself to be very knowledgeable about the assumptions underlying the philosophy of laissez-faire and their implications, but he also dealt very effectively with the critics of laissez-faire in nineteenth-century thought, especially the socialists.[41] Thus, by the time Leacock graduated from the University of Chicago, he had a good understanding of the two opposing creeds and perspectives of the industrial age – laissez-faire and socialism – which he set out succinctly in *The Unsolved Riddle of Social Justice*. He had also noted the dangers of putting too much faith in technology and of allowing the machine and the process of mechanization to dictate the values of the society. Such values were not always in the interest of society as a whole so much as in the interests of groups that held power and benefited the most from the dominant technology. Thus, Leacock had already come to appreciate the tension between technology and freedom – the technological imperative and the moral imperative – that has been a pervasive theme and a concern among Canadian theorists of technology.

While Lampman had used poetry to convey his message of social reform, Leacock occasionally used humorous novels to present his views. He saw humour as an effective way of conveying his message; it 'sweetened the medicine' while having the same remedial effect of offering a cure for the ills of society. In particular, humour became for Leacock a wonderful medium in which to poke fun at the attitudes and values – the intellectual component of the machine age – of the captains of industry, as well as the exponents of what he saw as the extreme theories on mechanization, namely, the individualists of the theory of laissez-faire and the socialists. Only when Leacock came to realize that his popularity as a writer of humour was undermining the willingness of his audience to take his criticisms of the social system and his proposed solutions seriously did he attempt to separate his serious and humorous writings, and even then with little success. He once informed a young friend of his problem: 'Do not ever try to be funny, it is a terrible curse. Here is a world going to pieces and I am worried. Yet when I stand up before an audience to deliver my serious thoughts they begin laughing. I have been advertised to them as funny and they refuse to accept me as anything else.'[42] However, the audience was not always to blame. It was as though Leacock found it difficult, even in his serious writing, not to be funny, probably because he believed that the general populace would not be interested in what he had to say unless it was entertaining. At least through humour he was able to reach a very wide audience. This exposure was important for Leacock the social reformer who had a strong mission to right the wrongs of the industrial age and who had a vision of a utopian world at the same time that he poked fun at other utopianists of his day with the same aspiration.[43]

Leacock combined humour and social reform most successfully in his first novel, *Arcadian Adventures with the Idle Rich* (1914). Literary critic Claude Bissell has described the novel as 'the fictional companion piece to Thorstein Veblen's *The Theory of the Leisure Class*.'[44] Indeed, the titles suggest a common preoccupation. In *Arcadian Adventures,* Leacock took great delight at poking fun at the pseudo-sophistication and pretentious cultural warp of the idle rich, who succeed in doing nothing useful and spending money lavishly, as Veblen's study of the 'pecuniary culture' of industrial society argued. Both Leacock and Veblen noted that the money of the wealthy class often came from inheritance or speculation rather than from honest work or useful manufacturing. So caught up in their own false world of high society, these 'drones' were oblivious to the poverty that surrounded them. In *Arcadian Adventures,* the daily world of the idle rich revolved around the Mausoleum

Club. Leacock juxtaposed the two worlds of rich and poor in the following description of the club:

> From Plutoria Avenue you see the tops of the skyscraping buildings in the big commercial streets, and can hear or almost hear the roar of the elevated railway, earning dividends. And beyond that again the City sinks lower, and is choked and crowded with the tangled streets and little houses of the slums.
>
> In fact, if you were to mount to the roof of the Mausoleum Club itself on Plutoria Avenue you could almost see the slums from there. But why should you? And on the other hand, if you never went up on the roof, but only dined inside among the palm-trees, you would never know that the slums existed – which is much better.[45]

Leacock was equally stinging in his criticism of and sardonic in his attack on churches and universities, the two institutions in society that he believed should have been most critical of this wealthy class but in fact had sold out to it. He ridiculed the church wardens who were more concerned with saving money than saving souls and maintaining status in society than upholding humanitarian concerns. University presidents came under attack as well for catering to the wishes of the business entrepreneurs who 'owned' the universities and demanded that the institutions be turned into factories – run on the basis of 'efficiency' and 'usefulness' in the industrial age – by turning out students in as mechanical a fashion as an assembly line in a production plant. As well, the university 'taught everything and did everything ... It offered such a vast variety of themes, topics, and subjects to the students, that there was nothing that a student was compelled to learn, while from its own presses in its own press-building it sent out a shower of bulletins and monographs like driven snow from a rotary plough.'[46]

Literary scholars of Leacock have debated whether his popular novel *Arcadian Adventures with the Idle Rich* had been set in Chicago or Montreal or in another North American city. Certainly, the contrast of rich and poor and the callous and indifferent attitude of the city's 'idle rich' in *Arcadian Adventures* could easily have been found in Chicago at the turn of the century but equally in Montreal – or in any other modern urban and industrial city – thus reinforcing Leacock's belief in *The Unsolved Riddle* that the machine age had pervaded every aspect of North American life, homogenizing societies everywhere into a socially stratified society of the wealthy class with their 'pecuniary culture' and the labouring class with their poverty.

Leacock wrote *Arcadian Adventures* on the eve of the First World War. The Great War accentuated for Leacock the good and bad aspects of the age and made clear what was wrong with the industrial system. He contributed two essays to the edited volume *The New Era in Canada* (published in 1917) entitled 'Democracy and Social Progress' and 'Our National Organization for the War,' which were the serious side of *Arcadian Adventures*. In the former article, which served as the introductory chapter of the book, Leacock set out the problem. He attacked the 'new government of the money power' as being soulless. 'The plutocrat, unfettered by responsibility,' he wrote, 'seemed as rapacious and remorseless as the machinery that has made him.'[47] He summed up the characteristics of the age as follows:

> The rise of the great trusts, the obvious and glaring fact of the money power, the shameless luxury of the rich ... – all this seemed to many an honest observer of humble place as but the handwriting on the wall that foretold the coming doom ... To many it seemed as if the country were falling under the rule of the great corporations, the railroads and the banks; as if our free democratic government, wrested after so many efforts from those who ruled us, had given us only the rule of the capitalist. For bread, a stone.[48]

Having set out the problem, Leacock offered his solution in his second contribution to *The New Era*, which constituted the concluding chapter to the volume. The chapter borrowed heavily from a paper he had written for government use, entitled 'Our National Organization for the War.' The title foretold his solution. He maintained that all men, women, and children must give their all for the war effort. He predicted that the Great War, once seen in its true light as a struggle for democracy, would have a moral cleansing effect on the Canadian soul. For one thing, it would bring to light the terrible waste of money and productive capacity of machines and human power on the production of superfluous goods for the idle rich, both of which could be used instead for war production. According to Leacock, then, the first step to a reformed Canada was to stop wasting national resources on useless and vulgar activities. 'Let there be no more luxuries, no wasted work, no drones to keep, out of the national production.'[49] The next step was for each individual to use only what was absolutely essential from industrial production to subsist and to give the rest to the government in the form of hard work, increased taxes, patriotic government loans, and volunteer service. In turn, the government needed to force manufacturers to produce only goods essential for the war effort. In this way, Leacock maintained, the entire nation could be galvanized for war. In essence, however, it meant a complete change of attitude from greed to philanthropy.

What Leacock proposed in 1917 was the opposite of the philosophy of laissez-faire that he had studied in depth in his doctoral dissertation. It was also a repudiation of the age of the idle rich that he had caricatured in *Arcadian Adventures*. Building on his knowledge from Thorstein Veblen's courses in which Veblen had set out the glaring weaknesses and moral discrepancy of the 'pecuniary culture,' and from his reading of the opponents of the philosophy of laissez-faire – the advocates of government intervention and those who empathized with the plight of the poor – along with his innate sympathy for the poor and downtrodden coming out of a belief that he too had been poverty-stricken as a child, Leacock began to go beyond diagnosing the problems of the 'age of machinery and power' to offer solutions. 'Our National Organization for the War' was his first foray in this direction; *The Unsolved Riddle of Social Justice* was the next and the most sustained attempt to find a solution to the 'riddle' of the machine age.[50]

In proposing solutions in *The Unsolved Riddle,* Leacock noted once again the importance of the Great War. He went further than he had in 'Our National Organization for the War' to argue that the great lesson of four years of fighting was the realization of the need for a new perspective on social questions and the importance of applying the right values to the industrial order. That lesson was most evident, he believed, in conscription. Leacock managed to find a positive perspective on this policy that had created such a deep fissure between French Canadians and English Canadians. He saw conscription as the crowning achievement of progressive democracy: 'the obligation of every man, according to his age and circumstance, to take up arms for his country and, if need be, to die for it.'[51] This obligation reinforced the right of individuals to expect in return for service to their country that the nation would care for them in times of need and offer 'the opportunity of a livelihood.'[52] Based on his perception of the great sacrifice the Canadian populace had made for the war effort, and the opportunity for a great national effort coming out of the crucible of war, Leacock constructed an extensive list of basic social rights that each individual could and should expect from government, including employment for all able-bodied individuals; social insurance for the sick, crippled, maimed, and aged; shorter hours of work; a minimum wage; and, most important of all, equal opportunity for all children. On the latter, Leacock was both passionate and adamant, and his justification is worth quoting at length:

> The children of the race should be the very blossom of its fondest hopes. Under the present order and with the present gloomy preconceptions they have been the least of its collective cares. Yet here – and here more than anywhere – is the point towards which social effort and social legislation may be directed immediately and successfully. The moment that we get away from the idea that the child is a mere appendage of the parent, bound to share good fortune and ill, wealth and starvation, according to the parent's lot, the moment we regard the child as itself a member of society – clothed in social rights – a burden for the moment but an asset for the future – we turn over a new leaf in the book of human development, we pass a new milestone on the upward path of progress.[53]

In essence, Leacock advocated a program of social welfare well ahead of its time. It was the embodiment of the ideal of technology being used for the betterment of society, a reconciliation of the technological imperative with the moral imperative. In the end, it was as utopian as that of the socialists Leacock attacked and as that of Mackenzie King in *Industry and Humanity*. Where the two social reformers differed was in their emphasis. King stressed the need for moral regeneration of the individual, while Leacock focused on the regeneration of society. However, both looked to technology as the means to achieve their end.

Leacock's *Unsolved Riddle of Social Justice* was the high-water mark of his optimism in the ability of technology to create a better world. It also proved to be his last major serious study of economic and social problems. From this point on, he devoted most of his time to his humorous writings. Still, the subject of technology occasionally appeared as a topic in both his serious and humorous writings. Leacock experienced the disillusionment the whole world experienced by the early 1920s, when the hoped-for better world coming out of the carnage of the Great War never materialized. Like others, he became pessimistic and cynical about events both in Canada and in the world in general. In particular, he began to question his faith in technology.

In 1921, Leacock published an article in *Collier's Magazine* entitled 'Disarmament and Common Sense,' in which he argued that the current Washington Conference on Disarmament was 'a turning point.' 'Civilization stands at the crossroads,'[54] he warned. It was in danger of collapsing under renewed international conflict resulting from an arms buildup. He blamed the situation on 'the industrial machinery' that he believed had become 'out of gear.' The age of machinery had not inaugurated a new era of peace and prosperity, as its advocates (including himself) had promised; quite the contrary, it had escalated conflict. Leacock wrote:

> We have taken it easily and freely for granted these two or three generations back that the thing we called progress had come to stay: that the machine inventions of the nineteenth century had conquered nature: that the great problems were solved, that the structure of society was bolted to the rock and that nothing remained but to fill in the upper details. To none did it occur that the world might move backward: that our civilization might be shattered: that even our progress ... might reach its 'peak load' and fall.[55]

Again, Leacock remonstrated that the problem lay in society's attitude toward technology. Now, however, the problem was not the wrong values underlying technology, as he had claimed was the case in *The Unsolved Riddle of Social Justice*. Rather, it was a blind faith in the ability of technology to solve social problems, the very faith Leacock himself had displayed. While he still differentiated between spiritualism and technology as he had done in the *Unsolved Riddle,* he no longer had the optimism that the former would prevail. The technological imperative seemed to be dominating society and dictating the values by which society operated. He saw those values as mechanical, materialistic, and based on power politics. In 'Disarmament and Common Sense,' he wrote:

> We looked about us at our achievements, the skyscraper, the aeroplane, and the Ford car. Wonderful indeed they seemed, and each decade added to the list. We were so lost in wonder at our progress that we did not realize that it might lack the one thing that makes progress worthwhile – the power of the spirit: that of themselves machinery and mechanical progress are nothing: that of itself machinery can obliterate neither hunger nor want nor cruelty nor war: that the machine that should be the servant may become a huge demon dominating its masters. So it comes that we stand in a bankrupt world littered with machinery and demanding in the form of naval armament more machinery still.[56]

It was a familiar warning but premised on the pessimistic belief now that humankind lacked the ability to control technology, that 'the machine that should be the servant' had become 'a huge demon dominating its masters,'[57] his version of the master/servant relationship. Nowhere was this more evident than in the armaments race, the very problem that had led to the last war and now threatened the world with another war. Nations vied for technological superiority in armaments and in so doing increased the possibility of Armageddon. The very survival of civilization lay in the hands of the 'demon' technology. Leacock's only ray of hope in an otherwise pessimistic article was his belief in an enlightened American public, aware of the stupidity of the arms race and willing to take a lead in ending it by reducing their nation's own buildup of battleships and war machinery. In other words, technology needed to be harnessed by humankind and utilized for public good.

In his satirical essay 'The Man in Asbestos: An Allegory of the Future,' Leacock doubted the ability of society to control technology. He even questioned the ideals of the machine age that he had so eloquently set forth in *The Unsolved Riddle of Social Justice*. Like Thomas Keefer, Leacock used the familiar literary device of falling into a deep slumber that lasted hundreds of years and then awakening in a dreamlike state in a world in which technology had achieved all that he had envisioned in *The Unsolved Riddle of Social Justice*. Hunger had been eliminated, no one lacked material comfort, work had become non-existent and non-essential, the changeable weather had been brought under control, even disease and death had been 'conquered,' discovered to be no more than 'a matter of germs.' All this came about thanks to technology. Humankind had discovered a way to master nature. The guide to the new society in Leacock's dream remarked, 'There came, probably almost two hundred years after your time, the Era of the Great Conquest of Nature, the final victory of Man and Machinery.'[58]

In his dreamlike state, Leacock became aware that this future society was no utopia, no earthly paradise, but rather a hellish wasteland. Noise, action, and even emotions had been eliminated. Leacock's guide kept telling him not to get excited, that he was not used to such outbursts of emotion. In this static, lifeless world, Leacock described his moment of revelation: 'We sat silent for a long time. I looked about me at the crumbling buildings, the monotone, unchanging sky, and the dreary, empty street. Here, then, was the fruit of the Conquest, here was the elimination of work, the end of hunger and of cold, the cessation of the hard struggle, the downfall of change and death – nay, the very millennium of happiness. And yet, somehow, there seemed something wrong with it all.'[59]

What was 'wrong,' he realized, was the dullness and monotony, the lack, ironically, of the very rapidity, frenzy, and chaos, of the psychological tension and restlessness, that he had found such an appalling aspect of the machine age. At the outset of the story, before he entered the dream state, he had set forth his view of the current society and condemned the very things he was now missing. 'I always had been, I still am, a passionate student of social problems. The world of to-day with its roaring machinery, the unceasing toil of its working classes, its strife, its poverty, its war, its cruelty, appals me as I look at it. I love to think of the time that must come some day when man will have conquered nature, and the toil-worn human race enter upon an era of peace.'[60]

In his dream, the time of which he wrote is realized, thanks to technology, and the juxtaposition of reality and dream is most poignant. Leacock had realized that the technology he himself had put such 'blind faith' in could turn out to be the curse, not the saviour, of civilization. Technology had the power to destroy civilization by undermining the values underpinning civilization. Technology dehumanized individuals, turning them into robots (Frankenstein-like monsters) that performed human functions but lacked human qualities, including a human spirit. For Leacock, the machine had entered the 'garden'; humankind could no longer return to a state of innocence. Ironically, one of Canada's great technophiles had become one of its leading technophobes.[61]

In this shift in perspective, Leacock mirrored the thought patterns of other Canadian intellectuals during the interwar years. Prior to the Great War, Canadian writers on technology in the era of material expansion and industrialization tended in general to dwell on the positive contributions of machines to society. After the war, Canadian intellectuals increasingly questioned the kind of society and civilization that technology was creating. Clearly, in terms of reflection on technology, the First World War can be seen as a watershed. Stephen Leacock reflected the changed perspective more than most Canadian thinkers on technology, because in such a brief period of time he had shifted his perspective from one of optimism to pessimism and even continued often within a single article to present contradictory positions on technology, to note its paradoxes or 'riddles.' In this respect, Leacock, King, George Sidney Brett, and Frederick Philip Grove reflected their times, when opposing views of technology emerged, based on differing perspectives on the issue of the relationship of the technological imperative and the moral imperative.

All four of these theorists of technology during the interwar years saw the mechanization of society as producing humans who were as mechanical in mind, spirit, and soul as they were in body, automatons whose mindset was one with the technological age. They had lost their soul and with it a concern for the less fortunate in the world. Each thinker advocated the introduction of spiritual values so as to transform the industrial age into a time that would serve the needs and fulfill the hopes of its citizens.

Part 3
Philosophizing the Imperative

7

Harold A. Innis and Eric Havelock: Technology as Power

Harold Innis was one of a small but significant group of Canadian thinkers in the mid-twentieth century who saw technology as a central component of modernity and therefore as inextricably linked to the fate of Western civilization. Other Canadian analysts who examined aspects of technology and modernity included Charles N. Cochrane, Eric Havelock, Marshall McLuhan, George P. Grant, and Northrop Frye. One of the distinguishing characteristics of this group of thinkers was their anti-modernist perspective,[1] in particular their critical assessment of the direction in which Western civilization was moving. In this respect, these Canadian intellectuals were themselves part of a larger contingent of Western thinkers who, during the interwar and post–Second World War eras, re-evaluated the nature of technology and critically assessed the virtues of modernity in terms of the future of Western civilization. Included in this group were such prominent thinkers as A.C. Kroeber, Oswald Spengler, Arnold Toynbee, Lewis Mumford, Siegfried Giedion, Martin Heidegger, Jacques Ellul, along with Theodore Adorno and Walter Benjamin, who were two members of the Frankfurt School, a group of German, predominantly Jewish, intellectuals committed to a radical critique of modernity. Central to the thinking of this eclectic group of European and Canadian intellectuals was the question of technology and how it related to modernity within Western civilization.[2]

Among the Canadian group of analysts of technology and modernity, Harold Innis is distinguished for offering a novel view of technology as the defining characteristic in the shaping of nation-states, empires, and civilizations. What Innis meant by technology changed and evolved over time, but central to his thinking at all times was the belief that technology was power: the power to shape human actions and human will by dictating the way people think – in other words, their mindset.[3] Innis and his colleague Eric Havelock, whose ideas on technology are also discussed in this chapter in the context of Innis's ideas, saw power as juxtaposed to and constantly in tension with freedom. The former – power – was a by-product of technology and the root of the technological imperative; the latter – freedom – was the root of the moral imperative. Havelock explored the source of this tension in Greek thought through a study of Aeschylus's play *Prometheus Bound,* which was the first attempt in Western thought, Havelock noted, to address the issue of power and freedom within the context of technology. The intellectual exchange between Innis and Havelock concerning technology, explored in the latter section of this chapter, greatly enriched and deepened the understanding of the relationship between the technological imperative and the moral imperative as power and freedom. Their ideas also raised questions about aims, aspirations, and visions within a society dominated by technological thought. Here is the fourth and final category of technology: as volition. Innis became aware of the challenge and difficulty of ensuring freedom in a world in which the technological imperative was so powerful. The challenge was all the more apparent with the realization that technology was not 'out there' but within the mind, shaping our very being. Technology was not 'making,' 'doing,' or 'using' but rather a way of 'thinking' shaped by the technology. Harold Innis was the first Canadian thinker to present this new perspective on technology, a perspective then picked up by Marshall McLuhan, Northrop Frye, and George Grant in their own unique views on the theme of technology as volition.

The tendency among analysts of Harold Innis is to associate technology only with his later 'communication studies' and not with his earlier 'staple studies.' There is also a belief that the question of technology was of secondary importance for Innis in his understanding of the nature of Canada – that is, the country's patterns of staple production and trade and the role of the Canadian Pacific Railway in Canada's history – and then later in analyzing the rise and fall of empires and civilizations. However, it is clear that the nature and impact of technology is the most important issue in Innis's writings. The subject of technology is evident in all of his writings, from his PhD dissertation, revised and published as *A History of the Canadian Pacific Railway,* through his staple studies, to his late work on communication studies. What changed over time was his understanding of and perspective on technology.

The evolution in Innis's thinking about technology reflects the way he approached the subject. As mentioned earlier, Innis was consistently interested in the concepts of space and time as determining factors in shaping the history and mentality of nation-states, empires, and civilizations. In his research, he followed his own interesting pattern of time and space. In terms of space, Innis began his research with the particular, immediate, and familiar locale of Canada in his first study, a history of the Canadian Pacific Railway, followed by his staple studies of furs and fish. He then moved outward, almost in a series of concentric circles of exploration, to North America, Europe, Western civilization, and, to a degree, the world. In terms of time, in his study of the CPR, Innis began with the advent of the industrial revolution in Canada in the late nineteenth century and then moved back in time to the beginnings of European exploration of North America in the era of the fur trade and cod fisheries. In his communication studies, he reached further back to the beginning of modern European civilization with the advent of the printing press and ultimately to the roots of Western civilization in Greece and Rome and to an understanding of other ancient civilizations. Throughout his diversity of interests and subjects, he focused consistently on the impact of technology on societies and civilizations, especially modern Western civilization, and particularly on the impact of technology on Canada's position within that civilization. As well, Innis realized that, at all times throughout history, technology has been a source of power, for both beneficial and destructive ends. Unfortunately, too often in the history of Western civilization, especially in the modern period, the destructive nature of technology dominated because of the power of space-binding media of communication.

Innis's thinking about technology is rooted in his past.[4] He was born on a farm near Otterville in Oxford County, Ontario, in 1894. The oldest of the four children of William and Mary Innis, he was also the fifth generation of his family to live in British North America/Canada. All four previous generations had been farmers. Carl Berger notes that growing up on a farm aroused within Innis 'an intense curiosity about the routine and details of farm activity, machinery, and the habits of animals, and an uncanny ability to recognize small differences in commonplace things.'[5] Mary Quayle Innis recalled that her husband 'kept in touch with the farm all his life: with the changing seasons he would often remark on what his family would then be doing.'[6] Certainly, farm life in central Ontario was undergoing rapid change at the turn of the century with increased mechanization, and Harold Innis's decision not to continue in the family tradition of farming was characteristic of many youth at the time (including Stephen Leacock) who were abandoning farm life for jobs and careers in towns and cities. An observant child, young Harold was undoubtedly aware of the impact of technology on farm life: it turned the family farm from a way of life into an occupation and transformed it into a business enterprise.

Innis's rural upbringing did more than simply arouse within him an awareness of the impact of technology on farm life. He never lost his feeling of indebtedness to his rural roots. He believed that new ideas emerged on the margin, and rural communities were on the margins of metropolitan centres. Rural communities also upheld the oral tradition that Innis believed was so important in order to achieve a balance between time-biased and space-biased societies. Throughout his life, he worried about the demise of rural communities, the kind that had helped shape his perspective on life.

The Innises and the Adamses, Harold's mother's side of the family, were devout Baptists. Yet he never made a public profession of faith or was baptized, as was customary in the Baptist tradition.[7] While he would abandon his Baptist faith during the First World War, it did instil in him a strong anti-authoritarian and anti-hierarchical strain – to challenge individuals in positions of power – which is evident in his later social science perspective.[8] His Baptist upbringing also encouraged Innis to question any belief system that appeared to be paramount and to appreciate how beliefs and values were products of their time. Innis would later apply such a perspective to his study of technology. As well, his Baptist faith instilled in him the importance of moral values and ethical standards. Judith Stamps notes that, in later life, Innis saw the decline of ethics in the social sciences as 'symptomatic of modern life, bereft of long-term, historical concerns and correspondingly overwhelmed by short-term or, as he put it, spatial ones.'[9]

Innis attended a one-room elementary school, the Otterville High School, and the Collegiate Institute in Woodstock before going off to McMaster University, the Baptist-affiliated university located at that time in Toronto. University opened up a new world of ideas. McMaster's Baptist affiliation meant that Innis could continue secure in his faith, but the university did not shy away from issues and beliefs that challenged church doctrine. During his undergraduate years, Innis would acquire a philosophical basis for his faith. He studied history and political economy, but he appeared to be most interested in his psychology course with James Ten Broeke, a Hegelian theorist with an interest in languages. Innis kept his rough lecture notes from Ten Broeke's class (although he did not keep his essays), something he did not do with other courses. As well, later in his Preface to *The Bias of Communication,* Innis claimed that this collection of essays was his attempt 'to answer an essay question in psychology which the late James Ten Broeke, Professor of Philosophy in McMaster University, was accustomed to set, "Why do we attend to the things to which we attend?"'[10] Innis's tentative answer to the question was the influence of communication technology on the thought patterns of a particular culture.

While a student at McMaster, Innis enlisted in the Canadian army in 1916. His reason for joining differed from the familiar ones given by most enlistees: others were joining for glory or to fight for God, king, and country. Instead, for him it was a moral issue: the choice between German autocracy and power or British democracy and freedom; it was, in essence, the need to choose between the technological imperative and the moral imperative. He saw the fate of Western civilization and all it stood for hanging in the balance in the Great War. In a letter home, in which he explained his decision to enlist, young Innis wrote, 'Germany started in this war by breaking a treaty, by breaking her sealed word [with Belgium]. Not only did she do that but she trampled over a helpless people with no warning and with no excuse.

If any nation and if any person can break their word with no notice, whatever, then, is the world coming to.'[11] This issue behind the Great War was one of power – whether Germany had the right to use its position of power to dominate lesser states of Europe such as Belgium. He joined as a private and became a signals expert, a telling position in light of his later fascination with communication technology. His war experience also made him aware of the diabolical side of technology, a technology that appeared to have its own imperative. The power of technology could dwarf the most optimistic of purposes in the struggle between ideals and reality.

Due to an injury, his time at the war front was brief, and he returned to civilian life by early 1918 and pursued his master's degree at McMaster, writing a thesis called 'The Returned Soldier.' As an offshoot of that study, he published an article in the *McMaster University Monthly* in which he no doubt used his own experience to generalize on the attitudes of most soldiers to army life, particularly the deference to authority and its debilitating effect on the freedom of the individual soldier: 'The hated subservience to officers, the detested persistence of obedience to orders, the monotony of the bugle, have all alike tended to crush the spirit of independence and individuality which have become as dear to them since he [the soldier] has been bereft of them.'[12] Innis occasionally referred to himself as a 'psychological as well as physical casualty of the Great War.' He did not use the term 'psychological casualty' lightly. Fighting in the trenches while fellow soldiers were maimed or killed all around traumatized him. Though he seldom talked about his war experience, the few times he did were to remind others that war brought out the worst in human nature, that veterans could never again have an optimistic perspective on the human race. He made an indirect reference to this changed perspective in his MA thesis: 'There remains the fact that no one who has been wounded, or no one for that matter who has seen a great deal of front line service, is physically or mentally better for the experience.'[13] War created a sense of camaraderie among its survivors, but at the same time it disconnected the soldier from the larger society. Innis's war experience contributed to his loss of faith, bouts of depression, and feelings of need to work constantly to the point of exhaustion to make up for those gifted youth who died on the battlefields of Europe before being able to make their mark in life. John Watson, Innis's biographer, attributes Innis's dark vision of himself and the world around him to his war experience. Innis's chronic sense of overwork and depression were 'the long-term result of the psychological torture of the trenches,' Watson writes. 'The intensity of these consequences of the guilt carried by Innis as a survivor of the First World War could well have led to his complete disintegration.'[14]

In the summer of 1918, Innis enrolled at the University of Chicago, another Baptist-affiliated university, where he pursued a higher degree in economics, as had King and Leacock before him. He took a variety of courses that exposed him to the economic theories of the day. Frank Knight, who taught economic theory and statistics, appeared to be the most influential of his professors. A skeptic of statistics, Knight reinforced in his students the belief, Innis recalled, that 'one could never again become lost in admiration of statistical compilations.'[15] Innis did his doctoral dissertation under Chester Wright, who had suggested a history of the Canadian Pacific Railway when Innis requested to do a Canadian topic of study.

At the time that Innis was at Chicago, it was known for its distinctive school of sociology – with such noted luminaries as Robert Ezra Park, George Herbert Mead, and John Dewey – that 'proposed an ecological framework based on the analysis of communication as the key to urban industrial life.' These sociologists saw the technology of communication as the means to create communities, institutions, and cultural forms.[16] Marshall McLuhan would later argue that Innis was particularly attracted to Robert Ezra Park's ideas – ideas that set Park apart from the other members of the Chicago School – so much so that '[they] seem[ed] to have appealed more to the mind of Harold Innis than to any other student of Robert Park.'[17] McLuhan also notes how Innis would have been particularly attracted to the following theme in one of Park's essays: '"I have gone into some detail in my description of the role and function of communication because it is so obviously fundamental to the social process, and because extensions and improvements which the physical sciences have made to the means of communications are so vital to the existence of society and particularly to that more rationally organized form of society we call civilization."'[18]

Thorstein Veblen had left the University of Chicago by the time Innis arrived, but his presence was still strongly felt. Innis read Veblen's works and participated in a Veblen study group while at the university.[19] The budding Canadian economist admired Veblen's 'tremendous irony' and iconoclastic approach to the study of economics. However, he saw more in Veblen's ideas than their debunking quality and regretted in later life that Veblen's economic theories had too often been dismissed for their iconoclasm rather than taken seriously. Innis noted that beneath the memorable bombastic 'Veblenian' expressions, such as 'conspicuous consumption' and 'pecuniary emulation,' was a solid analysis of the dynamic factors that underlay America's industrial growth in the post–Civil War period. They were an attempt to show 'the effects of machine industry and the industrial revolution' on American society and values, which Veblen formulated as the 'laws of growth and decay of institutions and associations.'[20] Fundamental to Veblen's thinking was the belief that advances in technology from handicrafts to machine industry altered social values from ones that emphasized the production of essential items to ones that accentuated the making of superfluous goods, because of the demand from the emerging class of nouveau riche for such goods. This shift in perspective resulted in an economy of waste and inefficiency. Veblen advocated a technocracy in which engineers would replace the existing capitalist class, because he believed that engineers would return the industrial system to its rightful goal of producing essential goods efficiently, systematically, and rationally. However, Innis disagreed with Veblen's faith in rule by engineers, putting his faith in an educated public with ideas coming from the university community, most notably the social scientists.

Yet Innis did agree with Veblen as to the importance of modern machine industry as a process that shaped the values of those who functioned within it, including their perspective on themselves, their society, and the world. He also found in Veblen's work a direction for his own research. In an article appraising Veblen's work, Innis noted its significance for future economic studies: 'The conflict between the economics of a long and highly industrialized country such as England and the economics of the recent industrialized new and borrowing countries will become less severe as the study of cyclonics is worked out and incorporated in a general survey of the effects of the industrial revolution such as Veblen had begun and such as will be worked out and revised by later students.'[21]

At the time, Innis pointed out some of the unique features of industrialization in new emerging industrialized countries like Canada compared to advanced countries like Britain and the role of the economist in studying these differences. In 'Notes and Comments' in *Contributions to Canadian Economics: University of Toronto Studies in History and Economics,* the forerunner to the *Canadian Journal of Economics and Political Science,* Innis noted, 'Heretofore Canadians have exhausted their energies in opening up the West, in developing mines, hydro-electric power, and pulp and paper mills of the Canadian Shield, in building transcontinental railways, grain elevators, and cities, and in taking a share in the war [First World War]. The rapidity and energy with which these gigantic tasks have been accomplished have only been possible with the technical advantages of modern industrialism and the concentrated efforts of a small population.' Innis noted that such rapid growth had put the engineer and the technician at the forefront in guiding this growth but had also led to a series of problems: 'Foremost among these problems is that of providing an adequate market for the raw materials which Canada has succeeded in developing so rapidly. Scarcely less important is that of protecting natural resources from rapid exploitation ... overproduction ... and exhaustion ... which follow from dependence on older, more highly industrialized areas.' Innis saw the economist as responsible for the 'development of ways and means by which this protection may be obtained ... The welfare of the nation for the future must depend to an increasing extent on the economist.'[22] Here was justification for a new scholarly series, 'Studies in History and Economics,' which Innis had initiated. Here too was

Innis's belief in the important role that Canadian economists should play in providing insights and guidance as Canada underwent its industrial revolution. In his mind, economics and technology were inexplicably linked.

In studying the role of technology in Canadian development, Innis became increasingly pessimistic, or at least alarmed, about its implications. He began referring to the 'Frankensteinian' nature of technology, as had King before him. A.F.W. Plumptre, one of Innis's students in the 1920s who would go on to become a noted economist in his own right, recalled the concern that came over Innis as he studied the increased power of technology in the Canadian economy:

> Innis had an undertone, if not of pessimism, certainly of deep misgivings at a time when we undergraduates really thought the world was going to be all right ... The League of Nations was getting going etc.
>
> Innis at that time was very much concerned with where industrialism was taking the world. I can remember him talking about something which in those days we'd never heard of – 'Frankenstein' – which later became famous, but at the time was still a novel by Mrs. Shelley ... The mechanical monster took over and ran its creator. [Innis] kept coming back to the worry that technology was going to do this to civilized man.[23]

Innis's doctoral dissertation at Chicago, later published as *A History of the Canadian Pacific Railway* (1920), established from the beginning both his writing style and the theme that would occupy him for the rest of his life. Given the upsurge in national feeling in Canada in the post–First World War era, it would have been logical for Innis to see the railway as a great national edifice, binding the country together and giving it a sense of identity and purpose. However, Innis resisted the obvious interpretation, as he would on many other occasions, and instead looked at the CPR as a form of technology that contributed to the spread of Western civilization in North America. He presented this theme only in his conclusion and in passing reference without developing the link between the CPR as 'technical equipment' and the expansion of Western civilization. He noted:

> The history of the Canadian Pacific Railroad is primarily the history of the spread of western civilization over the northern half of the North American continent. The addition of technical equipment described as physical property of the Canadian Pacific Railway Company was a cause and an effect of the strength and character of that civilization. The construction of the road was the result of the direction of energy to the conquest of geographic barriers. The effects of the road were measured to some extent by the changes in the strength and character of that civilization in the period following its construction ...

> The Canadian Pacific Railway, as a vital part of the technological equipment of western civilization, has increased to a very marked extent the productive capacity of that civilization. It is hypothetical to ask whether under other conditions production would have been increased or whether such production would have contributed more to the welfare of humanity.[24]

Innis saw technology, or 'technical equipment,' as more than machinery and industrialism, since he did not dwell on these aspects of the history of the railway. Instead, he emphasized in *A History of the Canadian Pacific Railway* the political and economic factors behind the decision to build a transcontinental railroad, dwelt in detail on the construction of the main line and the addition of spur lines, and cited statistics on passenger traffic, earnings from operations, expenses, capital, and profits. The implication was that the CPR was a form of technology that represented a mindset or *mentalité* within Western civilization that was itself 'technological.' That technological mentality measured everything in quantitative, mechanical, and mathematical terms – as profits, material values, and power – rather than in human and spiritual terms. Innis believed that it was this 'technological mentality,' characteristic of modern Western civilization, that spread across the North American continent through the CPR and enabled central Canada to dominate the west and Britain to dominate British North America. The implication was that the spread of Western civilization via technology had both positive and negative connotations: technology was both constructive and destructive. In looking at technology in this way, Innis was very much in tune with the perspectives of other Canadian theorists of technology in the post–First World War era such as George Brett, William Lyon Mackenzie King, and Stephen Leacock. This understanding enveloped an interpretation of technology as process, a theoretical stance that Innis would eventually supersede with his own intellectual positing of technology not just as process but also as volition, the interplay between power and freedom.

In the early 1920s, Innis took a number of research trips to the northern parts of Canada. He wrote an account of one of those trips in which he and his friend John Long, a Toronto high school teacher, canoed down the Mackenzie River for 2,000 miles and then returned on the Hudson's Bay Company steamer, the *Laird River*. Besides noting the geographical and geological features of the region, Innis commented on the impact of limited transportation technology on the cultural milieu of the region. 'Watches and clocks are delightfully haphazard,' he observed. 'No one cares about newspapers since all news is stale news. The wireless station erected this year at Simpson, under the direction of Lieut. Galbraith, will change this in part, but there will long remain the difference between inside and outside.'[25] Innis appeared to be suggesting that the limited transportation technology in the

North resulted in it being a region on the margin of the dominant centre of central Canada, a theme – margin and centre – that Innis would pursue in his later communication studies.

During the 1920s, Innis became interested in discovering the roots of the European dominance over the northern half of North America that was evident in the history of the CPR. He found those roots in the early fur trade and cod fishery. He discovered that many of the traits, patterns, and institutions in modern Canada had their origins in the nature and pattern of these early staple products. One of those patterns was the importance of transportation and communication technology in shaping the nature of Canada and its historical association with Britain and the United States. Once again, as in his history of the CPR, Innis only explored the association of transportation with technology and civilization in his Conclusion to *The Fur Trade in Canada*. He wrote, 'Fundamentally the civilization of North America is the civilization of Europe and the interest of this volume is primarily in the effects of a vast new land area on European civilization. The opening of a new continent distant from Europe has been responsible for the stress placed by modern students on the dissimilar features of what has been regarded as two separate civilizations. On the other hand communication and transportation facilities have always persisted between the two continents since the settlement of North America by Europeans, and have been subject to constant improvement.'[26] This awareness reinforced the observation Innis had made in his history of the CPR of the important role that transportation and communication technology had played in tying the country together and, more importantly, in linking Canada to the British Empire and, through the Empire, to Western civilization. Canada remained more closely tied to Britain and linked to European civilization than did the United States because of the importance of the communication and transportation technology that bound the two countries together. Innis realized that technology had shaped Canada's historical development to be different from that of the United States and assured that, on the northern half of the North American continent, a different civilization than that of the United States – a Canadian version of Western civilization – would exist.

That uniquely Canadian civilization emerged out of the complex interrelationship of European civilization with an indigenous Native civilization. In his history of the fur trade, Innis was more explicit than he had been in his history of the CPR as to the distinctive nature of the civilization that evolved in British North America.[27] He was also one of the first scholars to acknowledge that an indigenous culture and civilization existed in North America prior to European settlement, one that had assisted the early European settlers to adjust and survive in a harsh and unfamiliar terrain and climate. Unfortunately, Innis argued, this Native society could not withstand the onslaught of the Europeans with their superior technology. He noted:

> The history of the fur trade is the history of contact between two civilizations, the European and the North American, with especial reference to the northern portion of the continent. The limited cultural background of the North American hunting peoples provided an insatiable demand for the products of the more elaborate cultural development of Europeans. The supply of European goods, the product of a more advanced and specialized technology, enabled the Indians to gain a livelihood more easily – to obtain their supply of food, as in the case of the moose, more quickly, and to hunt the beaver more effectively ... [But] the new technology with its radical innovations brought about such a rapid shift in the prevailing Indian culture as to lead to a wholesale destruction of the peoples concerned by warfare and disease.[28]

What did Innis mean by the 'more advanced and specialized technology' of the Europeans? Clearly, he meant the superior tools such as guns, iron knives, and brass kettles that made life easier for the Native people. He saw these items as by-products of European industrialism that gave Europeans the advantage over the Aboriginal population. Once again, however, as in the case of his history of the Canadian Pacific Railway, Innis also saw such items as iron tools, along with the process of industrialism, as by-products of a mindset that was itself 'technological.' In the case of the fur trade, it was this aggressive technological mindset that was based on power, materialism, and exploitation that enabled the Europeans to exploit the First Nations, thus enabling the dynamics of centre and margin within Western civilization to work to the advantage of Europeans. Innis went on to note that industrial technology in North America developed first and to a greater extent in the United States, once again leaving Canada on the margin, only now dependent on the American as opposed to a European imperial centre of technology.

In *The Fur Trade in Canada,* Innis noted further that the pattern of a superior-inferior trade relationship, first established in the fur trade era, continued long after the fur trade had died out and dictated the pattern of trade of later staples. In his Conclusion, he observed that, within the history of Canada, the advancement and therefore superiority of industrialism in Great Britain and the United States vis-à-vis Canada shaped the nature of the staple trade between Canada and these two countries to Canada's disadvantage, making Canada a hinterland to industrial and urban centres outside its borders, thus retarding the growth of industrialism within the country. Innis pointed out:

> In Great Britain the nineteenth century was characterized by increasing industrialization with greater dependence on the staple products of new countries for raw material and on the population of these countries for a

> market ... In the United States the Civil War and railroad construction gave a direct stimulus to the iron and steel industry and hastened industrial and capitalistic growth ... Canada has participated in the industrial growth of the United States, becoming the gateway of that country to the markets of the British Empire. She has continued, however, chiefly as a producer of staples for the industrial centres of the United States even more than of Great Britain making her own contribution to the Industrial Revolution of North America and Europe and being in turn tremendously influenced thereby.[29]

Innis's next major study, *The Cod Fisheries: The History of an International Economy*, published in 1940, followed the pattern and general themes established in *The Fur Trade in Canada*. It was a detailed study of the 'geographical background and habits'[30] of the cod fish and their significance on Canadian-European-American relations. Again, as in the case of the fur trade, Innis stressed the interrelationship between transportation and communication technology and civilization in terms of its impact on Canada. It was the introduction of technical equipment to the fishing industry, he argued, such as the steamship, the railway, and the trawler, that changed the dynamics of the staple industry and led to a dramatic shift from a maritime to a continental economy. That shift in turn affected Canada in a multitude of ways, resulting in a demand for an extensive policy 'including government support for the Intercolonial Railway, the maritime freight-rates act, the West Indies agreement, coal subventions, extensive subsidies and aggressively active measures for Nova Scotia's fisheries, and the collapse of responsible government in Newfoundland. The transition from dependence on a maritime economy to dependence on a continental economy [was] slow, painful, and disastrous.'[31]

By the time Innis had completed his staple studies, he had discovered a pattern to explain Canada's historical evolution. He realized the importance of transportation technology in tying Canada to Britain and through Britain to Western civilization, thus setting it apart from that of the United States, whose historical evolution had been different. By the twentieth century, this British pull was facing a stronger American counter-pull that had revolutionary repercussions for Canada's future. However, the British connection through technology had assured the growth of a Canadian version of Western civilization that enabled the country to evolve independently of the United States but with a mindset that was 'technological,' in keeping with the technologically advanced society of Britain.

In the early 1940s, Innis underwent a significant shift in thinking. He abandoned his economic staple studies within the context of transportation technology as it shaped relations between the centre of civilization in

Europe and the margin in North America and embarked on a cultural study between centre and margin within the context of communication technology. Marshall McLuhan perceptively described Innis's changed perspective as being a shift 'from the trade-routes of the external world to the trade-routes of the mind.'[32] The latter encompassed a study of the various technologies of communication – stylus, papyrus, parchment, stone, clay tables, paper, and the printing press – that Innis believed had over time shaped the thought patterns of Western civilization. The study led him to examine the roles communication technologies played in the rise and fall of all the major centres of civilization in the West, from Mesopotamia up to and including Western Europe and the United States. He posed one of the questions that drew him to the topic: 'What is it that makes the condition of civilization a compelling preoccupation?'[33] It was his version of the question raised by James Ten Broeke, his psychology and philosophy professor at McMaster University: 'Why do we attend to the things to which we attend?' Innis's question was also a version of that raised by Charles Beard, the American iconoclastic historian of the interwar years, which became the title of one of Beard's books on technology and industrialism: *Whither Mankind?* Innis believed that, to know where Western civilization was going, it was necessary to understand from whence it had come. The question took him back into the recesses of history to study the roots of that civilization. The major theme underlying these communication studies continued to be the same as that pursued in his staple studies, namely, the relationship between technology and civilization. Indeed, this theme now became the focus of all his writings after 1940. Equally, the focus of his communication studies still remained on Canada, even when the articles dealt with ancient civilization or never mentioned Canada specifically. Innis's observation on Edward Gibbon's monumental study, *The Decline and Fall of the Roman Empire* – namely, that it was more a study of Gibbon's native Britain in the nineteenth century than it was of Rome – was equally true of Innis's communication studies of ancient civilizations.[34] They shed as much light on the decline of Western civilization in the twentieth century and on Canada's role as a nation on the margin of that civilization as they do about the ancient past.

Regarding his perspective on technology, Innis's communication studies resulted in a major shift. No longer did he see technology in terms of machines, mechanization, industrialism, or modes of transportation such as ships or railroads. Nor did he see staples of trade as the vital link between technology and civilization. Instead, he saw the material objects that create and bear the message – the technologies of communication – as decisive in shaping the history of civilizations. What made communication technology so important was the role it played in shaping the ideas and values of a civilization, for it was *ideas,* not material goods, that were the essence of a civilization. It was the *mentalité* or mindset of a civilization that dictated all aspects of its existence, and that mindset, he now argued, was shaped by the dominant technology of communication. Innis came to realize that it was communication technology alone that had the power to bring civilizations into existence and also to destroy them. Knowledge was power, and communication technology was the source of power to create the knowledge of an age. Machines, industrialism, mechanization, and even technical knowledge were all by-products of the modern world that began with the mechanization of print by the Gutenberg press; they became part of the mindset of the age of print and electronic media.

Innis's awareness of the power of ideas and values in shaping a civilization through the technology of communication led him away from economic history and into cultural history. He saw the shift in his thinking as characteristic of a general trend in the social sciences. He noted this trend in his Preface to *Political Economy in the Modern State,* a collection of essays spanning his writings from the late 1930s through the Second World War that more than anything else chronicle the transition in his thinking and research interests from his staple studies to his communication studies: 'Perhaps the most significant development in the social sciences in the past quarter century has been the interest in the study of civilization following Spengler, Toynbee, Kroeber and others. The importance of cultural growth as a reflection of the tenacity of civilizations compels an intense interest in the subject.'[35]

Within Innis's intellectual journey into an understanding of technology, his shift from economic history to cultural history reflected a shift from perceiving technology as process to seeing it as volition. Innis realized that technology was not external to the mind, a process that shaped economic patterns to which humans interacted, but rather within the mind, a way of thinking shaped by cultural patterns that human beings mentally created. Although these cultural patterns were shaped by the very technology the mind was attempting to comprehend, the 'technological mindset' nevertheless also gave humans freedom to shape different patterns of thought once they were aware of the power of the mind. Such awareness moved Innis's thinking on technology to a new level of consciousness: an awareness that technology as volition was a dynamic between power and freedom.

Analysts of Innis's thought have put forward a variety of explanations to account for the shift in his thinking and research interest during the early 1940s. Some see the shift in light of his earlier staple studies. Having completed studies of the major Canadian staples of fish, furs, and minerals, he was ready to embark on a study of the timber trade. In good Innisian fashion, Innis felt the need to go back and study the roots of the paper industry in the early history of printing and the mechanical press. This in turn forced him further and further back in time to study the beginning of the alphabet and to explore modes of communication other than paper and books.

A more convincing explanation has to do with the impact of the Second World War. As already noted, Innis had been adversely affected by the First World War. Throughout his life he carried the physical scars from his injuries and the mental scars from the carnage at the war front. He believed that the Great War had had a negative effect on the men and women who had fought in it and that it had changed their perspectives on life thereafter. For example, when in the midst of the Second World War Frank H. Underhill, Innis's colleague at the University of Toronto, was threatened with dismissal from the university because of disparaging remarks Underhill had made about the British Empire, Innis came to his defence against the political and university authorities even though he disagreed with Underhill's views. In a note to President Cody, in which Innis attempted to explain Underhill's actions in light of his involvement in the Great War, Innis revealed more about his own reaction to the war than Underhill's: 'It is possibly necessary to remember that any returned man who has faced the continued dangers of modern warfare has a point of view fundamentally different from anyone who has not. Again and again have we told each other or repeated to ourselves, nothing can hurt us after this. The psychic perils of civilization mean nothing to us.'[36]

With the outbreak of a second world war within the space of a quarter of a century, Innis felt compelled to understand the roots of such violence. He saw both world wars as power struggles between the major European powers of Britain and Germany as to which would be supreme. But he came to see such struggles from a longer perspective, a *longue durée,* as part of the decline of Western civilization, which he believed had begun during the peak period within that civilization in the mid-to-late nineteenth century. In a revealing address entitled 'This Has Killed That,' which Innis gave sometime during the Second World War, he noted the changed and charged power dynamics of the late nineteenth century that led up to the First World War: 'In attempting to suggest the background of this collapse of modern civilization [in the Second World War] we may well ask what has happened which brought to an end about a century of comparative peace from the end of the Napoleonic Wars to the outbreak of the first Great War. Are there any signs within the last twenty-five years before the outbreak of the first Great War which point to the dangers of collapse?'[37]

He attributed the cause of the world wars in the twentieth century to the rise of the popular press in the late nineteenth century. In turn, he linked the rise of the popular press to the changed technology of paper production and printing as well as to new technological inventions such as the typewriter, the linotype, and photographic reproduction processes. He noted that 'within about a quarter of a century before 1914 newsprint began to be manufactured on an enormous scale from wood, and its price declined sharply to about 1900.' The availability of large quantities of inexpensive wood for pulp and paper production along with cheaper means of production meant the availability of enormous quantities of newspapers at a cheap price that could be 'distributed at unprecedented rates of speed.' As well, news was placed 'in the hands of the reader in the form of an encyclopedia from which he could quickly pick out the information desired. The headline and the streamer began to [favour] the news which would attract the largest possible number of readers.'[38] War became a popular news item that sold newspapers. Innis implied that it was the newspaper magnates, such as Randolph Hearst in the United States and Lord Northcliffe in Great Britain, who fostered war through their papers to keep their newspaper empires thriving. 'The press, supported by the inventions of the late nineteenth century,' Innis wrote, 'had shifted to an entirely new level and was designed to reach lower levels of intelligence or literacy.'[39] The shift was reflected in a number of ways: with an emphasis on the present over the past or the future, in fickle public opinion that could easily be stirred up and inflamed, in politicians who resorted to slogans instead of reason in order to win public support, and in an obsession with nationalism instead of internationalism. In 'The Newspaper in Economic Development,' Innis went even further to point out some of the less obvious ways in which the rise of newspapers influenced thought patterns in the twentieth century: 'The concentration of the natural sciences on the problems of physics and chemistry concerned with speed reflects the influence of the newspapers. Education systems and literacy have been subject to their influence directly and indirectly. Speed in the collection, production and dissemination of information has been the essence of newspaper development.'[40] Innis had come to a better understanding of technology as a mindset that shapes who we are and what we think.

Through a historical study of earlier civilizations, Innis searched for a convincing explanation for the current decline of the West. He found a pattern in the way the dominant technology of communication shaped a civilization's social structure and cultural values by controlling the ideas or mentality of its populace. The technology of communication established a 'monopoly of knowledge' dominated by the elite who controlled the technology. The existing rigidity prevented any new forms of knowledge or creative thinking from emerging within the civilization to ensure that it continued to flourish and grow; hence, the civilization declined at the very peak of its cultural creativity. This was the point at which Minerva's owl took flight to a new centre of creativity on the margin of the dominant civilization, where a new, more powerful communication technology controlled by a new elite took hold. Innis first presented this schema, and the various communication technologies that predominated in the various stages of the evolution of Western civilization, in 'Minerva's Owl,' his presidential address to the Royal Society of Canada in 1947. He noted at the outset the dominant technologies of communication in the various periods of Western civilization:

> I have attempted to suggest that Western civilization has been profoundly influenced by communication and that marked changes in communications have had important implications. Briefly this address is divided into the following periods in relation to media of communication: clay, the stylus, and cuneiform script from the beginnings of civilization in Mesopotamia; papyrus, the brush, and hieroglyphics and hieratic to the Graeco-Roman period, and the reed pen and the alphabet to the retreat of the Empire from the west; parchment and the pen to the tenth century or the dark ages; and overlapping with paper, the latter becoming more important with the invention of printing; paper and the brush in China, and paper and the pen in Europe before the inventing of printing or the Renaissance; paper and the printing press under handicraft methods to the beginning of the nineteenth century, or from the Reformation to the French Revolution; paper produced by machinery and the application of power to the printing press since the beginning of the nineteenth century to paper manufactured from wood in the second half of the century; celluloid in the growth of the cinema; and finally the radio in the second quarter of the present century.[41]

What enabled a particular communication technology to take hold and dominate was its ability to control either time or space. Innis defined these technologies as either 'time-biased' or 'space-biased.' Communication technologies that were durable and difficult to transport, such as stone, clay, or parchment, were time-biased, whereas those that were light and easy to transport over long distances, such as paper and papyrus, were space-biased. By 'bias,' Innis meant much more than just a preference for time or space. He argued that time-biased and space-biased communication technologies created certain types of civilizations with a particular orientation of knowledge. Time-biased civilizations were culturally tradition-oriented and emphasized custom, community, continuity, the sacred, the moral, and the historical. Socially, they were hierarchical societies ruled by an elite group that controlled the technology of communication. Space-biased civilizations were culturally concerned with the present and the future over the past and with the technical and the secular over the moral and the sacred. Socially, they were ruled by secular political authorities through the power of the state, either of an authoritarian or democratic nature, again depending on the particular communication technology that dominated. Space-biased civilizations also tended toward the growth of empires in centres where the ruling class controlled the dominant communication technology, enabling such centres to subjugate marginal groups.[42]

Innis noted how the principal medium of communication favoured one particular group within the civilization – the group that controlled the technology of communication – which maintained its power by preventing the emergence of any alternative communication technology that could threaten it. The oral tradition, for example, enabled the Spartan oligarchy to prevail; writing on papyrus benefited the Roman imperial bureaucracy; and parchment allowed the medieval clergy and the Roman Catholic Church to monopolize knowledge in the Middle Ages. By contrast, Gutenberg's mechanical print fostered the vernacular and allowed the monarchs of nation-states to consolidate their power and create vast empires through the merchant class.[43] The modern newspaper, a hybrid of the printed word and electronic media, particularly the telegraph, came under the control of the press lords, who in turn were pressured by charismatic political leaders and totalitarian rulers to print what they dictated.

A year after his presidential address to the Royal Society of Canada, Innis gave a speech at a conference of Commonwealth universities at Oxford in which he presented a key point that he would dwell on in numerous papers and articles on the impact of communication technology in the modern world: the threat of the power of technology to individual freedom in the modern world. He presented his concern at the outset: 'The conditions of freedom of thought are in danger of being destroyed by science, technology, and the mechanization of knowledge, and with them, Western civilization.'[44] He made an appeal to return to the qualities of the oral tradition and in so doing noted the limitations of the written tradition, particularly the printed word. It substituted education for learning and made that education uniform and dull through the use of 'textbooks, visual aids, administration, and conferences of university administrators.'[45] The printed word also placed greater value on scientific knowledge over humanistic knowledge and turned universities into instruments of subjection to the state.

Innis argued that the problem confronting the Western world in its attempt to deal with the myriad issues arising out of mechanization of knowledge was its inability to get beyond its own bias in dealing with these issues. He noted this concern in an address given at the University of Michigan on April 18, 1949. He both reproduced the address and used its title as the title for his collection of essays, *The Bias of Communication*. He stated, 'We can do little more than urge that we must be continually alert to the implications of this bias and perhaps hope that consideration of the implications of other media to various civilizations may enable us to see more clearly the bias of our own. In any case we may become a little more humble as to the characteristics of our civilization.'[46]

He followed that address a day later, on April 19, 1949, with another, called 'Technology and Public Opinion in the United States.' In this speech, Innis traced the history of the newspaper in the United States from the pre-Revolutionary era to the interwar years, emphasizing the role that technological discoveries, along with changing societal expectations, had on the nature, composition, and quality of newspapers. He pointed out how the power of the press in the United States increased its monopoly of knowledge over time by ironically appealing to the concept of the freedom of the press. Quoting Clarence Darrow, Innis observed that '"the most profound irony [is that] ... our independent American press, with its untrammeled freedom to twist and misrepresent the news, is one of the barriers in the way of the American people achieving their freedom" ... Freedom of the press had been an essential ingredient of the monopoly for it obscured monopolistic characteristics. Technological inventions were developed and adapted to the conservative traditions of monopolies of communication with consequent disturbances to public opinion and to political organization.'[47] In concluding his speech, Innis warned of the dangers of increased monopolization of knowledge in the electronic age of the radio; it emphasized space over time with the resulting emphasis on 'bureaucracy, planning, and collectivism.'[48]

These two papers – 'The Bias of Communication,' which looked at the bias of communication technology on civilizations in the past, and 'Technology and Public Opinion in the United States,' which examined the influence of newspaper technology on American civilization in the twentieth century – were, as noted, on consecutive days in 1949 and yet do not follow each other in *The Bias of Communication*. Rather, they are separated by four other papers and some 100 pages. This makes it difficult when reading the essays in *The Bias of Communication* to see the evolution of Innis's thinking about communication technology. Furthermore, it can lead to a failure to realize how Innis's concerns about the decline of Western civilization in his day as a result of the impact of modern communication technology were shaping his view of the history of Western civilization and conversely how his historical perspective was shaping his perspective on his present day. Innis believed that one read present concerns back into history; this was part of the 'bias' of the current communication technology of mechanized print and electronic media. As he noted by way of introduction to 'Technology and Public Opinion in the United States,' 'few subjects are exposed to more pitfalls than those concerned with public opinion since the student is so completely influenced by the phenomena he attempts to describe. Objectivity may be improved by considering its development over a long period of time but even a description of this character must register the results of an astigma adjusted to present environment.'[49] Likewise, he believed that one used history to shed light on the present; it was the communication technology in vogue at the time that most biased the questions asked of the past as well as the answers to those questions.

Innis's concern with the bias of the current communication technology of print and electronic media and his fear for the future of Western civilization dominated his writings in the final years of his life. In 'A Plea for Time,' an address that was part of the sesquicentennial lectures at the University of New Brunswick in Fredericton in 1950, Innis showed the multifaceted nature and meaning of time in a historical continuum from the beginning of the concept of time to the present. He emphasized that time was only a mental concept, an idea, always shaped by the dominant technology of communication present in each civilization. He noted the possible ways of defining time, from astronomical time to social time, linear time, and historical time. Each concept of time prevailed at any given period, Innis argued, and then was replaced by a new meaning of time given by groups or regions on the fringe that challenged and supplanted the dominant paradigm at the centre through the use of a new technology of communication that perceived time in a different way. In this way, technology and thought became the warp and woof of the rise and fall of every civilization. It was Innis's concern about the neglect of time and history in the present age that drove him to study the concept of time in past ages. 'The economic historian,' he pointed out, 'must consider the role of time or the attitude towards time in periods which he attempts to study, and he may contribute to an escape from antiquarianism, from present-mindedness, and from the bogeys of stagnation and maturity. It is impossible for him to avoid the bias of the period in which he writes but he can point to its dangers by attempting to appraise the character of the time concept.'[50]

Innis came to associate the problem of bias with empires, particularly the British Empire. In his Introduction to *Empire and Communications,* he argued that one of the characteristics of Western civilization in the twentieth century was a concern for 'the role of economic considerations in the success or failure of empires.'[51] He associated this economic perspective with the modern interest in or obsession with the history of the British Empire: 'Recognition of the importance of economic considerations is perhaps characteristic of the British Empire and it will be part of our task to appraise their significance to the success or failure of the British Empire and in turn to the success or failure of Western civilization.'[52] He went on to note the difficulty of achieving historical objectivity for someone, like him, who was 'obsessed' with the economic importance of the British Empire:

> A citizen of one of the British Commonwealth of Nations, which has been profoundly influenced by the economic development of empires, who has

> been obsessed over a long period with an interest in the character of that influence, can hardly claim powers of objectivity adequate to the task in hand ... Obsession with economic considerations illustrates the dangers of monopolies of knowledge and suggests the necessity of appraising its limitations. Civilizations can survive only through a concern with their limitations and in turn through a concern with the limitations of their institutions, including empires.[53]

Thus, Innis saw the British Empire as an integral part of Western civilization whose importance was of an economic nature.

Here again was the link to technology in Innis's thinking. It was the 'superior' technology of the British after the Industrial Revolution that enabled Britain to amass the most powerful empire in the world, thus making it the imperial centre of Western civilization in the nineteenth century. Innis's *History of the Canadian Pacific Railway* and his staple studies had gone a long way in explaining how Britain had achieved such dominance through the superiority of its transportation technology. In his communication studies, he noted the importance of developments in communication technology, especially newspapers, in explaining Britain's initial dominance in the early twentieth century and then its precipitous decline by the mid-twentieth century, thus expanding on observations he had made at the outbreak of the Second World War.[54]

Innis maintained that the advent of newspapers in Britain resulted in political partisanship and corruption and the rise of advertising. The British Parliament responded by restricting the freedom of the press. As a result, newspaper editors fled Britain. One of those editors, Benjamin Harris, moved to the United States, where he started the first newspaper in America, *Publick Occurences*. The beginnings of newspapers in America coincided with an improvement in print technology, when pulpwood replaced rags as a better and more economical way to make paper. The cost of newsprint dramatically declined 'from 8½ cents a pound in 1875 to 1½ cents in 1897.'[55] In a sweeping statement, Innis noted, '"taxes on knowledge" in Great Britain prior to 1861 restricted the development of newspapers, favoured a monopoly of *The Times,* reduced the demand for rags, and accelerated development of newspapers in the United States.'[56] With cheaper newsprint came the rise of the penny press with its emphasis on sensationalism, especially war news. Equally, cheaper newsprint meant fiercer competition and the resulting dependency on advertising as a means to sell papers.

These changes benefited the United States because of the easy access American newspaper entrepreneurs had to sources of pulp and paper, especially in northern Ontario and Quebec. As well, the American newspaper industry benefited from efficient railway transportation, which crisscrossed the United States for quick, easy, and wide distribution across the continent, telegraph service for news transmission, and, by the twentieth century, the radio as a source of information.

Innis realized, however, that rapid changes in communication technology in the modern age meant that Minerva's owl had no sooner set down in the United States when this 'newspaper civilization' entered its denouement in the 1920s. Part of the explanation was the rise of a new communication technology, the radio, that was supplanting the newspaper. More importantly, however, in keeping with his wider understanding of the pattern behind the rise and fall of civilizations throughout time, Innis noted that the newspaper's decline was due to the monopoly of knowledge newspapers had engendered, which prevented any new ideas that went counter to those of the newspaper lords and their political lackeys, who dominated the thinking of the time.

In his final years, Innis sketched out the implications of the decline of the British Empire for Canada and its relations with the United States. For example, he noted that the shift from the supremacy of the British to the American Empire altered Canada's economy such that demand for staples such as timber and wheat in Britain was being replaced by demand for staples such as pulp and paper and minerals desired by the United States. Since the transportation system in Canada in the early nineteenth century had been geared to trade with Britain, the rise of American dominance of the Canadian economy, beginning in the late nineteenth century, necessitated a fundamental shift in Canada's transportation network. Also, the new staples in demand for the American market were concentrated in Ontario and Quebec, which led to conflict between central Canada and the hinterlands of the west and the Maritimes to a significant degree. The advent of radio aided regional politicians and thus contributed to increased regionalism in Canada. Furthermore, Canadian dependency on the American economy, and the fact that the country came within the American communication orbit, resulted in Canadian subservience to American foreign policy too. Innis appealed to Canadians to 'call in the Old World to redress the balance of the New,'[57] seeing this action as their only hope in offsetting American dominance.

Innis held out hope that Canada might be the new centre of creativity where Minerva's owl would find its next resting place. He pointed out that the country's entire history, from the time of European exploration and settlement, had been one of marginality within Western civilization, first under the French Empire, then the British, and, more recently, the American. While Canada could not maintain its marginal position vis-à-vis the centre of imperial dominance in the United States during American imperial dominance in the twentieth century to the same extent that it had during French and British dominance in the eighteenth and nineteenth centuries, still, through a long tradition of anti-Americanism, an association and identification with Britain to offset the American influence, and a political ideology that incorporated both conservative and liberal values – in other words, values that were both time-biased and space-biased – Canada had created a society and culture different from that of the United States.

In his last publication, a booklet entitled *The Strategy of Culture,* Innis set out his concern for Canadian creativity and survival in the face of American imperial power. He also suggested that Canada, a country on the margin, had something positive to contribute to the modern world:

> The dangers to national existence warrant an energetic programme to offset them. In the new technological developments Canadians can escape American influence in communication media other than those affected by appeals to the 'freedom of the press.' The Canadian Press had emphasized Canadian news but American influence is powerful. In the radio, on the other hand, the Canadian government in the Canadian Broadcasting Corporation has undertaken an active role in offsetting the influence of American broadcasters. It may be hoped that its role will be even more active in television. The Film Board has been set up and designed to weaken the pressure of American films. The appointment and the report of the Royal Commission on National Development in the Arts and Sciences imply a determination to strengthen our position ...
>
> We are fighting for our lives. The pernicious influence of American advertising reflected especially in the periodical press and the powerful persistent impact of commercialism have been evident in all the ramifications of Canadian life. The jackals of communication systems are constantly on the alert to destroy every vestige of sentiment toward Great Britain holding it of no advantage if it threatens the omnipotence of American commercialism. This is to strike at the heart of cultural life in Canada. The pride taken in improving our status in the British Commonwealth of Nations has made it difficult for us to realize that our status on the North American continent is on the verge of disappearing. Continentalism assisted in the achievement of autonomy and has consequently become more dangerous. We can only survive by taking persistent action at strategic points against American imperialism in all its attractive guises.[58]

Innis's strident anti-Americanism in his latter years mystified his colleagues and has been a point of contention ever since. Yet it was in keeping with his analysis of the role of communication technologies in shaping mindsets within civilizations. Innis was concerned with the power of modern communication media, both mechanical and electronic, to control public opinion over vast spaces to such an extent that it was almost impossible for a counter-balancing time-biased media to emerge. The ability of the United States to extend its influence over distant parts of the world accounted for the unprecedented power of the American empire compared to all previous empires. It also made it all the more difficult, yet essential, for new centres of creativity, which could offer a counter-perspective to the dominant American paradigm, to emerge on the margin. As Innis stressed in *Empire and Communications,* 'The limitations of mechanization of the printed and the spoken word must be emphasized and determined efforts to recapture the vitality of the oral tradition must be made.'[59] The United States represented not only the latest of imperial dominance but also the worst, because it lacked a balance between time-binding and space-binding media and therefore a necessary check on its power. Marshall McLuhan perceptively noted that the roots of Innis's anti-Americanism came out of his aversion to power, which in turn was a result of his Canadianism:

> His [Innis's] hostility to power may well be a clue to his essential Canadianism, because power for Canadians has at all times been absentee and irresponsible power. At first Europe and England and now the United States – these are the centres where the decisions concerning us were and are made. The effects of this situation on national psychology have been according to basic temperaments. In the man of 'inner direction' it has bred an extremely independent attitude and a distaste for every kind of authority and social hierarchy. In the 'outer directed' man it has fostered an inclination to rigid bureaucratic structures. So that Canadians, on one hand, distrust Ottawa or have no interest in it, and on the other look to centralized government action to achieve the most ordinary local results. In neither case is there any vestige of American acceptance of power as essentially local and amenable to personal intervention.[60]

Eric Havelock, Innis's colleague from the Department of Classics at the University of Toronto, also dwelt on the issue of power and freedom in the context of technology. Havelock's views greatly assisted Innis in his reflections on the subject in his communication studies; conversely, Innis's views assisted Havelock in formulating his ideas. Of special importance to Innis was Havelock's lengthy Introduction, entitled 'The Crucifixion of Modern Man,' to his translation of Aeschylus's play *Prometheus Bound,* published in 1950. Havelock emphasized the danger to the freedom of the individual, particularly academic freedom, in a world of power politics reinforced by technology. Havelock's work also reminded Innis that the roots of the tension between freedom and power lay in ancient Greece. In his acknowledgments in *The Strategy of Culture,* Innis noted the importance of Havelock's work for his own views:

> The general argument [in *The Strategy of Culture*] has been powerfully developed in Aeschylus, *Prometheus Bound* as outlined in E.A. Havelock, *The Crucifixion of Intellectual Man* (Boston, 1950). Intellectual man of the nineteenth century was the first to estimate absolute nullity in time. The present – real, insistent, complex, and treated as an independent system, the foreshortening of practical prevision in the field of human action – has penetrated the most vulnerable areas of public policy. War has become a result and a cause of the limitations placed on the forethinker. Power and its assistant force, the natural enemies of intelligence, have become more serious since 'the mental processes activated in the pursuit and consolidating of power are essentially short range' (p. 99). But it will not do to join the great chorus of those who create a crisis by saying there is a crisis.[61]

In *The Crucifixion of Intellectual Man,* Havelock offered his own unique perspective on technology but one that was in keeping with Innis's perspective that the moral imperative was constantly being challenged and compromised by the emergence of the technological imperative. He argued that the myth of Prometheus, the immortal who stole fire (representing technology) from Zeus, the god of power, to give to mankind, was central to Greek thought and therefore to Western thought. He noted that the myth originated 'in the age of metals, when the use of fire for smelting has been taken as the hallmark of a whole culture.'[62] In *Prometheus Bound,* Havelock pointed out, fire represented 'the technological flame' and was responsible 'for the introduction of a whole array of those tools and skills which build the structure of a civilized life.'[63] Thus, fire became associated in the Greek mind with the applied sciences and the scientific mindset: 'The figure of the Fire-Stealer and Fire-Giver is made over into the figure of the great inventor and teacher and thinker.'[64] Havelock went on to explain that central to the Western scientific mindset had been the belief in the power of the human intellect, through technology, to shape one's own destiny. That idea was first enunciated in Greek thought. He wrote, 'The acceleration of industry and technology, with all its quantitative results in the broadcasting of material advantages, has had its own effect upon the changes in the mind of man ... The eyes and minds of men must themselves learn to read, and their tongues to speak, the vocabulary of technology.'[65]

According to Havelock, this scientific mindset, along with its technological by-products, had been responsible for some of the finest values of Western civilization: 'the extension of literacy to the masses, the rise of humanitarianism with its high regard for the individual life, and the forms of political freedom.'[66] However, Western civilization had paid a high price for such triumphs in the form of primitivism: the demonic side of human nature most poignantly expressed in wars and human genocide. Havelock noted that this duality – rationalism and primitivism – had been familiar to the Greeks, and they associated it with technology. This realization made *Prometheus Bound* an enduring play with a universal message for all ages. In essence, that message was 'technological man at the mercy of dictatorship.'[67]

Havelock saw the theme of the play as the intellectual always in opposition to the power brokers in society. In the play, Prometheus the 'Forethinker' had both wisdom derived from knowledge, including the 'power of technical invention and science applied to the creation of civilization'[68] and the gift of foresight, essential in order for knowledge to be used for the good of humanity. Pitted against Prometheus were the figures of Might (the Controller) and his assistant, Force (the Executive). Havelock argued that Prometheus represented 'the intelligence of man,' while 'the enemy he confronts bears a close resemblance to that other spiritual force, man's will to power. It is an influence equally operative along with technology and philanthropy in the history of human societies.'[69]

According to Havelock's interpretation of *Prometheus Bound,* the blame for the human predicament did not lie with technology: 'If change for the worse has occurred in certain areas of great public policy ... the trouble cannot lie in the technological habit of mind.'[70] Havelock noted that the Greeks saw technology in positive terms, as instruments 'to achieve what is "useful" and "beneficial" to the human race.'[71] *Techne* for the Greeks was a term that identified the intelligence of humans as tool users, both in theoretical and in practical ways, and as scientists *and* artists. 'The only "culture,"' he wrote, 'whether plastic or literary art, was a culture of the intellect.' What was to be feared was technology in the hands of power brokers intent on using it for purposes of dominating others – the will to power. Havelock warned, 'The bitter dialectic of the *Prometheus* seems to pursue us still. As the intellectual powers of man realize themselves in technology, with all its possibilities for human leisure and freedom, there seems to be raised up against them the force of a reckless dominating will, a compulsion of political and military power which says: "All else must yield to necessity. All intricate and subtle discussions must be laid aside. Man must be subjected afresh to agonizing pain and death."'[72]

Havelock argued that the role of the intellectual was to challenge those in positions of power by bringing his or her wisdom and foresight into play, for true foresight must of necessity be moral and philanthropic: 'Man cannot prethink evil, but only good.'[73] Havelock reminded his readers: 'This quality of intellectual prevision is close kin to the scientific imagination, and it needs the patience, the precision and the analysis of science to accomplish the stages of forethought; it calls for the discipline of measurement and a large dose of experimental courage.'[74]

Havelock concluded that *Prometheus Bound* offered no utopia, no redemption, only the challenge of the intellectual to resist the will to power within both the human dynamic and society. Here alone, the Greeks believed, lay the hope for humankind. Havelock wrote, 'If he [each human being] reinforce [sic] the courage of his intellect, he may yet achieve a better reconciliation between his will to power and his scientific vision. This is the race he must run, the price he must pay, for a just and lasting peace within his own soul, and among all nations.'[75]

In later studies, Havelock explored the roots of the Greek technological mindset as set out in *Prometheus Bound* with its 'powers of prevision' but also its 'habits of exact prognosis, scientific calculation, and rational exposition.'[76] However, the Greeks saw humans more as tool users than thinkers or, more precisely, as a fusion of the two modes of thought. He wrote, 'The Greek originally drew no formal distinction between the sciences of use and those of leisure, between technology and the fine arts.'[77] Havelock would thus have sided with Heidegger over Mumford on the issue as to whether humans were first tool makers or creative thinkers.

It was technology, or more precisely *techne,* Havelock argued, that lay at the heart of Greek thought and, consequently, Western thought. In *The Literate Revolution in Greece and Its Cultural Consequences,* he noted, 'Technology at varying levels supplies the necessary basis for all cultures modern and ancient. We have detected its presence in the non-literate period of Greek culture as applied to architecture, metallurgy and navigation.'[78] But the greatest technological feat of the Greeks, and possibly of Western civilization, was the discovery of the written word through the invention of the alphabet. 'Homo sapiens sapiens – the prototype of ourselves – achieved that condition of culture which we can identify as societal and humane only after his achievement and mastery of language.'[79] Noting Marshall McLuhan's emphasis on the revolutionary psychological and intellectual effects of the printing press in *The Gutenberg Galaxy,* Havelock argued that the true revolution in Western thought occurred with a much earlier technological invention in communication media than the printing press: the Greek alphabet. Script enabled the 'storage' of language, 'a body of "useful" oral information. Equally, through this same alphabetic instrument, there was discovered a new means of storage infinitely more efficient than the oral kind of which it had put into the record.'[80] Havelock noted, 'The use of vision directed to the recall of what had been spoken (Homer) was replaced by its use to invent a textual discourse (Thucydides, Plato) which seemed to make orality obsolete. Here was a paradox indeed of dialectical process, of transformational change. The singing Muse translates herself into a writer: she who had required men to listen now invites them to read.'[81]

Equally significant, coming out of the technological invention of the Greek alphabet, was the emergence of the 'concept of selfhood and the soul.' The emergence of self-identity occurred at a particular historical moment and 'was inspired by a technological change, as the inscribed language and thought and the person who spoke it became separated from each other, leading to a new focus on the personality of the speaker.'[82]

Eric Havelock's study *Prometheus Bound: The Crucifixion of Intellectual Man* assisted Innis in formulating his ideas on the power of technology to shape thought. In his article 'Harold Innis and Classical Scholarship,' John Watson suggests that Innis's concept of monopolies of force and knowledge in his communication studies was the equivalent of Havelock's classical dialectic of power and intelligence. Watson also suggests that Innis's dialectic of force and knowledge formed the basis of his time-binding and space-binding civilizations in that the former valued and enhanced knowledge, while the latter emphasized force. Therefore, in his search for balance or stability within civilizations, Innis was looking for a means to reconcile power and knowledge akin to that of the Greeks as presented in *Prometheus Bound*. What he discovered was 'the paradox of empire,' Watson contends, 'that power and intelligence were antithetical, and yet, at the same time, both had to be associated in the articulation of a successful imperial project.'[83] But what distressed Innis was the lack of balance between power and knowledge in the modern West, and Havelock's study of Greek thought reinforced in Innis's mind that this was an age-old problem, part of the human predicament.

Havelock equally benefited from Innis's insights into the role of communication technology in shaping Western civilization in his analysis of Greek mythology. The work of Innis and McLuhan helped Havelock discover the roots of modern Western thought in the technological invention of the alphabet, the likes of which enabled the power of the written word not only to record what was said but also to invent a textual discourse that formed the basis for the emergence of the reflective self by separating the person speaking from the inscribed language and thought, thus enabling a new focus on the personality of the speaker to emerge.

In *Harold Innis: A Memoir*, an insightful account of Innis's personality revealed in his thought, Havelock offered his own critique of Innis and noted where the two of them differed on the subject of technology. Havelock believed that Innis had too romantic a perspective on the oral tradition:

> It is all very well to stress the oral component in Greek culture, but after all, it was mainly the alphabet which released the energies of this culture into history. Without this technology, how much would the Romans, not to mention ourselves, have known of the Greek mind? They had more of the physical monuments than we have, but would these have been enough? There is the communication which is spoken in converse, the truly oral word. But there is also the communication which is placed in storage so that it can be consulted, re-used, and enlarged. In an oral culture this service

> can be performed by poetry which is orally memorized. But script adds a new, a fantastically enlarged dimension to this capacity, and it is this kind of stored communication which becomes the support of advanced civilizations.[84]

In the context of his analysis of Innis's study of modern technologies of communication, Havelock offered an explanation for Innis's pessimism and anti-Americanism in later life:

> As he [Innis] brings his survey into the nineteenth and twentieth centuries, one detects an underlying vein of pessimism. As invention has increased the rapidity with which information is processed for consumption and multiplied in its amount, the mind of modern man becomes preoccupied with instantaneous experience, at the expense of long range calculation. It is being deluged with ephemeral, at the expense of retrospection over the past and forethought for the future ... For him [Innis], the mass media are not ushering in any brave new world. In terms of the historical dialectic, the necessary balance between control of space and control of time is being destroyed. This conclusion colours a great deal of what he says about the United States and its history.[85]

Havelock's and Innis's perspectives on technology greatly advanced the understanding of the nature of technology and its impact on the rise of Western civilization and, in the case of Innis, on the predicted fall of that civilization due to its heavy dependency on the very technology that had enabled it to be so dominant. Clearly, these two intellectuals, and especially Innis, moved the discussion of technology well beyond the confines of external objects that had an impact on human thought to an exploration of the 'technological mindset' alluded to by earlier analysts of technology but never developed by them, where a tension between power and freedom was always present. For Innis and for subsequent theorists of technology in the post–Second World War era, it was the nature of this inner technological mindset – and the tension that occurred between the technological imperative and the moral imperative – that fascinated them and became the focus of their theories on technology.

8

Marshall McLuhan: Making Sense(s) of Technology

Marshall McLuhan was one of the first theorists of technology to thoroughly examine the nature, significance, and impact of the new electronic technologies of communication (the telegraph, telephone, radio, cinema, television, and computers) on the human senses, society and culture, and the history of Western civilization. While influenced by Innis's theories on communication technology, and while using the same elliptical writing style, McLuhan did not so much build on Innis's work as depart from it and take it in a new direction. He focused on a minor idea in Innis's work, namely, the impact of communication technologies on the human senses, and built his own unique theories on the significance of communication technology in shaping the modern mind of Western civilization.

McLuhan defined technology as any medium or artifact that was an extension of the human body and mind through our senses; this could be anything from clothing to computers: 'I think of technologies as extensions of our own bodies, of our own faculties, whether clothing, housing, and the more familiar kinds of technologies like wheels, stirrups, extensions of various parts of the body. The need to amplify the human powers in order to cope with various environments brings on these extensions, whether of tools or furniture. These amplifications of our powers, sorts of deifications of man, I think of as technologies.'[1] McLuhan believed that each new technology created a new human environment and thus a new way of thinking. However, since he saw humans as essentially communicative animals, it was the media, or technologies, of communication that were primary in shaping who we were, what we thought, and how we acted. As analyst James Carey notes, 'Technologies of communication were [for McLuhan] principally things to think with, molders of minds, shapers of thought.'[2] McLuhan saw each technology of communication as an externalizing or 'outing' of one or more of the senses that then dictated the nature of the sensorium ratio and thus altered human character and changed the way a person saw their natural environment, the structure and dynamics of their society and culture, the world, and knowledge itself. In *The Gutenberg Galaxy,* McLuhan wrote, 'When technology extends *one* of our senses, a new translation of culture occurs as swiftly as the new technology is interiorized.'[3] In keeping with Innis's analysis of staples of production and their impact on shaping the Canadian economy, McLuhan compared the media of communication to staples of natural resources in their ability to reshape a society and culture. 'Technological media are staples or natural resources, exactly as are coal and cotton and oil. Anybody will concede that [a] society whose economy is dependent upon one or two major staples like cotton, or grain, or lumber, or fish, or cattle is going to have some obvious social patterns of organization as a result ... Cotton and oil, like radio and TV, become "fixed charges" on the entire psychic life of the community. And this pervasive fact creates the unique cultural flavor of any society.'[4]

McLuhan believed that the impetus for change came from the external, dominant communication technology that reacted on, rather than through, the senses and that then caused a reconceptualizing and re-orchestrating of the thought patterns and therefore the mindset of the age. Thus, McLuhan noted, 'We shape our tools and afterwards our tools shape us.' McLuhan was less interested in understanding the roots and origins of new technologies or the social conditions that explained their emergence (as Innis was) than in explaining how they affected the senses and as a result influenced thought, creating with each new communication technology a new 'technological mindset.'

What was significant, then, McLuhan argued, was not the content, or the message, conveyed by the communication media but the means, or the medium, by which it was conveyed. Hence his famous aphorism 'The medium is the message.' McLuhan used the light bulb to explain his point in the context of the electric age. The light bulb could be used in a variety of ways which then became its 'message.' But behind 'the message' was a power or force, a medium, in the technology of the electric light that shaped time and space in a new and radically different way than was used in the industrial age. McLuhan explained, 'The message of the electric light is like the message of electric power in industry, totally radical, pervasive, and decentralized. For electric light and power are separate from their uses, yet they eliminate time and space factors in human association exactly as do radio, telegraph, telephone, and TV, creating involvement in depth.'[5]

The medium was also 'the *massage,*' a subliminal technological massaging of our senses without our awareness of its impact on our psychic and social relationships. As McLuhan noted in *The Medium Is the Massage,* 'All media work us over completely. They are so pervasive in their personal, political, economic, aesthetic, psychological, moral, ethical, and social consequences that they leave no part of us untouched, unaffected, unaltered. The medium is the massage. Any understanding of social and cultural change is impossible without a knowledge of the way media work as environments.'[6] The medium also created the massage, a radical new age of mass culture that referred less to the size than to the simultaneous interaction of individuals.[7] In 'Playboy Interview,' McLuhan expressed his frustration at the failure of people to realize the manipulative role of communication technologies on our senses and our mindsets: 'The new extensions of man and the environment they [communication technologies] generate are the central manifestations of the evolutionary process, and yet we still cannot free ourselves of the delusion that it is how a medium is used that counts, rather than what it does to us and with us. That is the zombie stance of the technological idiot.'[8]

McLuhan claimed that the impact of communication technology on the human senses had been the primary factor in shaping societies throughout history, from the earlier tribal societies, through the introduction of the phonetic alphabet, writing, and mechanized print, to the modern electronic communication technologies. Each new communication technology had had a major impact, but in the overall history of communication technology within Western civilization there had been three truly revolutionary changes. The first was a shift from an oral to a predominantly written media with the advent of the phonetic alphabet; the second was the introduction of mechanical print with the invention of the Gutenberg press; and the third was the beginning of the electronic age with the creation of the electric telegraph in 1844. Each of these revolutionary changes in technology had a greater impact than the previous one. McLuhan wrote, 'If the phonetic alphabet fell like a bombshell on tribal man, the printing press hit him like a 100-megaton H-bomb.'[9] The introduction of electronic technology was having an even more dramatic impact, resulting in a retribalizing of man and the creation of a global village, because it was the 'outing' of all the senses, being an extension of the human nervous system rather than just one (the visual) as in the case of phonetic writing and the printing press. In affecting all of the senses, the electronic media would ultimately restore a balance to the senses. There was an optimism, tempered by caution, in McLuhan's perspective on electronic technology in the modern age and its impact on Western civilization that was absent in the other major post–Second World War Canadian theorists of technology.[10]

While McLuhan believed that technologies of communication reacted on our senses often in subliminal ways, thus making it difficult for individuals to be aware of the impact or capable of reacting to it until after the fact and a new technological force had taken hold, nevertheless the whole thrust of his thoughts on technology was to make people aware of its impact. He became the guru of the technological imperative, providing guidance to the uninitiated on the belief that knowledge of the technological imperative was the first step to dealing with it. Once aware, individuals would then be free to decide what they accepted as good in new technology and what they rejected as harmful. McLuhan verged on being a technological determinist, but he never entirely abandoned his faith in the power of choice in the way people could respond to technology. Even to accept what was happening and to 'go with the flow,' as the sailor in Edgar Allan Poe's story did when caught in a 'whirlpool' – a story McLuhan delighted in telling as justification for his own acceptance of the new electronic technology – was a conscious choice. Thus, McLuhan identified with the concept of technology as volition.

McLuhan was so convinced of the ultimate benefits of the new electronic technologies of communication, especially the most recent, television, that he seldom questioned or doubted their virtues. If there were limitations or pains, they were in the means of getting to what he believed would ultimately be a technological utopia, not the end itself. Still, he realized that the new society he envisioned emerging out of the electronic technological revolution would be so radically different as to 'unravel the entire fabric of our society.' A part of him wanted to hang on to what was being lost, to maintain the moral imperative of his Catholic faith while enjoying the freedom that the technological imperative promised. That was the furthest he went in attempting to reconcile the moral and technological imperatives.

The roots of McLuhan's thinking on technology lay in his background and early education and can be seen in the evolution of his thought on the subject as reflected in his first two important studies, *The Mechanical Bride* and *The Gutenberg Galaxy,* before his major study on the subject, *Understanding Media: The Extensions of Man*. Herbert Marshall McLuhan was born in Edmonton, Alberta, on July 21, 1911.[11] He had one sibling, a younger brother, Maurice, born in 1913. His grandparents of both sides of the family had been the first generation to homestead on the Prairies. His great-grandparents on the McLuhan side had come from Northern Ireland and migrated to Ontario. Marshall McLuhan's grandfather, James McLuhan, and his son Herbert, Marshall's father, came west in 1907, settling on land near Miniburn, Alberta, east of Edmonton. In 1909, Herbert laid claim to his own 160-acre homestead in the area. There he met his future wife, Elsie Hall.

The Halls originated in England, but the family had resided in Nova Scotia since 1800. In 1906, Marshall's maternal grandfather, Henry Seldon Hall, moved west with his wife and two sons to homestead near the town of Minniville, some 100 miles east of Edmonton. Their daughter, Elsie, joined them in 1907, after completing her teacher training at Acadia University, in Wolfville, Nova Scotia. She taught in districts neighbouring her parents' homestead at Minniville. At one time she boarded with Marshall McLuhan's parents. It was there that she first met Herbert.

The couple married in 1909. They homesteaded for one year before moving to Edmonton, where Herbert became a real estate agent, forming a company with three other men during the boom year of 1912. By 1914, the company had gone bankrupt, a result of the economic downturn of 1913-14. In 1915, Herbert joined the army and served as a recruiting officer, but he was discharged within the year as a result of influenza. The McLuhan family then moved to Winnipeg, where Herbert became a salesman for the North America Life Assurance Company. Elsie joined the Alice Leone Mitibell School of Expression to begin a lifelong career as an elocutionist and monologist.

Marshall McLuhan, while born in Edmonton, spent most of his childhood and youth in Winnipeg. He claimed in later life that his Prairie upbringing had provided him with 'a kind of natural "counterenvironment" to the great centers of civilization.'[12] According to Philip Marchand, McLuhan's biographer, McLuhan 'felt he had the advantage that any bright outsider brings with him from the boondocks when he comes to the big city: a freshness of outlook that often enables him to see overall patterns missed by the inhabitants who have been molded by those patterns. It was the advantage that he felt accrued to Canadians in general vis-à-vis the United States and Europe.'[13] The perspective was not unlike that of McLuhan's mentor, Harold Innis, who also believed that Canadians had the advantage of a different perspective on Western civilization by living on the margin of that civilization in Canada. They were able, in a sense, 'to look in at it,' unlike those intellectuals who were immersed in the centre of that civilization without the advantage of distance or a counter-perspective. Being a Canadian had its advantages for a critic. Not only Innis and McLuhan came to appreciate this 'Canadian advantage'; so did two of McLuhan's contemporaries, Northrop Frye and George Grant.

Marshall McLuhan attended the University of Manitoba from 1928 to 1934, where he received his BA and MA in English. Being a precocious student, he was never impressed by the calibre of his Manitoba professors. The two who did stand out were only temporarily at the university as visiting professors from England: R.C. Lodge, a professor of classics from Oxford who introduced McLuhan to Plato and an appreciation of classical philosophy in general, and Noel Fieldhouse, a history professor from Cambridge, who introduced his impressionable young student to the writings of the great British historian Thomas Babington Macaulay. Macaulay became one of McLuhan's intellectual heroes and the subject of his first publication, an article in *The Manitoban,* the student newspaper, in 1930. Although McLuhan enjoyed history and classics, he decided to major in English (even though he was unimpressed with anyone in the department), a decision he would later come to regret. He wrote his MA thesis on George Meredith as a poet and dramatic novelist.

In the summer of 1932, between the completion of his BA degree and the beginning of his MA program, McLuhan and his friend Tom Easterbrook (who became a student of Harold Innis's and a well-known economist in his own right) made a trip to England. McLuhan fell in love with England, with its rich history and equally stimulating literary tradition. He recalled in later life purchasing a copy of Palgrave's *Golden Treasury,* an anthology of great poetry. 'Its verse,' Philip Marchand maintains, 'added a mellow haze to every ruined abbey and meadow he [McLuhan] encountered.' Marchand noted, 'He [McLuhan] was twenty-one years old, and he saw in poetry and literature a noble protest of the human soul against the mechanical and the commercial, against the vulgarities of modern life. Modern life wholly dominated Winnipeg, as it did all of North America; but in England one could still find echoes of something cultivated and literary that had once set the tone of civilization. "It was like going home," McLuhan later said. "The place I had grown up imagining was my headquarters."'[14]

McLuhan expressed this viewpoint in his second article for *The Manitoban,* entitled 'Tomorrow and Tomorrow?' He described the modern world as 'hopelessly sunk in corruption.' He cited with approval such moral critics of modernity, with its emphasis on machines, as Thomas Carlyle, John Ruskin, and William Morris, all of whom had looked back on the Middle Ages as a golden era in European civilization. According to Marchand, McLuhan told a colleague in later life that as a youth he had rejected the twentieth century as 'totally unfit for human habitation.'[15]

From Manitoba, McLuhan went to Cambridge University, where he did a second BA and also a PhD. His doctoral dissertation was on the Elizabethan writer Thomas Nashe, in which he used Nashe's writings to examine 'the modern biases in logic and science that culminated in the triumph of the Newtonian worldview, and its clockwork mechanical perfection, in 19th-century Europe.'[16] Being a 'colonial,' McLuhan, like so many other Canadians who studied in England, felt like an outsider. As Marshall's friend John Wain, the English writer, noted, 'Marshall was a country boy from the prairies, and in a way his obsession at being up-to-date, his delight at being able to manipulate the modern world, always contained a bit of the rustic.'[17]

However, McLuhan was impressed with the intellectual calibre of his mentor I.A. Richards, who taught him to see a poem as 'simply a supreme form of human communication. A reader analyzes a poem to see how it is able to achieve its effects – that is, to communicate an experience. English studies are themselves nothing but a study of the process of communication.'[18] Richards and his colleague William Empson would become the 'godfathers' of the New Criticism in English literature. In later years, McLuhan would boast that he was the only student of their work 'to perceive its usefulness in understanding electronic media.'[19] However, according to

Marchand, it would be a while before McLuhan would come to appreciate the core idea of the New Criticism: that 'if words were ambiguous and best studied not in terms of their "content" (i.e., dictionary meaning) but in terms of their effects in a given context, and if those effects were often subliminal, the same might well be true of other human artifacts – the wheel, the printing press, and so on.'[20] As well, Empson's comment that 'the process of getting to understand a poet is precisely that of constructing his poem in one's own mind' became the basis for McLuhan's later point that the 'content' of any poem is the reader of that poem. Marchand notes, 'McLuhan extended the insight to mean that the content of any medium or technology is its user.'

F.R. Leavis, a former student of Richards and also in the Department of English at Cambridge, suggested to McLuhan that the New Criticism could be a useful approach to studies of the entire human environment. Leavis's *Culture and Environment* showed McLuhan the way. It was a lament for the passing of what Leavis called the 'organized community,' 'in which people were educated in folk traditions, crafts, and ways of life based on the soil and on cottage industries.'[21] In the 1930s, McLuhan agreed with this perspective, but he would later come to reject it. What he retained from Leavis's study was the ability to apply the New Criticism to other areas of study besides literature, like advertising. Leavis also introduced McLuhan to the twentieth-century poets Ezra Pound and T.S. Eliot and showed him how to 'read a human artifact,' be it a poem or a machine, not by what it said or did but in terms of its impact on the thinking process.

After Cambridge, McLuhan acquired a one-year teaching assistant position in the English department of the University of Wisconsin. During that year he converted to Roman Catholicism. Wanting now to teach at a Catholic university, he got the opportunity when St. Louis University, a Jesuit institution, hired him as an instructor in their English department in 1937. According to Marchand, St. Louis University 'completed McLuhan's education in the history and development of Western civilization by immersing him in the thought of the Middle Ages and the Renaissance.'[22] The institution also reinforced his dislike of the modern age. In an article for the university's literary magazine, the *Fleur de Lis,* entitled 'Peter or Peter Pan,' McLuhan condemned advertising, industrialism, big business, and Marxism as modern evils that undermined the family and the church. They forced Western man to choose between two 'Peters': Peter, being Saint Peter, representing the Roman Catholic Church as its first leader, or Peter Pan, representative of the fantasy world of modern Western civilization.[23]

While at St. Louis, McLuhan familiarized himself with the ideas of three individuals whose writings on technology would influence his own thinking on the subject. One person was Walter Ong, his MA student, who wrote his thesis on Gerard Manley Hopkins. McLuhan also directed Ong to an obscure

Renaissance theologian by the name of Peter Ramus for his PhD study. Out of Ong's writings would come an idea that would infuse McLuhan's *The Gutenberg Galaxy,* namely, that 'Western culture in the Renaissance had shifted from a primarily auditory mode of apprehending reality to a primarily visual mode and that the vehicle for this shift was the invention of printing.'[24]

Lewis Mumford and Siegfried Giedion were the other two individuals whose works on technology influenced McLuhan at this time. Mumford's major work, *Technics and Civilization,* posited a radical distinction between the first stage of industrial civilization, based on steam power and highly mechanical in character, and a second stage based on electricity and organic in nature. McLuhan would also come to see this transition from the mechanical age to the electronic age as revolutionary, but it was in terms of its impact on the human senses, and through them, on the nature of Western society and thought. Mumford saw the second stage as beginning with the discovery of the telegraph and then the telephone in the mid-to-late nineteenth century, a point that McLuhan would also make in his writings on the electronic age. As well, McLuhan would agree with Mumford that this new electronic age held the potential of creating a worldwide communication community, a 'global village,' although Mumford did not use that phrase that has come to be associated with McLuhan. The electronic age could decentralize society and thus reverse the process of building up huge cities and large factories and return society to its old rural, artisan, and community-based way of life. While at that point in time McLuhan would dismiss Mumford's utopian vision as simplistic and Rousseau-like, he did agree with Mumford's argument that it was the printing press that had inaugurated the modern standardized and conglomerate society of mechanistic artifacts that in turn had contributed to the '"lost balance between the sensuous and the intellectual, between image and sound, between the concrete and the abstract" that had been achieved, to some extent, by medieval civilization.'[25]

McLuhan had met Siegfried Giedion in St. Louis at the beginning of the Second World War. Giedion's *Mechanization Takes Command: A Contribution to Anonymous History* (1948), published after McLuhan had left St. Louis, remained a valuable resource for McLuhan throughout his life. The book showed how everyday objects or activities could serve as 'artifacts' to reveal the values of a society, especially the increasing mechanization of human life. Part of that mechanization process, Giedion pointed out and McLuhan concurred, was the severing of thought from feeling. McLuhan came to believe, however, that what was true of the mechanical age could be undone in the electronic age, where intellect and emotions could be reintegrated by creating a 'new' (i.e., prosthetic) human being.

In 1944, McLuhan accepted a teaching post at Assumption College (now the University of Windsor) in Windsor, Ontario. It was clearly a step down in prestige from St. Louis University. McLuhan described the University of

Windsor as '"a little backwater in a stagnant stream" (the "stagnant stream" being Canada).'[26] He remained there for only two years. In 1946, St. Michael's College at the University of Toronto offered McLuhan a position, which he readily accepted. The position ended his days as a wandering scholar. He remained at the University of Toronto for the rest of his academic career with the exception of one year, 1967-68, when he held the Schweitzer Chair at Fordham University in New York City.

McLuhan often used slides of advertisements as a teaching tool to encourage students to 'read' these ads for their insight into the mores and milieu of the modern age. At one such series of lectures at Assumption College, he labelled the present era 'the Age of the Mechanical Bride,' an aphorism that stuck. He decided to write a book on the subject in which he would reproduce an advertisement or comic strip as an exhibit or cultural artifact that he would then analyze and reflect on in an accompanying short essay. The book, published in 1951 by Vanguard Press, was called *The Mechanical Bride: The Folklore of Industrial Man*.[27] It was meant to capture the impact of mechanical technology with its accompanying industrialism, mechanization, dialectical thinking, and capitalism at the point in time when this industrial or mechanical age was being supplanted by a new and radically different electrical age, and thus, he commented jokingly, the 'Mechanical Bride' would become the 'Electronic Bride.' In a sense, *The Mechanical Bride* became McLuhan's farewell discussion of the machine age of Western civilization, since in later publications (with the exception of parts of *The Gutenberg Galaxy*) he turned to a study of the new electronic age.

The Mechanical Bride was an attempt to alert the public to the power of technology over their lives through the manipulation of their minds by advertisements.[28] McLuhan believed that in the modern world of technology – the machine age – the aim was the total mechanization of the human personality. McLuhan's aim was to use these advertisements 'to enlighten [their] intended prey' by showing people what they were designed to do. As he noted in his Preface, 'Where visual symbols have been employed in an effort to paralyze the mind, they are here used as a means of energizing it.'[29] He compared his task to that of Edgar Allan Poe's sailor in 'A Descent into the Maelstrom,' who saves himself by studying the action of the whirlpool and by going with the current rather than against it. McLuhan believed that the industrial age had too much influence on humankind to resist, so that the best that one could do was study the effect of technology so as to deal with it more effectively.

While McLuhan would later claim that his task was only to offer commentary and criticism on technology – 'probes,' as he liked to call his aphorisms – rather than moralize and condemn, his disgust at the manipulation of the human mind through advertisements so as to create the 'mechanical man' – his version of the modern Frankenstein 'man-machine' – was clearly evident. Indeed, McLuhan compared the humans in these advertisements to machines, ones resembling living organisms, comparable to the vision of such nineteenth-century critics of technology as Samuel Butler and Mary Shelley. He noted, for example, that, 'as early as 1872, Samuel Butler's *Erewhon* explored the curious ways in which machines were coming to resemble organisms not only in the way they obtained power by digestion of fuel but in their capacity to evolve ever new types of themselves with the help of the machine tenders. This organic character of the machines, he saw, was more than matched by the speed with which people who minded them were taking on the rigidity and thoughtless behaviorism of the machine.'[30] Equally, the robotic response of mechanized individuals reminded him of Frankenstein's monster: 'Many of the Frankenstein fantasies depend on the horror of a synthetic robot running amok in revenge for its lack of a "soul." Is this not merely a symbolic way of expressing the actual fact that many people have become so mechanized that they feel a dim resentment at being deprived of full human status?'[31]

He warned that, in the modern world of technology, machines were as threatening to the well-being of human life as wild animals were to prehistoric man. Both required taking on the guise of the enemy – prehistoric man donning animal skins; modern man 'propagating the behavior mechanisms of the machines' – in order to survive. Thus, by the early 1950s already, McLuhan saw the power of technology in shaping modern thought by creating a mindset necessary for mechanization to thrive.

At this point in his intellectual journey, McLuhan discovered the works of Harold Innis. Innis occasionally joined a coterie of McLuhan's colleagues, who met most days at 4:00 p.m. at the coffee shop at the Royal Ontario Museum. McLuhan acknowledged Innis's contribution to his own thinking about communication technology. On one occasion he noted, 'Until the work of Harold Innis, I have been unable to discover any epistemology of experience, as opposed to epistemology of knowledge.'[32] In *The Gutenberg Galaxy,* he acknowledged that 'Harold Innis was the first person to hit upon the *process* of change as implicit in the *forms* of media technology. The present book is a footnote of explanation to his work.'[33] Later on, in addressing the subject of nationalism and print, McLuhan acknowledged that 'the present volume to this point might be regarded as a gloss on a single text of Harold Innis: "The effect of the discovery of printing was evident in the savage religious wars of the sixteenth and seventeenth centuries. Application of power to communication industries hastened the consolidation of vernaculars, the rise of nationalism, revolution, and new outbreaks of savagery in the twentieth century."'[34] In his Introduction to a reprint of Harold Innis's *The Bias of Communications,* McLuhan once again noted, 'I am pleased to think of my own book *The Gutenberg Galaxy* (University of Toronto Press,

1962) as a footnote to the observations of Innis on the subject of the psychic and social consequences, first of writing and then of printing.'[35] In a less charitable acknowledgment, McLuhan once commented that he could not fathom how this 'hick Baptist' ever came up 'with this amazing method of studying the effects of technology.'[36]

However, the two theorists strongly disagreed on the nature of radio as a communication technology. Innis saw the radio as a space-biased medium, an extension of the power of the written word, whereas McLuhan saw it as a radical break from the age of the printed word, as one of the electronic communications media that would liberate humankind from the sensual limitations of linear print, return the world to the age of oral culture, and in so doing bring about the global village. 'It had not occurred to Innis,' McLuhan wrote, 'that electricity is in effect an extension of the nervous system as a kind of global membrane. As an economic historian he had such a rich experience of the technological extensions of the bodily powers that it is not surprising that he failed to note the character of this most recent and surprising of human extensions.'[37]

The Gutenberg Galaxy, McLuhan's second book, was an attempt to understand the world of print technology with its mechanical environment when it was disappearing with the advent of the new electronic age. 'In our time,' McLuhan wrote as a synopsis of the book, 'the sudden shift from the mechanical technology of the wheel to the technology of electric circuitry represents one of the major shifts of all historical time ... *The Gutenberg Galaxy* is intended to trace the ways in which the *forms* of experience and of mental outlook and expression have been modified, first by the phonetic alphabet and then by printing.'[38] In essence, the aim of the book was to study the current electrical age by looking at it from the perspective of the preceding print age so that the former unfolded as the latter was revealed.

McLuhan claimed that *The Gutenberg Galaxy* attempted to answer the question raised by Peter Drucker in his article 'The Technological Revolution': '"What happened to bring about the basic change in attitudes, beliefs, and values which released it [the technological revolution]? ... How responsible was the great change in world outlook which, a century earlier, had brought about the great Scientific Revolution?" *The Gutenberg Galaxy* at least attempts to supply the "one thing we do not know."'[39] McLuhan found the answer to be the technology of communication – particularly Gutenberg's printing press – which re-orchestrated the human senses with an emphasis on seeing. Print was a form of technology that extended the eye and emphasized the visual over the oral. It isolated the sense of seeing from the other senses and thus hypnotized society. 'The formula for hypnosis is "one sense at a time." And new technology possesses the power to hypnotize because it isolates the senses.' Then, quoting William Blake, McLuhan noted that "they become what they beheld."[40] *The Gutenberg Galaxy* attempted to explain 'what they beheld' and why they beheld it.

When asked why others had not studied 'the obvious,' McLuhan replied that the introduction of a new communication technology had a revolutionary effect that was jarring for the individual. The nervous system responded by instituting 'a self-protecting *numbing* of the affected area, insulating and anesthetizing it from conscious awareness of what's happening to it.' McLuhan called it 'Narcissus' from the Greek word *narcosis,* 'a syndrome whereby man remains as unaware of the psychic and social effects of his new technology as a fish of the water it swims in. As a result, precisely at the point where a new media-induced environment becomes all-pervasive and transmogrifies our sensory balance, it also becomes invisible.'[41] McLuhan believed that it was only when a new technology and environment took hold and transformed the old environment that both environments were made visible by contrast. Thus, he argued that the age of mechanized print, with its resulting mechanization of thought and society as a whole, was revealed and made comprehensible only with the inauguration of the age of electric circuitry. At the same time, a study of the mechanical age shed light on the electrical age by means of contrast.

McLuhan saw the Gutenberg age of print as 'fragmented, individualistic, explicit, logical, specialized and detached.'[42] The whole man became disjointed and alienated because the emphasis on seeing, at the expense of the other senses, focused on only one aspect of his character. Dennis Duffy notes that, for McLuhan, 'printing with its continuous and uniform production established a fixed point of view on a psychic level.'[43] Society broke down into individual units, and the world operated on linear time and Euclidean space. McLuhan argued that a dualism emerged between sight and sound, head and heart, man and nature that ultimately resulted in 'atrophy,' since the vitality and imagination that were offshoots of a balanced sensorium were lost. Furthermore, the technology of typography created 'an analytic sequence of step-by-step processes ... by producing the first uniformly repeatable commodity'[44] that could be mechanically applied to all other functions. Thus, according to McLuhan, the printing press 'created Henry Ford, the first assembly line, and the first mass production. Movable type was archetype and prototype for all subsequent industrial developments. Without phonetic literacy and the printing press, modern industrialism would be impossible.'[45] In this respect, McLuhan argued, Gutenberg's printing press laid the groundwork for the Industrial Revolution. But it did much more. By producing books in infinite numbers, it assured universal literacy and enabled individuals to pursue knowledge independently and in isolation. Also, since print matter could be easily disseminated across Europe, 'vernacular regional languages of the day' were turned into 'uniform closed systems of national languages.'[46] McLuhan therefore saw a direct correlation between the printing press and the rise of such disparate phenomena as 'nationalism, the Reformation, the assembly line and its offspring, the Industrial Revolution, the whole concept of causality, Cartesian and Newtonian concepts of the universe, perspective in art, narrative chronology in literature and a psychological mode of introspection or inner direction that greatly intensified the tendencies toward indivi-dualism [sic] and specialization engendered 2000 years before by phonetic literacy.'[47] Mechanical print was the *sine qua non* of the mechanical age with all its psychic, social, economic, and political ramifications. He noted in *The Gutenberg Galaxy* that 'the modern phases of this process are the theme of *Mechanization Takes Command* by Siegfried Giedion.'[48]

McLuhan believed that the technology of the printing press inhibited the ability of modern human beings to face the new technology of electric circuitry while ironically preparing the way for the new age by providing the means for it to emerge. He claimed that each new technology built on the previous one while at the same time making it redundant. McLuhan used as his metaphor the concept of humans as the 'sex organs' of a new technology, the agents of 'cross-fertilization' that brought it into being.[49] He described such periods of transition as new and exciting but also frightening and painful, since the old was replaced by the new. In *The Gutenberg Galaxy,* he noted two major eras of transition. One was the sixteenth-century Renaissance as 'an age on the frontier between two thousand years of the alphabetic and manuscript culture, on the one hand, and the new mechanism of repeatability and quantification, on the other.'[50] The other was the new electronic age with its media of film, radio, and television. McLuhan wrote, '[Here] the actual *transfer of learning* and change in mental process and attitude of mind that has occurred has been almost entirely subliminal. What we acquire as a system of sensibility by our mother tongue will affect our ability in learning other languages, verbal or symbolic. That is perhaps why the highly literate Westerner steeped in the lineal and homogeneous modes of print culture has much trouble with the non-visual world of modern mathematics and physics. The "backward" or audile-tactile countries have a great advantage here.'[51] He commented on the general significance of these transition periods: 'An age in rapid transition is one which exists on the frontier between two cultures and between conflicting technologies. Every moment of its consciousness is an act of translation of each of these cultures into the other.'[52] His perspective was reminiscent of Innis's concept of new communication technologies emerging on the margin and eventually supplanting and eliminating the old technology and the group who controlled it at the centre by creating a new monopoly of knowledge.

Understanding Media: The Extensions of Man was McLuhan's attempt to analyze the significance of the new age of electric communication technology. As cultural analyst Lewis H. Lapham notes, 'Despite its title, the book was never easy to understand. By turns brilliant and opaque, McLuhan's thought meets the specifications of the epistemology that he ascribes to the electronic media – non-lineal, repetitive, discontinuous, intuitive, proceeding by analogy instead of sequential argument.'[53] The book consists of an Introduction, seven chapters on the nature of the media in general, and twenty-six chapters on specific media, including human speech, print, clocks, money, and of course television. Philip Marchand perceptively notes that 'the number of chapters appeared to be cabalistic in design – seven introductory chapters followed by a number of chapters equal to the number of letters in the alphabet, in illustration of McLuhan's thesis that the phonetic alphabet was the real source of all subsequent Western technology.'[54] McLuhan began by noting the revolutionary significance of this new age:

> After three thousand years of explosion, by means of fragmentary and mechanical technologies, the Western world is imploding. During the mechanical ages we had extended our bodies in space. Today, after more than a century of electric technology, we have extended our central nervous system itself in a global embrace, abolishing both space and time as far as our planet is concerned. Rapidly, we approach the final phase of the extensions of man – the technological simulation of consciousness, when the creative process of knowing will be collectively and corporately extended to the whole of human society, much as we have already extended our senses and our nerves by the various media.[55]

McLuhan believed that the significance of electric technology was its 'total-field awareness,' extending not just one of the senses but the entire human body through an extension of the central nervous system. Equally, however, McLuhan emphasized that the world contracted inward into the human psyche – it 'imploded' – altering our thought patterns and the way we react to the external world. He compared the body's nervous system to electric circuitry, thus reinforcing the analogy. And just as electric circuitry made information instant, so too did it make humans instantly and fully conscious of what was happening around them and through them, thus enabling them 'to win back control of [their] own destinies.'[56] It also meant that modern human beings responded instantly and were more involved in the world around them. They became 're-tribalized' in 'the global village.'

McLuhan saw this instant response to stimuli and events as one of the major differences between 'mechanical' and 'electrical man.' He wrote:

> Western man acquired from the technology of literacy the power to act without reacting ... We acquired the act of carrying out the most dangerous social operations with complete detachment. But our detachment was a

> posture of noninvolvement. In the electric age, when our central nervous system is technologically extended to involve us in the whole of mankind and to incorporate the whole of mankind in us, we necessarily participate, in depth, in the consequences of our every action. It is no longer possible to adopt the aloof and dissociated role of the literate Westerner ... The aspiration of our time for wholeness, empathy and depth of awareness is a natural adjunct of electric technology.'[57]

In 'Playboy Interview,' he noted that the electronic media brought us 'back in touch with ourselves as well as with one another.'[58] If the essence of machine technology was to fragment, compartmentalize, and isolate, then the essence of automation technology was to unify, integrate, and reunite. It made everything instant and whole. McLuhan claimed that myth was 'the instant vision of a complex process that ordinarily extends over a long period.'[59] With information expanding at a rapid rate in the present electric age, humans lived mythically, even if they still thought fragmentally.

However, such unity and involvement varied according to the different electronic technologies used; it depended whether it was a 'hot' medium or a 'cool' medium. No aspect of McLuhan's theories created more controversy than his discussion of hot and cool media with their accompanying 'high definition' and 'low definition.' McLuhan argued that each new electric technology was either hot or cool according to the dominant sense or senses that it brought into play. A hot medium was one that extended one single sense in high definition, which he defined as 'a state of being well filled with data.' By contrast, a cool medium was one that extended a sense or number of senses in low definition, meaning it provided little information and thus required greater participation on the part of the individual to acquire or add the additional information. The telephone was a cool medium, McLuhan argued, since the ear was given a meagre amount of information, whereas radio and film were hot media with high definition, requiring very little participation on the part of the listener or the observer to receive the message. He noted that 'hot media are, therefore, low in participation, and cool media are high in participation or completion by the audience.'[60]

McLuhan argued that television was a cool medium with low definition – little detail and a low degree of information – that, like the telephone, required greater participation of the observer. McLuhan also claimed that television was the most significant of the electric media and the one that, more than any of the other electric media, ended 'the visual supremacy that characterized all mechanical technology,'[61] because, contrary to popular belief, it was not primarily a visual media but rather a *tactile* one and thus involved maximal interplay of all the senses. 'With TV, the viewer is the screen,' McLuhan noted. 'He is bombarded with light impulses that James Joyce called the

"Charge of the Light Brigade" that imbues his "soulskin with subconscious inklings."' McLuhan attempted to explain why this was so:

> The TV image is not a still shot. It is not photo in any sense, but a ceaselessly forming contour of things limned by the scanning-finger. The resulting plastic contour appears by light through, not light *on,* and the image so formed has the quality of sculpture and icon, rather than of picture. The TV image offers some three million dots per second to the receiver. From these he accepts only a few dozen each instant, from which to make an image ... The TV image requires each instant that we 'close' the spaces in the mesh by a convulsive sensuous participation that is profoundly kinetic and tactile, because tactility is the interplay of the senses, rather than the isolated contact of skin and object.[62]

Hence, he concluded that the TV image 'demands participation and involvement in depth of the whole being, as does the sense of touch.'[63] According to McLuhan, this accounted for the social activism of the youth of the 1960s; they were the first generation of children to be raised on television. 'The young people who have experienced a decade of TV,' he wrote in 1964 in *Understanding Media,* 'have naturally imbibed an urge toward involvement in depth that makes all the remote visualized goals of usual culture seem not only unreal but irrelevant, and not only irrelevant but anemic.'[64] The observation appeared to hold true for the 'flower culture' of the 1960s as the postwar baby boomers became involved in protest both on and off university campuses, although the explanation McLuhan gave for such involvement did not convince his critics.

Here was the basis for McLuhan's concept of the 'global village.' In the new electric age, the old distinction of centre and margin that was characteristic of the mechanical age no longer applied. The speed of information created centres everywhere and anywhere, making margins obsolete. In visual space a distance existed between the seer and the object seen, but in acoustic space that distance disappeared. Sounds bombarded the individual and filled the space around that individual. Acoustic space, McLuhan argued, 'is organic and integral, perceived through the simultaneous interplay of all the senses.'[65] It was the space in which tribal man lived, and thus the new electric age with its emphasis on sound over sight and ear over eye was returning modern man to the tribal age in the mix of sensorium, creating the modern global village reminiscent of the tribal village.

Instant communication meant the possibility of universal communication. McLuhan pointed out that modern computers held the possibility of instant translation of any code or language into any other code or language. It was therefore but a short step to eliminate language altogether and create 'an integral cosmic unconsciousness somewhat similar to the collective unconscious envisioned by Bergson,'[66] a collective world soul. 'The computer, in short,' McLuhan euphorically wrote, 'promises by technology a Pentecostal condition of universal understanding and unity.'[67] Computers had the ability to program the sensory life of whole populations by programming 'the media to determine the given messages a people should hear in terms of their overall needs, creating a total media experience absorbed and patterned by all the senses.'[68]

This was the world of cybernetics or automation that Norbert Wiener sketched out in *Cybernetics: Control and Communication in the Animal and the Machine* and that McLuhan saw as pertinent to his vision of the electronic age. Cybernetics or automation was a way of thinking as much as a way of doing, he claimed, by looking at the problem not as a segmented sequence of individual, respectable parts but as 'an integrated system of information handling.'[69] McLuhan argued that instant synchronization of numerous operations created an 'organic unity' that was the essence of 'mass production' and 'mass media.' In the electric age, these terms referred not to size but to 'an instant inclusive embrace' because everything occurred simultaneously or everyone got involved at the same time. Therefore, McLuhan believed that cybernetics required a more liberal education, one that could integrate knowledge as opposed to the compartmentalization of knowledge that existed in the age of specialization.

Understanding Media was the high point of McLuhan's thinking about technology, his only attempt to provide a comprehensive and overarching account of the nature and impact of electric technology. Later books, usually written in collaboration with others, addressed only a component or elaborated on a segment of his ideas presented in *Understanding Media*. Claude Bissell, president of the University of Toronto during McLuhan's heyday in the 1960s and a long-time McLuhan enthusiast, claimed in a reflective article on McLuhan that *Understanding Media* 'takes its place in that wide channel of cultural criticism of the twentieth century that includes writers like T.S. Eliot, Oswald Spengler, D.H. Lawrence, F.R. Leavis, David Riesman, Hannah Arendt.'[70]

Like Innis, McLuhan had come to see communication technology as the decisive form of technology. For him, its decisiveness was due to its ability to orchestrate the human senses in such a way as to affect what people in different technological ages perceived to be 'reality' and thus the very way people thought. Even in the mechanical age, technology was more than machines; it was a mechanistic mindset that caused Western civilization to emphasize rationalism, logic, causality, and science because it was a print-oriented society, and print – first in the form of the phonetic alphabet and then, more significantly, in the form of the printing press – led to individualism and rationalism as the sense of sight and the importance of the eye dominated. This age came to an end as the new media technology such as the telegraph, telephone, radio, and television came to dominate communications. This new electronic technology, especially those (like the telephone and the television) that were 'cool' media, involved all the senses. Thus, Western society was embarking on a new era – the electronic age – the nature of which society was only beginning to understand. McLuhan's mission, as he saw it, was to sketch out the perimeters of that new society. In doing so, in the very 'whirlpool' that he, like Edgar Allan Poe's sailor, was trying to understand so as to conquer it, he had flashes of brilliance and insight amid thoughts that to many appeared simply foolhardy if not downright wrong. Certainly, his arcane thought stupefied most people, and they were never sure which of McLuhan's ideas were brilliant and which were insane or silly.

His esoteric ideas made him a celebrity of the very electric communication technology he was attempting to analyze and that, in the case of television, he found suspect and even at times menacing. He was sought out for interviews and in demand as a speaker, who could in turn demand a high price for his speaking engagements.[71] Fellow media critics eagerly wanted to collaborate with him in writing books. McLuhan equally sought out collaborators, since he was never good at following through on projects of his own. Many of these collaborative efforts never saw the light of day in terms of publication; almost all ended in frustration for both parties as McLuhan's erratic behaviour led to tension.

In 1967-68, along with Quentin Fiore and Jerome Agel, he published two popular books: *The Medium Is the Massage,* a 'McLuhan Made Easy' book, which his biographer notes sold 'nearly a million copies worldwide'[72] (clearly McLuhan's success story in terms of sales), and *War and Peace in the Global Village,* in which he argued that the present can only be seen when it has become the past, describing this as 'the rearview mirror phenomenon.'[73] In the book, he also offered his explanation for wars: 'I suppose that one could even produce a theory of war to say that when a certain amount of technological changes happens very quickly to a whole community, they are so lost about who they are that they want a basic war to find out.'[74]

In 1968, along with Harley Parker, McLuhan published *Through the Vanishing Point: Space in Poetry and Painting,* in which he elaborated further his idea of the impact of media technology on a person's sense of space, such as visual space and acoustic space. A year later, he and Parker did *COUNTERBLAST,* McLuhan's version of *BLAST,* the book by the famous British painter Wyndham Lewis published on the eve of the First World War. Just as *BLAST* was a non-traditional book in format and appearance, so too was *COUNTERBLAST,* a 'counter-book' in that it was full of aphorisms, non-sequiturs, and psychedelic print. McLuhan explained his rationale for the book in the Introduction: 'The term COUNTERBLAST does not imply any attempt to erode or explode *BLAST*. Rather it indicates the need for a counter-environment as a means of perceiving the dominant one. Today we live invested with an electric information environment that is quite as imperceptible to us as water is to a fish. At the beginning of his work, Pavlov found that the conditioning of his dogs depended on a previous conditioning. He placed one environment within another one. Such is *COUNTERBLAST*.'[75] The original environment into which McLuhan wanted to fit the new electronic environment was of course that of print, which he claimed, using the metaphor of the 'goose quill,' 'put an end to talk, abolished mystery, gave us enclosed space and towns, brought roads and armies and bureaucracies. It was the basic metaphor with which the cycle of *CIVILIZATION* began, the step from the dark into the light of the mind. The hand that filled a paper built a city.'[76]

The book further explored the concept of visual and acoustic space that the old print and the new electronic technology created respectively and argued that the Western civilization the technology of the printing press had created was no longer in existence or at least was on the way out in the new electric age. What would replace it no one knew, McLuhan claimed, but *COUNTERBLAST* was designed to offer some thoughts in the form of 'probes' on the 'brave new technological world' he saw emerging. As McLuhan boldly asserted:

> By surpassing writing, we have regained our sensorial WHOLENESS, not on a national or cultural plane, but on a cosmic plane. We have evoked a super-civilized sub-primitive man.
>
> NOBODY yet knows the languages inherent in the new technological culture; we are all technological idiots in terms of the new situation. Our most impressive words and thoughts betray us by referring to the previously existent, not to the present.
>
> *Movies and TV* complete the cycle of mechanization of the human sensorium. With the omnipresent ear and moving eye, we have abolished the dynamics of Western civilization.
>
> We begin again to structure the primordial feelings and emotions from which 3000 years of literacy divorced us. We begin again to live a myth.[77]

McLuhan's last book, *Laws of Media: The New Science*, written in collaboration with his son Eric, was published posthumously. It attempts to provide a scientific basis to media studies by applying 'laws' in the form of four questions – a tetrad – to any human artifact, information, or idea. These questions were: What does it enhance or intensify? What does it render obsolete or displace? What does it retrieve that was previously obsolesced? And what does it produce or become when pressed to an extreme? In an attempt to set the context for an understanding of such laws, the McLuhans claimed that the new electronic technology was returning society to the acoustic space of a pre-print and pre-alphabetic age in early Greece, one in which 'figure' and 'ground' were reunited and in harmony. The severing of 'figure' and 'ground,' characteristic of visual space, began in the post-alphabetic period in Greek thought and reached its zenith in the Gutenberg and post-Euclidean age of the Renaissance. The McLuhans maintained that only two great thinkers challenged the wisdom of this perspective – Francis Bacon in *The New Organon; Or, True Directions concerning the Interpretation of Nature* and Giambattista Vico in *Scienza Nuova* – before the advent of the electric age forced scientists in the twentieth century to challenge and undo the theories on which science had operated for centuries. *Laws of Media* sought to explain the revolutionary change engendered by the transition from visual space to acoustic space, both in terms of scientific thought but more importantly in terms of popular thought, by outlining the impact on the thought process, or mindset, of the modern age. Artifacts become 'a word, a metaphor that translates experience from one form into another.'[78] That translation occurred spontaneously and simultaneously through an acoustic space that was 'spherical, discontinuous, non-homogeneous, resonant, and dynamic.'[79] In so doing, the technology transformed modern thought and therefore the modern world in which humans lived.

McLuhan appeared ambivalent about the new electric age. When pressed to express his personal views, he would usually rejoinder that his job was only to observe and reflect, not to judge or moralize. Yet, as a number of critics have pointed out, McLuhan did make judgments about this new age. Such judgments showed just how ambivalent he was. In the space of a single interview, McLuhan presented the electric age in both euphoric and frightening terms. He claimed in 'Playboy Interview' that electricity was ushering in a world of harmony, 'a rich and creative retribalized society.'[80] When the interviewer stated that McLuhan's envisioned tribal society in the new electronic age appeared to him to be 'a rigidly conformist hive world in which the individual is totally subordinate to the group and personal freedom is unknown,'[81] McLuhan retorted:

> Literate man is alienated, impoverished man; retribalized man can lead a far richer and more fulfilling life – not the life of a mindless drone but of the participant in a seamless web of interdependence and harmony. The implosion of electric technology is transmogrifying literate, fragmented man into a complex and depth-structured human being with a deep emotional awareness of his complete interdependence with all of humanity. The old 'individualistic' print society was one where the individual was 'free' only to be alienated and dissociated, a rootless outsider bereft of tribal dreams; our new electronic environment compels commitment and participation, and fulfills man's psychic and social needs at profound levels.[82]

At other times, he viewed the electronic age with trepidation and fear. He noted in *Understanding Media* that it was not simply coincidental that Søren Kierkegaard's *The Concept of Dread* should appear in the same year, 1844, as the invention of the telegraph. 'For with the telegraph,' McLuhan explained, 'man had initiated that outering or extension of his central nervous system that is now approaching an extension of consciousness with satellite broadcasting. To put one's nerves outside, and one's physical organs inside the nervous system, or the brain, is to initiate a situation – if not a concept – of dread.'[83] When pressed in the same interview in which he described the electric age as 'euphoric' to express his personal thoughts and feelings about the changes occurring in the transition to the electronic age, McLuhan the moralist revealed just how 'fearful' and 'full of dread' he was about such changes:

> I view such upheavals with total personal dislike and dissatisfaction. I do see the prospect of a rich and creative retribalized society – free of the fragmentation and alienation of the mechanical age – emerging from this traumatic period of cultural clash; but I have nothing but distaste for the *process* of change. As a man molded within the literate Western tradition, I do not personally cheer the dissolution of that tradition through the electric involvement of all the senses: I don't enjoy the destruction of neighborhoods by high-rises or revel in the pain of identity quest. No one could be less enthusiastic about these radical changes than myself. I am not, by temperament or conviction, a revolutionary; I would prefer a stable, changeless environment of modest services and human scale. TV and all the electric media are unraveling the entire fabric of our society, and as a man who is forced by circumstances to live within that society, I do not take delight in its disintegration.[84]

Still, McLuhan believed such change was inevitable, and he remained optimistic to the end that out of the ashes of the destruction and disintegration of the mechanical age would come a rebirth. He ended his interview on an upbeat note: 'Personally, I have a great faith in the resiliency and adaptability of man, and I tend to look to our tomorrows with a surge of excitement and hope. I feel that we're standing on the threshold of a liberating and exhilarating world in which the human tribe can become truly one family and man's consciousness can be freed from the shackles of mechanical culture and enabled to roam the cosmos ... We live in a transitional era of profound pain and tragic identity quest, but the agony of our age is the labor pain of rebirth.'[85]

McLuhan's belief that much of what happens is 'inevitable' – a familiar term in his writings – raises the question as to whether he was a 'technological determinist.' McLuhan did argue that new communication technologies emerge independently of our knowledge or actions, even if they are in some way an extension of our senses. Such communication technologies also interplay with our senses so as to alter our perspective, knowledge, and perception of the external world; this too, he claimed, would happen no matter what we might do to try to prevent it. Yet he did appear to believe and hold out hope that, through an understanding of what was happening, humankind could have some control over the process. To an extent, such understanding meant giving in to the new technology and seeing where it took one: the frequent reference in McLuhan's writings to Edgar Allan Poe's sailor in the maelstrom who survives by going with the current. Resistance was to turn a blind eye and negate any chance of understanding so necessary for positive action. Yet again, McLuhan also talked a great deal about the role of the artist, who was better able to see what was happening subliminally by setting up a 'counter-movement' or 'counter-culture' to the existing paradigm, in essence to resist by refusing to see that which was. Others could do the same, McLuhan maintained, but only after the fact, after a new media technology had taken hold, at which time its new sensory mix 'liberates' the mind from the old perspective, thus allowing it to see what was and (by implication and contrast) what is. Thus, there appears to be a strange mixture of technological determinism and human freedom in McLuhan's thinking that leaves open the possibility that in the end, despite the pervasiveness of technology, it could be manipulated to allow for human creativity and to usher in a better world. As he concluded in 'Playboy Interview' when confronted by the question of determinism, the choice is action through awareness or inaction (atrophy) through ignorance: 'If we understand the revolutionary transformations caused by new media, we can anticipate and control them; but if we continue in our self-induced subliminal trance, we will be their slaves.'[86] Technology as volition was premised for McLuhan on the belief that human beings had to be aware of the technology that was acting on their senses and thus shaping their thoughts.

Harold Innis and Marshall McLuhan had a similar perspective on technology, one that was characteristic of the thinking of post–Second World War theorists of technology, including Northrop Frye and George Grant. Both Innis and McLuhan asserted that the technology of communication that conveyed information and ideas shaped the mind itself. For both men, the importance of communication technology was not in the ideas conveyed, the content or message, but in the means, or medium, by which it conveyed ideas and information. The media of communication technology governed what one perceived as valid and important ideas and thus shaped the thought process of the age. Furthermore, both theorists ultimately came to see technology not as objects, such as machines or even the process of industrialism to which the individual reacted, but as a relationship between things created in the mind, what McLuhan referred to as 'ground' as opposed to 'figure' and Innis meant in part by the term 'bias,' a way of thinking that was in line with quantum logic. As William Kuhns notes, 'In an electronic age, these men [Innis and McLuhan] are saying, we must think of environments afresh. We should look not simply to what is "out there," but also to the profound and unexpected processes between the "out there" and the "in here." Man does not relate to his environment like a ball being bounced against the walls of a closed room. He is constantly creating a new environment, and shaping it with biases that will mirror and thus redefine himself in many inexplicable ways.'[87]

However, one important difference between Innis and McLuhan lies in the focus of their analyses of media technology on individuals and society. Innis focused on the message *senders* and hence on the source of *power* behind the medium of technology. Innis also sought to understand how the technology shaped the society and culture of the sender – the context – and thus was interested in the political economy of technologies. McLuhan was more interested in the impact of media technology on the message *recipients*, analyzing the effect of technology on the senses of individuals. Thus, he was less concerned with issues of sources of power and control of technology and not overly interested in the impact of media technology on society except as experienced by individuals within the society through their senses. Therefore, McLuhan did not focus on issues related to political economy but rather in the area of social psychology.[88]

Innis's and McLuhan's ideas moved the debate on technology among Canadian intellectuals in a significantly new direction from that of the interwar years. In that period, only the beginning of an awareness of the pervasiveness of technology was apparent, in a realization that technology was a dominant force that seemed beyond the reach and control of human beings and thus threatened to dominate and consume humankind. Theorists of technology in the interwar years viewed technology – whether identified as objects, machines, industrialism, or mechanization – as external entities that impinged on human beings for good or evil. The Second World War and the novelty of Canadian intellectuals to look at technology as a psychic phenomena – that is, ideas and thought patterns – contributed to the perspective of technology as a mental process, shaped by the media of communication. A true understanding of technology required an examination of the inner technological mindset that technology shaped. This awareness in turn raised questions about technology as volition and gave new meaning and concern to the age-old tension between technology and freedom or the technological imperative and the moral imperative. This concern also surfaced in Northrop Frye's and George Grant's writings on technology.

9
Northrop Frye and E.J. Pratt: Technology as Mythology

Northrop Frye, professor of English at the University of Toronto from 1937 until his death in 1991, had a shared interest with Harold Innis and Marshall McLuhan in understanding the impact of the technology of transportation and communication on Western civilization in general and on Canadian society in particular. In a CBC interview, Frye pointed out how he was part of 'a Canadian tradition.' One of the unique and shaping features of the Canadian environment is its obsession with communication, 'which took itself out in building bridges and railways and canals in the nineteenth century and in developing very comprehensive theories of communication, like Innis's and McLuhan's in the twentieth century. I suppose I belong to some extent in that category.'[1] When pressed by his interviewer to explain his link to these other Canadian theorists of communication technology, Frye replied:

> Well, Innis started out with a Lawrentian [sic] theory of Canadian expansion, with the fact that coming to Canada from Europe is a totally different experience than coming to the United States and with the fact that you live in what is, for practical purposes, a kind of one-dimensional country. It's just a long line of river and Great Lakes and railways from one ocean to the other. The difficulties within communication, within the very act of communicating at all, and the fact that the settlements in Canada are isolated from each other in geographical ways – these things, I think, have brought about certain affinities among people who have talked about communication in Canada.[2]

Frye noted the ways that his views differed from those of McLuhan, the other 'guru' at the University of Toronto in the 1960s and 1970s. He pointed out, for example, that, whereas McLuhan saw the introduction of the printed word and especially the mass production of books as having had a negative effect on communication by isolating individuals and forcing them to think in a one-dimensional, linear, and rational fashion, a way of thinking that McLuhan predicted was about to disappear with the introduction of new electronic technology, he (Frye) saw the book as 'the most efficient technological instrument ever devised in learning,'[3] much more efficient than even the computer. The book 'is a model of patience because it keeps saying the same thing no matter how often you consult it. It's a stable thing. It's a growing point of learning in a way that I don't quite see the computer being.'[4] He went on to expound his views on the computer:

> There are many superstitions about the computer I don't share. I think it is being looked at on a kind of phony Cartesian basis, a view in which the mind is opposed by a body which is a mechanism. Nobody worries about the fact that the automobile runs faster than the human being because only the body, the mechanism, is involved there. But as soon as it becomes obvious that the computer can calculate and in many respects think faster than a human being, people get the jitters. They think that something peculiar to the human animal is being invaded by a rival kind of alien, science-fiction being. I don't think that is true. The computer is a tool, an instrument. The difficulty is in thinking of man as conscious and of having a mechanism attached to that consciousness, whereas man is primarily a conscious will. The machine has no will to do anything. It depends on being plugged in or turned on. Only when it develops an autonomous will can it become a fully conscious being.[5]

Frye also maintained that each new technology of communication, including electronic technology, made humans introverted, rather than extroverted, as McLuhan argued. The book, the movie, television, and the most recent and therefore unknown electronic technology, the computer, ironically caused people to turn inward and thus lessen communication with others in a social context, a context Frye believed was an essential component of good and meaningful communication. He maintained that this introversion need not be inevitable, but it took some imaginative action on the part of individuals to 'get control of it.'[6] 'In the course of time,' Frye believed, 'the machine becomes more and more what it ought to be, an extension of a personality and not an independent personality set over against you.'[7]

Frye therefore disagreed with those individuals who saw technology as controlling humankind. He pointed out that this had been a well-established perspective since earliest times when writers used the wheel, for example, to talk about 'a wheel of fate or a wheel of fortune or a wheel as a cosmological force which is alienating him [man] from himself.' In the same way, humans invented the book and then started talking about 'the book of life, in which all our sins are recorded.'[8] These, Frye argued, are examples of how the scientific-technological view of the world gets translated into a mythology within literature that becomes part of our psyche unconsciously, at least until we are able to see such views as mythical, as inventions of the mind.

Frye also disagreed with McLuhan's interpretation of print culture as being 'linear' and 'time-bound' and oral culture as being 'simultaneous' and 'mosaic.'[9] He saw the linear and simultaneous as 'not a difference between two kinds of media, but a difference between two mental operations within all media,' that there was always a linear response followed by a simultaneous one whatever the medium. 'For words, the document, the written or printed record, is the technical device that makes the critical or simultaneous response possible,' Frye argued. 'The document is the model of all teaching, because it is infinitely patient, repeating the same words however often one consults it, and the spatial focus it provides makes it possible to return on the experience, a repletion of the kind that underlies all genuine education.'[10] Frye then went on to explain how print as a technological phenomenon was really the basis of freedom in Western civilization, the root of democracy; it was his counter-argument to Innis, who argued that the printed word, especially mechanized print, enabled empires to develop, and McLuhan, who maintained that print restricted the senses and thus limited human vision, so to speak, to a one-dimensional perspective. Frye reasoned:

> The document is ... the focus of a community of readers, and while this community may be restricted to one group for centuries, its natural tendency is to expand over the community as a whole. Thus, it is only writing that makes democracy technically possible. It is significant that our symbolic term for a tyrant is "dictator," that is, an uninterrupted oral speaker.
>
> The domination of print in Western society, then, has not simply made possible the technical and engineering efficiency of that society, as McLuhan emphasizes; it has also created all the conditions of freedom within that society: democratic government, universal education, tolerance of dissent, and (because the book individualizes its audience) the sense of the importance of privacy, leisure, and freedom of movement.[11]

Frye also challenged McLuhan's theory that 'the medium is the message,' that it is the communication medium of technology used that shapes and governs the message. Frye pointed out that different media give the same message in different societies, be it in the Soviet Union or the United States. 'This is because the real communication media are still, as they always have been, words, images, and rhythms, not the electronic gadgets that convey them. The differences among the gadgets, whether they are of high or low definition and the like, must be of great technical interest, especially to those working with them, but they are clearly of limited social importance.'[12]

Thus, in the ongoing debate among Canadian intellectuals between technology and ideas, form and content, the medium and the message, Frye clearly came down on the side of ideas. Ideas are the 'words' that make up the 'imaginative vision' of the individual, which acts as a counterforce to the technological materialism of the social mythology. Technologies may shape ideas, but it is what people do with those ideas that will explain their importance, even if they should turn those ideas into a belief that they are technologically induced, as Innis and McLuhan argued. To Frye, this was just one possible interpretation or myth among many possible myths.

Frye's interest in technology, particularly the technology of communications, came about indirectly out of his main interest, which was the study of Western mythology coming from the Bible.[13] He noted that two myths had dominated Western thought up to that point: the Christian myth until the seventeenth or eighteenth century and the scientific-technological myth since that time. The latter was premised on the belief that the world was like a machine that ran according to predictable scientific laws. Once humans knew these laws, they could control nature by means of technology. Frye claimed that this perspective did not necessarily lead to a world doomed to technological dominance, as George Grant warned, or to a possible utopia, as Marshall McLuhan predicted for the electronic age; it was simply one possible way of constructing reality, one myth in a variety of possible mythologies. Once humans achieved this perspective, they were already liberated from the myth that technology was all-pervasive.

For Frye, then, the technological imperative was not the only imperative, although he agreed that in our modern world it was the most pervasive at the time. What might dislodge it was the ability of human imagination to think outside the technological paradigm. This would be the means by which humans could liberate themselves from the modern scientific-technological myth, just as in the seventeenth and eighteenth centuries European thinkers had been able to liberate themselves from the earlier Christian myth. Myths were ways of constructing a view of the world and the place of humans in that world that gave meaning to 'reality.' Such myths came out of the human imagination: the ability to imagine the world to be different and better. This was the realm of the moral imperative, to provide the moral values by which the ideal world would come into being and upon which it would function. For Frye, this ideal world had to be one of concern: that society was greater than the individual alone, that the good of society transcended the freedom of the individual. Thus, the modern world of technology needed to be looked at in terms of the extent to which it served the needs of society as a whole.

Frye's former teacher, colleague, and friend, E.J. (Ned) Pratt, was the Canadian poet who had most successfully constructed a Canadian mythology based on the role of technology in shaping Canadian society and thought.

In his epic poems, especially *The Titanic* and *Towards the Last Spike,* the latter being an account of the building of the Canadian Pacific Railway, Pratt showed how these technological wonders revealed the marvel of human ingenuity but equally the weakness of human hubris. Something was invariably lost in the relentless pursuit of technological progress, Pratt suggested. What was lost were human lives and moral values so essential for the survival of civilization. Pratt's poetry presented his perspective on the tension between the technological imperative and the moral imperative.

Frye's study of the modern scientific-technological myth stressed the interplay and tension between imagination, coming out of the moral imperative, and 'reality,' constructed on the basis of the technological imperative. He withheld judgment on the ability of the modern world of technology to create our 'imagined' ideal world. However, he believed that this ideal would have to be a harmonious relationship between the moral imperative and the technological imperative. This, he believed, would be a social utopia, by which he meant a society where the fundamental concerns of freedom, health, equality, happiness, and love were dominant. This needed to be an open mythology. While his fear was that the technological imperative would make it into a closed mythology, his faith was in the ability of human imagination to create an open mythology, a harmonious relationship between the myth of concern and the myth of freedom. An understanding of how Frye arrived at his views of the modern scientific-technological myth and his vision for a reconciliation of the moral imperative and the technological imperative requires knowledge of his intellectual journey.

Northrop Frye was born on July 14, 1912, in Sherbrooke, Quebec.[14] He was the fourth child of Herman and Cassie Frye. Their first son, Erastus Howard, was born in 1899 but died in the First World War. Erastus Howard was very athletic and popular, and Northrop had to carry the familial expectation of living up to his older brother, whom he hardly knew. A second son had been stillborn or died soon after birth; the parents had intended to call him Northrop after Herman's mother's family. Thus, Northrop Frye carried an onerous family responsibility as the only son to survive to adulthood. A daughter, Vera Victoria, was born on Christmas Day, 1900. She and Northrop were close throughout their lives, despite the significant age difference of twelve years between them.

Northrop Frye wove his own mythology concerning his family origins. He maintained that his first New World ancestor, John Frie, had been a preacher from Andover, Hampshire, when in fact he was a wheelwright. John's son, Samuel Frie, produced a line of militia colonels and mandarins. One of Samuel's sons, Peter Frye, Northrop's great-great-grandfather, was a Tory in the American Revolutionary War who settled near Sherbrooke, Lower Canada, after the revolution. Peter's son Herman and his grandsons Abbott and Samuel Frye (the latter was Northrop Frye's grandfather) farmed in the region. Samuel married Sarah Ann Northrop, also of Puritan descent, in 1861. They had six children, three daughters – Etta, Carie, and Ella, and three sons – Herman (Northrop's father), Alfred, and Austin. They lived in Lowell, Massachusetts, and it was there that Herman brought his wife, Cassie Howard, to live after their marriage in 1897.

The Howards were descendants of Methodist ministers. Northrop's maternal grandfather, Eratus Seth Howard, was an itinerant minister who was posted to Windsor Mills, near Sherbrooke, in the 1880s. The Howards were staunch Loyalists, and Cassie perpetuated the feelings of anti-Americanism among the Frye-Howard clan. Indeed, soon after moving to Lowell with Herman, she became disgruntled with the United States and finally in 1904 persuaded her husband to return to Canada.

Herman Frye attended Belleville Business College and then worked as a hardware clerk for J.S. Mitchell in Sherbrooke, Quebec. In 1915, he set up his own hardware business to compete with Mitchell's. The wartime economy kept the store going, but it went under in 1919. At fifty years old, Herman became a commercial traveller. The family moved frequently in the Sherbrooke and Lennoxville areas of Quebec before settling in Moncton, New Brunswick, a Maritime railway centre. Northrop spent many of his boyhood and teenage years there.

Frye was a precocious child and learned to read and play the piano at the age of four. Family lore had it that 'Northrop carried around the copy of *Pilgrim's Progress* like a teddy bear and would read from it with little prompting to his twin cousins who were three years older than he.'[15] As he grew older, Northrop's mother added other literature, as well as history (Charles Dickens's *A Child's History of England* was one of Frye's favourites) and the Bible, the mainstay of the Methodist faith.

Northrop found school boring. In later life, he viewed his early schooling 'as a form of penal servitude.'[16] Nevertheless, he excelled in literary subjects but did poorly in mathematics. His most memorable school experience occurred outside of the classroom. During one of his early morning walks to school, he had what amounted to a parody of the Methodist conversion experience. He recalled the experience in later life: 'That whole shitty and smelly garment (of fundamentalist teaching I had all my life) just dropped off into the sewers and stayed there. It was like the Bunyan feeling, about the burden of sin falling off his back only with me it was a burden of anxiety.'[17] Thus, from an early age, Frye saw religion as something to take or to leave according to personal conviction. As he recalled in later life, 'I think I decided very early, without realizing it at the time, that I was going to accept out of religion only what made sense to me as a human being.'[18]

Meanwhile, he won a scholarship in stenographic training at the Success Business College for standing first in English in his high school graduating year. His facility at the piano made him an excellent typist. It is significant and interesting to note that the scholar who in later life became fascinated with 'words with power' and with the power of technology began his illustrious career by mastering the typewriter, the mechanical creator of words. In the spring of 1929, he entered the National Typing Contest run by the Underwood Company in Toronto. He used the opportunity of being in the city to familiarize himself with Victoria College, the Methodist-affiliated school at the University of Toronto, to which he applied and was accepted.

'Vic' was everything young Northrop Frye dreamed it would be – that is, his ticket to the life of an intellectual. As his biographer notes, 'He [Frye] felt that he had finally found a community which represented his own interests, that was willing to accept him for what he was. Instead of dilettantes he faced interesting minds.'[19] As well, university opened up a wealth of scholarly books on subjects of great interest to him. He described an undergraduate's life as 'a precious time of remaking individuals, separating them from mob mentality and providing for them '"a poise and culture resulting from moving in an intellectually stimulating society."'[20] He did the philosophy (English and history) program, which was Toronto's version of Oxford's 'Great Moderns.' At Vic, he met three English professors who profoundly influenced his views on literature: John Robins, Pelham Edgar, and E.J. Pratt. He also discovered Oswald Spengler's book *The Decline of the West: An Interpretative History*. Later, in a 1955 radio talk, he described the book as 'a vision rather than a theory or a philosophy, and a vision of haunting imaginative power. Its truth is the truth of poetry or prophecy, not of science ... If it were nothing else *The Decline of the West* would still be one of the world's great romantic poems.'[21] Later still, in a CBC interview, Frye noted the importance of *The Decline of the West* on his thinking during his impressionable student days: 'I suddenly got a vision of coherence. That's the only way I can describe it. Things began to form patterns and make sense.'[22] Spengler's 'vision' had also captivated Harold Innis, who used the study to understand the rise and fall of civilizations, including the West, as part of the cyclical nature of human existence. Innis, like Spengler, saw this cyclical nature in a historical context, whereas Frye would apply it to literature as patterns of comedy, romance, satire, irony, and tragedy.

After Vic, Frye attended Emmanuel College at the University of Toronto, the divinity school of the United Church, acquired a theology degree, and became an ordained minister. After a gruelling summer as a student minister, Frye knew he did not want to spend his life as a minister. His real love lay in theology studies, particularly the ideas of William Blake. He took a graduate course on Blake from Herbert Davis toward an MA degree in English. While working on a seminar paper, Frye mapped out the ideas that would form the basis of his seminal work, *Fearful Symmetry*. He realized how intricately connected were comparative theology, the Bible, and Western literature and that a study of Blake was the best means to understand the interconnection. As he wrote at the time:

> No great poet, with the very doubtful exception of Shakespeare, has, since the rise of Christianity, been able to write without the inspiration of the Christian religion behind him, and consequently he is forced to give expression to the deepest religious impulse of his age. Hence in every period there is one supreme poet who expresses the very essence of that period's attitude to Christianity ... Blake is peculiarly interesting because he lived in the time when the French and the Industrial Revolution with all that they signified, were shattering the unity of Christian civilization, and he, and Goethe, are the two great artist intellects who represent the final and culminating effort of Western culture to express that unity.[23]

Frye's interest in Blake took him to Oxford. He completed a second BA at Merton College in 1939, the year of the outbreak of the Second World War, and received an MA degree in 1940. While he was at Oxford, Victoria College offered him a full-time teaching position, so he and his wife of three years, Helen Kemp, settled in Toronto. As a clergyman, he received an automatic dispensation from serving in the war.

Frye used his early years at Victoria to perfect his teaching style. To students, his lectures appeared to be spontaneous, since he gave them without notes, but in fact this was Frye's way of working through the ideas in his head before putting them down on paper. He confessed in later life that he always wrote his lectures *after* he had given them and that these lectures then often became, without much editing, his published works. The approach made him a dynamic and engaging teacher.

Frye first broached the question of what role technology and industrialism had played in Canadian poetry in his article 'The Narrative Tradition in English-Canadian Poetry,' published in 1946. He noted the impossibility of Canadian poets doing something distinctive in poetry writing because the language used, whether French or English, already bound the poet to certain literary conventions and because Canada had always been part of a European socio-cultural tradition that dictated patterns of thought. Canada, he argued, was not 'a "young" country in the sense that its industrial conditions, its political issues, or the general level of its civilization, are significantly different from contemporary Europe.'[24] Still, there existed distinctive aspects of the settlement process in Canada that differentiated it from the European pattern of settlement due to the country's unique geography and environment. On one occasion, Frye noted, 'A Canadian village, unlike an English one[,] sprawls awkwardly along a highway or railway line, less an inhabited centre than an episode of communication. Its buildings express an arrogant defiance of the landscape; its roads and telephone wires and machinery twist and strangle and loop.'[25]

Therefore, a good Canadian poet needed to be able to capture the unique qualities of the Canadian version of the modern industrial-technological society. Similar to McLuhan, Frye maintained that the poet, as artist, was a barometer of his time especially with regard to the impact of technology. He noted that poets tended to be particularly sensitive to the mechanical dehumanizing of life that resulted from technology. Literary critic Jonathan Hart notes, 'When technology and mechanics dominate society, the poet becomes socially alienated ... Poets, who since Blake have inveighed against "dark Satanic mills," have protested against mechanization.'[26] In *The Critical Path,* a study of the Bible within a literary context, Frye observed that, 'among the poets, in particular, a strong movement takes shape, of the type that [Robert] Snow ... calls a "Luddite tendency." It is really a protest against the mechanized dehumanizing of life ... and it only looks reactionary when its opposite is assumed to be beneficent.'[27] Frye saw Lampman's 'The City of the End of Things' as one of the few great Canadian poems that succeeded in capturing the dehumanizing nature of industrialization. In it, Frye noted, was 'a vision of the Machine Age slowly freezing into idiocy and despair.'[28]

Frye maintained that a poet in the twentieth century was required to move beyond the problems of a pioneering country to a new 'vision of Canada as a settled and civilized country, part of an international order, in which men confront the social and spiritual problems of men.'[29] In dealing with this new theme, he advocated the continued use of the narrative form of poetry, because there was something 'Canadian' and 'modern' in that form: 'In an age when new contacts between a poet and his public are opening up through radio, the narrative, as a form peculiarly well adapted for public reading, may play an important role in reawakening a public respect for and response to poetry.'[30] Frye noted how the modern technology of communication dictated the best means by which the poet could develop his or her ideas; it was an idea analogous to Harold Innis's belief that the dominant technology of communication in any age shaped its social and cultural matrix and Marshall McLuhan's belief that different technologies of communication orchestrated the human senses in different ways and thus altered human perception of the external world.

From the outset, Frye had a balanced perspective on technology. He associated technology with machines and industrialism that could be either a source of oppression or a means for creative expression. In his 'Preface to an Uncollected Anthology' – his wish list of poems he would include in an ideal Canadian poetry anthology – Frye noted how technology frequently generated 'imaginative energy'[31] that led to creative literature. Unfortunately, in the case of the United States, this was not necessarily so:

> It is often said that a pioneering country is interested in material rather than spiritual or cultural values. This is a cliché, and it has become a cliché because it is not really true, as seventeenth-century Massachusetts indicates. What is true is that the imaginative energy of an expanding economy is likely to be mainly technological ... Successful nations usually express a restraint or a matter-of-fact realism in their culture and keep their exuberance for their engineering. If we are looking for imaginative exuberance in American life, we shall find it not in its fiction but in its advertising: not in Broadway drama but in Broadway skyscrapers; not in the good movies but in the vista-visioned and technicoloured silly ones.[32]

Within Canada, Frye observed, 'imaginative energy' through technology expressed itself in feats of conquest over the obstacles to communication and transportation. He cited E.J. Pratt's poem *Towards the Last Spike* – a recounting of the history of the Canadian Pacific Railway – as an especially notable example of the application of the creative imagination in poetry to the study of technology and its impact in the context of Canadian mythology.

E.J. Pratt, one of Canada's greatest poets, taught Frye at the University of Toronto and became one of Frye's lifelong friends. Pratt was fascinated with technology, which he weaved into a mythology in his poetry. In a number of his poems, he raised many of the questions about technology that Frye was addressing in his literary criticism. Thus, together these two literary giants advanced our understanding of the nature of technology as a defining myth in nineteenth- and twentieth-century Western literature in general and Canadian literature in particular.

Pratt's friend and fellow poet Louis Dudek once described Pratt as 'the Poet of the Machine Age.' Dudek explained, 'The poetry of E.J. Pratt strikes everyone who reads it by its peculiar kind of energy, a physical drive, a mechanical exuberance, unlike that of any other poetry we know.'[33] Pratt acknowledged his fascination with machines and other technological wonders while reflecting on his poem 'Behind the Log,' an account of the use of radar by Allied convoy ships during the Second World War to detect enemy submarines. 'I have an interest which has almost a cruel fascination for me, as I know it must for countless others, that is, the role that physical science is playing today in the construction and destruction of life, its relation to the bare essential fact of existence ... We are learning only too well how any formula, the product of man's mind, or any instrument, the product of his hand, can be used with equal readiness to save or to kill, and I know nothing which, as material, offers a sterner challenge to drama and poetry.'[34] Pratt's awareness of the constructive and destructive capacity of technology reflected the dualistic and balanced perspective Frye had with regard to technology.

However, both men focused on the destructive aspect of technology because they saw this aspect as the dominant one in the twentieth century. While Pratt could not find any more answers or resolutions to the riddles of life in the 'culture of industrialism' than he could in his religious training or in the study of human nature, he felt compelled to address in his poetry the issue of the role of technology in shaping and governing an indifferent and mechanistic world in the modern age.

Such a perspective may have come out of Pratt's experience of the First World War as a Newfoundlander. While he did not fight in the war, his brother and many of his friends did, and he was greatly affected by the impact of the war on them. The most mechanistic of wars to that time, it was also the most destructive of human lives. Among the millions who died in the Great War were many Newfoundlanders, including over 700 in the Newfoundland battalion – three-quarters of the regiment – in the Battle of Beaumont-Hamel, part of the ill-conceived Battle of the Somme. Such large-scale and useless slaughter revealed for Pratt a world gone mad on technology and a mechanistic universe without purpose. His unpublished play *Clay,* written at the end of the First World War, raised questions that agonized him; for example, 'How can man believe that the universe has order and that God is beneficent when death, the ravages of nature, and war point to an uncaring deity?'[35] He also questioned whether humankind was evolving or devolving, progressing to a higher level of civilization or reverting back 'from steel to stone' and a more primitive state of being. Literary critic Sandra Djwa suggests that Pratt's views of the machine as part of the human evolutionary process may have come from a combination of reading Samuel Butler's *Erewhon* (1872), in which Butler 'satirically presented the machine (and mechanism) as simply a higher stage of the evolutionary development of animal to man,' and the works of the psychologist Wilhelm Wundt, who 'had identified the human nervous response as a mechanism.' He may also have been influenced by Freudian psychology, which was 'mechanistic in its view of human aggression.'[36]

In a number of his poems, Pratt depicted technology as a major source of power for humans in their relentless struggle to survive in a world that appeared indifferent to the well-being of humankind. (In his perspective on technology as power, Pratt was following in the footsteps of his University of Toronto colleague Harold Innis, who, as noted, held a similar perspective.) A common theme in a number of Pratt's poems is the struggle between the world of nature, depicted as mechanical, cruel, and indifferent, and that of humans, themselves mechanical beings as a result of their reliance on technology as their chief 'weapon' in this epic battle. Pratt noted the optimistic view of technology that prevailed at the end of the nineteenth century, the belief that technical knowledge would enable humankind to end the vagaries of nature and bring it under human control. As a result of this perspective, a false sense of the superiority of humans over nature emerged, based on the belief that humans enjoyed a decided advantage, thanks to technology. Thus, there arose a sense of technological hubris, which Pratt presented most effectively in his poem *The Titanic* (1935).

The poem recounted the sinking of the most technologically advanced ship of its day, touted by its creators as being unsinkable, by nature's mightier 'weapon,' an iceberg. Pratt accentuated the hubristic perspective on technology that lay behind the building of this mighty vessel by noting in his reflections on the poem the technological sophistication of the ship:

> The claims put up for the Titanic are now a matter of public knowledge. She was the biggest ship in the world, one thousand tons larger than her sister, the Olympic. She would furnish the most comfort in travel and would be the steadiest boat in the sea ...
>
> But the outstanding presumption was that of safety. She was advertised as the unsinkable ship. Under the laws of hydrostatics she couldn't go down. Back of it all was the confidence in the construction as an engineering triumph. The mechanization of the ship was so perfect that all possibilities of trouble were believed to be anticipated. So automatic was the running and control that little was left to the human judgement ...
>
> Moreover, the ship was launched at a time when the Marconi discovery, scarcely ten years old, was still inflating the mind with its dramatic performances. Those who remember the announcement of the discovery may recall the sense of conquest over nature that visited the hearts of men, the trust in science for the prevention of the grosser human calamities.[37]

In the poem, technological hubris blinded the captain, his crew, and the passengers to the dangers of ice all around them, dangers of which they had been forewarned by other ships in the vicinity. It also blinded them into believing that this ship could never sink, even as it slowly sank into the ocean. The band played on, people chose to stay on board, and the captain refused to send out a message of alert to the passengers or even an appeal for help to neighbouring ships. Pratt mocked this naive faith in technology that blinded generations to the reality of technological fallibility. The inability of this mighty ship to withstand the onslaught of an iceberg was as unthinkable as the sinking of the ship. By juxtaposing a panegyric to this great technological wonder with an elegy to its demise, Pratt both undermined the faith in technology that prevailed in the pre–First World War era and captured the paradoxical and ambiguous view of technology that prevailed in the interwar years, evident in the writings of George Sidney Brett, William Lyon Mackenzie King, Stephen Leacock, and Frederick Philip Grove, as noted in this study.

In the post–Second World War period, Pratt shifted his perspective on technology in his poetry from technology as machines or a mode of mechanization to that of a mindset shaped from within the human psyche, which was itself technologically induced. This altered perspective is in keeping with the views of Frye's contemporaries, Innis, McLuhan, and Grant; it also fitted in with Frye's belief that a scientific-technological perspective on the world was mythological, shaped within the mind. Pratt presented this mythological perspective on technology most effectively in his epic poem *Towards the Last Spike,* an account of the building of the transcontinental railway by the Canadian Pacific Railway Company, seen at the time as a triumph of technology over nature. Whereas in his earlier poem *The Titanic* technology was the mighty ship – a machine – that came up against nature in a titanic struggle, in *Towards the Last Spike* the struggle was within the minds of politicians and the Canadian public as to the wisdom of putting such faith in the technology of railways. Prime Minister Sir John A. Macdonald and the chief engineer, William Van Horne, needed to 'sell' the wisdom of building the railway in the same way that Thomas Keefer had to sell railways to the population of the Canadas in his *Philosophy of Railroads*. The challenge was to create this technological mindset.

Thus, in *Towards the Last Spike,* Pratt re-created the debate between Edward Blake, the Liberal MP who doubted (the doubting Thomas) the feasibility of building an all-Canadian railway and who lacked faith in technology, and John A. Macdonald, whose faith (the biblical faith of Paul) in railway technology never wavered. As Pratt's fellow poet James Reaney perceptively notes in his reflections on Pratt's poem:

> In this poem Pratt is always careful to show the struggle against Nature, against the Rockies and the North Shore of Lake Superior as a struggle against something inside the human mind as well as something outside it. Sir John A. Macdonald and Van Horne are always struggling with their own inability to express their vision in words so that other minds will be persuaded of its truth. This is the real struggle, the attempt to make words conquer all that is unknown. The transcontinental railroad has to be built painfully in people's minds as well as in their country.'[38]

This building of the railway 'in people's minds' – the establishment of the credibility of technology by creating a mythology of technology – was as important at the time as the physical construction of the railway and has continued to be important in identifying the railway as a symbol of nation building, the triumph of technological nationalism.[39] Frye noted the importance of the theme of the technology of communication in Pratt's poetry: 'Pratt's poetry is full of his fascination with means of communication, not simply the physical means of great ships and locomotives, though he is one of the best of all poets on such subjects.'[40]

Frye's writings had from the beginning focused on the myths underlying Western thought. In *Fearful Symmetry* (1947), he had analyzed the ideas of the noted English poet William Blake to show that Blake wrote within a mythological framework that was strongly biblical in nature and Christian in orientation. In his later writings, Frye was more interested in understanding the scientific-technological myth that by the seventeenth or eighteenth century had supplanted, or at least rivalled, the Christian myth for dominance in Western thought.[41] He noted that this new myth also grew out of the Bible and Christian thought and was best expressed in the concept of progress, points that George Grant was also making at the same time Frye was writing about this modern mythology. But whereas Grant was critical of this technological perspective because he saw it as all-encompassing, enveloping the world, Frye was optimistic, realizing through his study of Western literature that this myth was only one perspective that could easily be replaced as the earlier Christian myth had been. For Frye, there was relativity to the mythology of technology that held hope for the future, so long as it was an open rather than closed myth.[42] To Frye, this meant recognizing the scientific-technological myth as only one possible view of the world in which there were many possible myths or viewpoints.

Frye was vague as to when the scientific-technological mythology emerged. At different times, he placed it in the Renaissance era, in the Cartesian view of the world of the seventeenth century, and in the great revolutionary age of the eighteenth century, embodied in the American, French, and Industrial Revolutions. However, he consistently argued that this mythology reached its full flowering in the Romantic period of English literature, most notably in the works of William Blake. For Frye believed that ideas became mythologies only when creative artists, most notably poets, took the imagery implicit in the new worldview and weaved it into literature, at which time the ideas become a popularized mythology. In his view as literary critic, that occurred with the Romantic poets and novelists. In his Introduction to *Fables of Identity* (1963), Frye revealed the importance of the Romantic movement in English literature for the modern view of the world:

> Slowly but steadily the doctrine of the divine creation of the human order fades out, not perhaps as a religious conception, but as a historical and literal fact taking place at a specific point in past time. Man thus comes to be thought of as the architect of his own order, a conception which instantly puts the creative arts in the very centre of human culture ... The conception of nature as a divine artefact also fades out, and nature is thought of, not so much as a structure or system presented objectively to man, but rather

> as a total creative process in which man, the creation of man, and the creation of man's art, are all involved ... If man has created his own order, he is in a position to judge of his own achievement, and to measure that achievement against the kind of ideals his imagination suggests.[43]

There arose, therefore, the belief, the modern myth, that the truly creative act was not God's creation of the world but man's *re*-creation of the world through the power of God within. This was the realm of the creative imagination, which expressed itself as 'the educated imagination' since it was based on one's knowledge of the world of imagination that education, especially a study of literature, provided. It was the inner world of vision. Frye maintained that the Bible became the book upon which to construct this imaginative universe, because it was 'the Word,' the act of creation through the telling of the story, or mythology, as envisioned in the mind's eye.

This re-created world of the imagination stood juxtaposed against the created world of fallen nature and alienated man symbolized in the biblical story of the fall of Adam and Eve, which Frye associated at times with the technological world. Thus, the tension Frye saw working itself out in Western literature was between the free individual (the myth of freedom) and the social norms of society (the myth of concern) – the tension between the moral imperative and the technological imperative – that, unless equalized, ended up either as a world of chaos, the primordial wilderness in which man was alienated from nature and in which freedom became licence, or a world of authority and external power, a totalitarian state in which freedom was suppressed through the power of technology. Somewhere in between was the ideal world, the imagined world, where the individual was free but chose to co-operate with other free individuals in a socially harmonious and socially conscientious society.

In his public lecture series *The Educated Imagination* (1963), part of the Massey Lecture series for the CBC, Frye explored the nature of the creative imagination. He pointed out that the role of the creative imagination was to create the vision of the ideal world. Such a vision invariably came up against the real world. Thus, a clash occurred between the ideal world and the real world, the visionary and the practical worlds, the world of ideas (words) and the world of technology (objects). In his Conclusion to *The Educated Imagination,* Frye related the tension between opposites to the biblical story of the Tower of Babel and in so doing reinforced the theme implicit throughout his lecture series of the tension in Western civilization between ideas and technology, freedom and power, a theme that this study notes as being pervasive throughout Canadian thought:

> The civilization we live in at present is a gigantic technological structure, a skyscraper almost high enough to reach the moon. It looks like a single

> world-wide effort, but it's really a deadlock of rivalries; it looks very impressive, except that it has no genuine human dignity. For all its wonderful machinery, we know it's really a crazy ramshackle building, and at any time may crash around our ears. What the myth tells us is that the Tower of Babel is a work of human imagination, that its main elements are words, and that what will make it collapse is a confusion of tongues. All had originally one language, the myth says. That language is not English or Russian or Chinese or any common ancestor, if there was one. It is the language of human nature, the language that makes both Shakespeare and Pushkin authentic poets, that gives a social vision to both Lincoln and Gandhi. It never speaks unless we take the time to listen in leisure, and it speaks only in a voice too quiet for panic to hear. And then all it has to tell us, when we look over the edge of our leaning tower, is that we are not getting any nearer heaven, and that it is time to return to the earth.[44]

In 1967, Frye sketched out the modern scientific-technological myth in the Whidden Lecture series, the Canadian centennial lectures at McMaster University. He entitled the lectures *The Modern Century* and claimed that this was the century in which 'Canada came of age.'[45] He identified self-consciousness as one of the defining characteristics of this 'modern century.' The self-reflection of the modern age revealed a world moving at a phenomenal speed. 'One very obvious feature of our age,' Frye wrote, 'is the speeding up of process: it is an age of revolution and metamorphosis, where one lives through changes that formally took centuries in a matter of a few years.'[46] Frye noted the value assumptions underlying this world of change that became translated into the myth of progress: 'The dynamic is better than the static, process better than product, the organic and vital better than the mechanical and fixed, and so on.' Yet he also observed that the positive perspective implicit in the myth of progress turned out to be the opposite: 'the alienation of progress.' Such alienation manifested itself in the feeling that 'man has lost control, if he ever had it, over his own destiny.' It was advertising, with its false illusions and hyperbole that, according to Frye, created this feeling of helplessness. 'Something has happened to atrophy one's responses when the most soporific words one can use are such words as "challenge," "crisis," "demand," and "endeavour."'[47]

Frye identified the source of this feeling of alienation in the technology of communication. 'The quickening of the pace of news, with telegraph and submarine cable, helped to dramatize a sense of a world in visible motion, with every day bringing new scenes and episodes of a passing show.' He compared the experience to being in a fast-moving vehicle: 'In a swift movement we are dependent on a vehicle and not on ourselves, and the proportion of exhilaration to apprehensiveness depends on whether we are driving it or merely riding in it. All progressive machines turn out to be things ridden in, with an unknown driver.'[48]

In the nineteenth century, the sense of alienation within the individual had been seen as externally induced, the result of an ugly landscape scarred by industrialization or of a social and spiritual alienation by a society that had put a value on materialism and capitalistic exploitation over spiritualism and the good of society. In the twentieth century, Frye noted, the sense of alienation became inner directed, the result of electronic communication media capable of captivating and regulating the mind. As Frye explained:

> It is quite possible ... for communications media, especially the newer electronic ones, to break down the associative structures of the inner mind and replace them by the prefabricated structures of the media. A society entirely controlled by their slogans and exhortations would be introverted, because nobody would be saying anything: there would only be echo, and Echo was the mistress of Narcissus. It would also be without privacy, because it would frustrate the effort of the healthy mind to develop a view of the world which is private but not introverted, accommodating itself to opposing views. The triumph of communication is the death of communication: where communication forms a total environment, there is nothing to be communicated.[49]

In *The Modern Century,* Frye showed how this new scientific-technological mythology was evident in the arts since the Romantic era. He noted, for example, the tendency of artists in the Romantic period *not* to look *at* nature but to identify with its processes and power, 'the creative forces symbolized by the swirling colours, the dissolving shapes, and the expanding perspective where we seem to see everything at once, as though the eye were surrounding the picture ... Impressionism portrays, not a separated objective world that man contemplates, but a world of power and force and movement which is in man also, and emerges in the consciousness of the painter.' In literature, modernism became associated with 'streamlining': 'the sense of exhilaration of mechanical movement'; in poetry, it was expressed as a revolt against continuity. 'Modern poetry,' Frye wrote, 'tends to be discontinuous, to break the hypnotic continuity of a settled metre, an organizing narrative, or a line of thought.' All of these qualities of the modern age, he maintained, were associated with technology, for 'the modern world began with the industrial revolution.'[50]

The new scientific-technological mythology also turned inward rather than outward, orientated toward the human sciences as opposed to the physical sciences. As such it was a 'myth of concern,' 'an expression of man's concern about himself, about his place in the scheme of things, about his relation to society and God, about the ultimate origin and ultimate fate, either of himself or of the human species generally.'[51]

Frye pointed out that the modern myth of concern could take two forms: (1) the ideal form of a benevolent world in which the good of society transcended the needs of the individual – what Frye called, after William Blake, the world of the lamb – 'the vision of the world as created by a benevolent and intelligent power,' or (2) the 'real' form of a world of power struggles where the concerns of the individual took precedence over the needs of society – Blake's world of the tiger – 'the subhuman world of nature, a world of law and of power but not of intelligence or design,'[52] another perspective on the tension between the moral and technological imperatives.

Frye concluded these Canadian centennial lectures by returning to a discussion of Canada's place in the scientific-technological mythology of 'the modern century.' He suggested that, in their hesitancy to wholeheartedly accept technology as the Americans had done, to be more skeptical of what technology had to offer the world, and to be more sympathetic to the myth of concern over the myth of freedom while being aware of the need for both in the modern technological world, Canadians had brought their country closer to being the world of the lamb, unlike the United States, which symbolized the world of the tiger. As Frye so aptly expressed in his concluding remarks:

> One of the derivations proposed for the word Canada is a Portuguese phrase meaning 'nobody here.' The etymology of the word Utopia is very similar, and perhaps the real Canada is an ideal with nobody in it. The Canada to which we really do owe loyalty is the Canada that we have failed to create. In a year bound to be full of discussions of our identity, I should like to suggest that our identity, like the real identity of all nations, is the one that we have failed to achieve. It is expressed in our culture, but not attained in our life, just as Blake's new Jerusalem to be built in England's green and pleasant land is no less a genuine ideal for not having been built there. What there is left of the Canadian nation may well be destroyed by the kind of sectarian bickering which is so much more interesting to many people than genuine human life. But, as we enter a second century contemplating a world where power and success express themselves so much in stentorian lying, hypnotized leadership, and panic-stricken suppression of freedom and criticism, the uncreated identity of Canada may be after all not so bad a heritage to take with us.[53]

In the 1970s and 1980s, Frye realized the importance of the *historical roots* of Canadian thought as opposed to geographical location in understanding the uniqueness of the Canadian experience within Western civilization. The difference in emphasis is evident in Frye's 'Conclusion' in the two editions of the *Literary History of Canada,* the first published in 1965, the second in 1976. In 1965, he emphasized Canadian *geography* or the country's physical environment as the decisive factor in shaping a different literary tradition in Canada from that of the United States. To make his point, he used the example of two immigrants from Europe, one entering the United States and the other entering Canada. The immigrant to the United States entered via a distinct eastern seaboard, whereas the immigrant to Canada entered by way of a great gulf with invisible land all around him or her, 'like a tiny Jonah entering an inconceivably large whale.'[54] Then, too, Canadian settlement occurred in pockets in the different geographical regions of the country at much the same period of time but isolated from one another, whereas Americans systematically moved across the continent bringing settlement and an established culture and civilization with them. According to Frye, this pattern of pockets of settlements in Canada was a contributing factor to the regional nature of Canadian literature. As well, he emphasized in the first edition the physical proximity of most Canadian towns and cities to the American border, which resulted in a strong American influence in Canadian literature that early on supplanted the British influence. He observed that, in the search for a Canadian identity within the nation's literature, the focus was not on the question 'Who am I? (which is or at least can be seen as a more historically oriented question of identity) but 'Where is here?' – situating oneself in a physical space, or geography, and environment. This awareness of the importance of geography in shaping the nature of Canadian society and culture also explained the emphasis in Canadian literature on the technology of transportation and communication as the means to unite a country that was vast in size, divided geographically, and sparsely populated.

In his 'Conclusion' to the second edition of the *Literary History of Canada* (1976), Frye focused on the different *historical* experiences between Canada and the United States as the distinguishing feature of Canadian and American literature. With regard to the United States, he stated that the country 'found its identity in the eighteenth century, the age of rationalism and enlightenment.' Frye wrote, 'The eighteenth-century cultural pattern took on a revolutionary, and therefore deductive, shape, provided with a manifesto of independence and a written constitution. This in turn developed a rational attitude to the continuity of life in time, and this attitude seems to me the central principle of the American way of life.'[55] He compared the American perspective to an 'express train.' He explained, 'It is a conception of progress, but of progress defined by mechanical rather than organic metaphors, and hence the affinity with the eighteenth century.'[56]

By contrast, Canada bypassed the Enlightenment, going 'directly from the Baroque expansion of the seventeenth century to the Romantic expansion of the nineteenth.'[57] The Loyalists who had come to British North America had rejected the 'enlightened ideas' underlying the American Revolution, while the French Canadians had turned their backs on the 'enlightened beliefs' of the French Revolution. As a result, Frye noted, 'identity in Canada has always had something about it of a centrifugal movement into far distance, of clothes on a growing giant coming apart at the seams, of an elastic about to snap ... This expanding movement has to be counterbalanced by a sense of having constantly to stay together by making tremendous voluntary efforts at intercommunication, whether of building the CPR or of holding federal-provincial conferences.'[58] Thus, from a historical perspective, as from a geographical one, the emphasis for Canadians had been on the technology of communication and transportation as a way of pulling a disparate people together, something that had not concerned Americans to the same extent.

What Canada and the United States did share from a historical perspective and what made them both 'North American' in outlook was a common attitude toward nature that came out of a seventeenth-century mathematical view of the world. In the essays in *Divisions on a Ground* (1982) and in the National Film Board film *Journey without Arrival,* narrated by Frye, he emphasized the mathematical and therefore abstract, mechanical, and technological view of the world coming out of what he defined as the 'Baroque age.' It was Frye's version of a point that George Grant made in *Time as History,* that both Canada and the United States had no history before the age of progress (Frye's age of mathematics) with the exception of their indigenous Native populations, whose views unfortunately had little impact in shaping a North American perspective of nature. As a result, the non-Aboriginal population had no sense of the land as 'autochthonous' or being possessed by the spirits or God. Instead, the early European settlers saw 'an immense gap' between themselves qua humans and 'all other living creatures, who belonged primarily in a world of mechanism.'[59] Frye wrote:

> The newly discovered power of mathematics ... was at its clearest at the exploring and pioneering periphery of Baroque culture, in the astrolabes and compasses that guided the explorer, in the grid-patterns that eventually were imposed on city and country alike. Every 'improvement' in communication since then, in railway, highway, or airway, has meant a shorter and straighter path through nature until, with the plane, the sense of moving through nature practically disappears. What does not disappear is the attitude of arrogant ascendancy over nature. For the white conquerors of this continent, creation does not begin with an earth-mother who is the womb and tomb of all created things, but with a sky-father who planned and ordered and made the world, in a tour de force of technology.[60]

While both Americans and Canadians shared this perspective, Frye argued that it played out differently in the mythology of the two countries. In his analysis, he focused only on its impact on Canadian mythology. He pointed out that this mechanical and technological view of the world was evident in the Canadian scientific and coldly calculating perspective on both the non-human and human world. It was evident, for example, in the emphasis on and almost obsession with the science of measurement and numbers, or 'quantitative relationships,' such as in the fur traders' ledgers, in which they had kept a careful record of the number of furs bought and sold, and in the missionaries' recordings in *The Jesuit Relations* of the number of Indians who converted to Christianity.[61] But it was equally evident in the humanities, with its 'verbal technology, the use of words for practical or useful purposes.' He elaborated further: 'The two words practical and useful do not of course mean quite the same thing: some forms of verbal technology, like preaching, may be useful without always being practical; others, like advertising, may be practical without always being useful. Many of the university's professional schools – law, theology, education – are concerned with verbal technology, and so is every area of human knowledge that employs words as well as numbers, metaphors as well as equations, definition as well as measurement.'[62]

This technological perspective was also evident in our modern view of God. In the new scientific-technological age of the post-Newtonian and post-Cartesian world, God became an 'engineer' who had started the world operating in a mechanical fashion that then left it to run on its own with clockwork precision. Today, Frye noted in the 1980s, God was a 'force' or 'energy' in the same way that quantum physicists think of the natural world as being made up of force and energy – that is, ever-changing and non-descriptive in the old mechanistic sense. Words too have become not objects but forces unto themselves: 'Words with Power' (the title of Frye's second major book on the Bible and literature). As Frye explained in his CBC interviews, published in 1992 as *Northrop Frye in Conversation:*

> One of the most seminal books I have read is Buber's *I and Thou.* Buber says we are all born into a world of 'its,' and if we meet other human beings we turn them into 'its.' In this view, everything is a solid block, a thing. Consequently, when we think of God, we think of a grammatical noun. But you have to get used to the notion that there is no such thing as God, because God is not a thing. He is a process fulfilling itself. That's how he defines himself: "I will be what I will be." Similarly, I am more and more drawn to thinking in terms of a great swirling of processes and powers rather than a world of blocks and things. A text, for example, is a conflict of powers. That's why the Derrida people can pursue a logic of supplement. They can extract

> one force and set it against another. But the text is not a thing any more. A picture is not a thing. It's a focus of forces.[63]

Frye also argued that the centrality of technology in the modern scientific-technological view of the world challenged the earlier Christian myth of man as a creature of God, subservient to him. Frye pointed out that technology in the Bible acted as a counterfoil to the goodness of man. It became associated with iron weapons or idols made from iron, both of which were linked to warfare. John Milton was therefore following a well-established view of his time, Frye stated, when in *Paradise Lost* he presented the devil as inventing gunpowder and artillery during the war in heaven. Frye wrote, 'The release of titanic powers in man through invention is feared and dreaded at every stage in history [within the biblical context], partly because of the intertwining of the titanic and the demonic, and partly because technology, having no will of its own, is readily projected as a mysterious, external and sinister force.'[64] As a result, within the Christian view of the world, technology came to be associated with warfare and was thus a deterrent to the vision of the biblical peaceable kingdom.[65]

In the 1980s as he brought his study of literary mythology up to the present, Frye observed that even in the twentieth century technology continued to have this negative connotation because of its use for destructive rather than creative ends. It was evident in the modern literary genre of dystopia novels, 'the Utopian satire or ironic social vision, represented in this century by Orwell's *1984* and Solzhenitsyn's *First Circle*.' According to Frye, the alternative to a dystopia was not a utopia, 'which is so often a contrived and anxiety-ridden form,' but rather a 'sense of social norm,'[66] a vision of a better society not in a future or past state as in a utopia but in the present state. It is a vision in which the fundamental concerns of freedom, health, equality, happiness, love are dominant over secondary concerns, which Frye described as being ideological. If and when technology is used to bring the fundamental concerns to all people without the limitations imposed by an ideology, including the Christian one, of exclusion, then the Christian prophecy of 'heaven on Earth' will be achieved. However, such a world would not be given by God to human beings as in the 'first' or original creation story but rather by human beings to God as they 're-created' their world in terms of their imaginary vision of a better world. As Frye expressed it so eloquently in his concluding section of *Words with Power,* 'The Biblical perspective of divine initiative and human response passes into its opposite, where the initiative is human, and where a divine response ... is guaranteed.'[67] Although this vision was apocalyptical, technology could be the practical means to bring it about.

Frye thus ended where he began half a century earlier, with a balanced perspective on technology: negative, based on past experience, but also positive in terms of future potential, the using of technology as creative arts. Once achieved, technology would cease to be a myth associated with a particular ideology, be it capitalism, Marxism, or socialism, or any other ideology, and would become a 'creative reality.'

10
George Grant and Dennis Lee: Technology as Being

George Grant was the Canadian theorist of technology who went furthest in exploring the implications of the concept of technology as a mindset that shapes our very being as humans. 'Technique is ourselves,'[1] he concluded at one point; it is our being, our end, and our morality. On another occasion he noted, 'Technology is the metaphysics of our age, it is the way being appears to us.'[2] In countering the optimism about technology in Canadian thought in the pre–First World War era and in the perspective of his contemporary Marshall McLuhan, Grant has often been dismissed as only a pessimist, a negative soothsayer who longed for a past age when the moral influence of his ancestors had prevailed. Yet such glib condemnations of the man and his ideas ignore the deep and agonizing thinking on the subject of technology and the hope of salvation from the technological imperative that underlay his warnings and predictions of technological doom. Even in the midst of his most negative thinking about the imperial dominance of technology over the entire world as evidenced by the Vietnam War, Grant could write an article entitled 'A Platitude,' in which he began, 'We can hold in our minds the enormous benefits of technological society,'[3] as a reminder that, despite its destructiveness and devastation, technology held wonders for the modern age. For Grant, it was not simply a matter of negating technology but of balancing and evaluating its strengths and weaknesses so as to better understand the magnitude and complexity of the subject under discussion. He reminded Canadians and anyone else willing to listen that the benefits of technology might be the 'fruit of the tree of knowledge' that hid the rot within. He demanded that the modern age confront technology to understand its nature and its dominance and pervasiveness. Grant believed it was important to define technology, to name it, as the best means to understand it. His thinking on the subject was like a spiral. He began in the outer reaches, initially accepting much of the conventional thinking on the subject of technology and then slowly, and painfully, discarding the prevailing beliefs as he narrowed in on the core and essence of technology. In the end, he found technology to be all consuming, so pervasive that it shaped the very essence of being.

Grant's thinking on the subject of technology follows in many ways the evolution of thought on the subject as it is presented in this book. He began as a proponent of technology, one of its enthusiasts sharing the perspective of Keefer, Haliburton, Fleming, Bell, and Bovey. Then he began to have nagging doubts about the potentiality of technology and the ability of humans to control technology through reason while still holding on to a faith in technology's power for creating the good society and upholding the moral imperative. He would have identified with the perspectives on technology of Brett, King, Grove, and Leacock. Doubts gave way to despair as Grant explored the pervasiveness and subliminal dominance of technology in our lives and, more significantly, in our thought processes. He came to understand technology as the *sine qua non* of modern thought and therefore believed it virtually impossible to think outside the technological paradigm; he described this awareness as 'hell,' an 'abyss,' a 'madness.' Yet the very fact that Grant could plumb the depths of the technological imperative and still surface to write about his experience for the benefit of others who might be experiencing a similar moral crisis is testimony to his belief in technology as volition.

The Canadian poet Dennis Lee was one of Grant's followers who went through a moral crisis over the ability of the United States to dominate Canada and ultimately the world as a result of technology. On reading Grant's *Lament for a Nation,* Lee wrote his epic poem *Civil Elegies,* in which he explored themes of dominance, colonial space, absence, and silence, which he came to associate with the technological imperative, as his way to come to terms with the tension between the technological imperative and the moral imperative.

Grant's response to the technological imperative was to ultimately realize that technology's hold may be suffocating at times but never final. The one avenue of escape was faith in a higher order that he identified as God. For him, faith in God was a fundamental belief underlying the moral imperative. His faith prevented him from committing suicide as he wrestled with the 'demon technology.' To appreciate how Grant came to this understanding about technology, one requires knowledge of his intellectual journey over a lifetime of thinking about the subject.

George Parkin Grant was born into distinguished Canadian families: his father's side, the Grants, and his mother's side, the Parkins, were of high intellectual and social status.[4] George remained acutely aware of his intellectual and social heritage as he evolved as a philosopher. James Grant, George's great paternal grandfather, had left Scotland in 1826 to farm in Pictou County, Nova Scotia. In 1831, he married Mary Monro. Their son,

George Monro Grant, was born on December 22, 1835. Young 'Geordie,' as he was known, lost the use of one of his hands in a farm accident, thus limiting his ability to carry on the family farm. Instead, James encouraged his bright young son to attend university in Glasgow, Scotland, in preparation for the Presbyterian ministry. Geordie followed his father's advice and was called to be the minister of St. Matthew's Church in Halifax in 1863. While in Halifax, he assisted in the revival of Dalhousie University, work that prepared him well for his next position as president of Queen's University in Kingston, Ontario, from 1877 until his death in 1902. During that time, he turned a fledgling and financially strapped institution into one of the finest Canadian universities. He modelled Queen's after the German universities, which emphasized specialization and research, and based its curriculum on the eminent Ivy League universities in the United States.

George Monro Grant was also one of the distinguished leaders of the social gospel movement in Canada at the turn of the twentieth century, devoted to turning Protestantism toward social reform so as to create a better society, a veritable 'Kingdom of God on Earth.' His grandson George Parkin Grant came to associate this form of Protestantism with liberalism and liberalism in turn with a faith in unlimited human freedom. He also came to identify this liberal Protestantism as the major philosophy underlying the modern technological age. Thus, in criticizing that philosophy in later life, he realized he was criticizing and rejecting his family's intellectual heritage. But he also came to appreciate the role that his heritage played in enabling him to both understand and ultimately reject the modern technological mindset.[5]

In May 1867, two months before Canadian Confederation, Reverend George Monro Grant married Jesse Lawson, the eldest daughter of a prosperous West Indies merchant, whose family had resided in Halifax since its beginning as a military base in 1749. Their son William Grant, George's father, was born in 1872. William attended Queen's and then Balliol College, Oxford, where he received a First in the challenging Greats (Classics) program. He was the first Canadian to achieve such a distinction. In 1898, he took a position as a history teacher at Upper Canada College, where George Parkin, his future father-in-law, was principal. William remained there until 1902, at which time he moved to another private boy's school, St. Andrew's College. In 1906, William was invited back to Oxford as Beit lecturer in colonial history. He remained at Oxford until taking up his new position as a history professor at his father's university in Kingston, Ontario. It was while back in England that William resumed his acquaintance with Maude Parkin, whom he had met at Principal Parkin's residence at Upper Canada College.

The Parkin family had equally humble beginnings, but by the time of William's marriage to Maude in 1911 they had achieved greater social status.

George Parkin, George Parkin Grant's maternal grandfather, was born in Albert County, New Brunswick, in 1846, the thirteenth child of a Yorkshire Methodist yeoman farmer. George Parkin was headmaster of the collegiate school in Fredericton, New Brunswick, until 1889. In that year Lord Milner, one of England's great imperialists, conscripted George Parkin to give lectures on imperial federation in Britain and throughout the dominions. Parkin had already joined the Imperial Federation League in Canada, founded in 1887. While serving the Empire, he wrote three notable books on the subject of imperial unity. His most famous, *Imperial Federation: The Problem of National Unity,* was published in 1892, along with *Round the Empire,* while *The Great Dominion: Studies of Canada* came out in 1895. He envisioned the British Empire as having a moral responsibility to spread justice and civilization around the world. As a result of his religious fervour for the Empire, one contemporary dubbed him 'the prince of imperialists and their first missionary.' Parkin preferred his own description of himself as 'a wandering Evangelist of Empire.'[6] In 1895, he gave up this exhausting job to become headmaster of Upper Canada College. In 1902 came Parkin's finest moment, when he was invited to implement and administer the Rhodes Scholarships, set up by Cecil Rhodes of South Africa to send promising young men from the dominions, as well as the United States and Germany, to Oxford, the seat of higher learning in the Empire and its intellectual guardian. Parkin performed this duty until his retirement in 1920, at which time he was knighted Sir George Parkin in recognition of his outstanding contribution to the Empire.

In 1878, Parkin married Annie Fisher, the daughter of a prominent New Brunswick family. Maude was the second oldest, born to George and Annie in 1880, of a family of four girls and one boy. She carried on the honour of the family name and tradition and passed it on to her son, George, who never forgot throughout his life that he was George Parkin's grandson. Maude was educated at McGill University. Upon graduation, she was offered the position of dean of women (warden) in a residence at the University of Manchester. She was still teaching there when William Grant proposed to her in August of 1910. The couple were married on June 1, 1911, in the village of Goring-on-Thames, before departing for Canada so that William could take up his new position at Queen's.

Like most English Canadians, William felt pressured to come to the aid of Mother England in the Great War, and he enlisted in 1915. He saw active service with the 59th Battalion and then with the 20th as major. He was seriously wounded in the summer of 1916 while serving in France when his horse threw him and then landed on top of him, resulting in a 'fractured rib, a concussion of the back and a wound of the head.' During convalescence, he realized that he would never be well enough to serve again and so decided to accept an offer that had been made to him just before he left to become principal of Upper Canada College and thus follow in his father-in-law's footsteps. In resigning his Queen's position, he explained his decision to accept the principalship as 'an opportunity to set the tone for education in the province as a whole.'[7] William remained the affable principal of Upper Canada College until his death in 1934 from pneumonia complicated by war wounds.

George Parkin Grant was convinced that the Great War had destroyed his father as it had so many other young Canadians. In an interview in later life, he explained, 'My father was a Nova Scotian, who'd grown up in Kingston, Ontario. He was essentially a very gentle, strong scholar, who I think, above all, was ruined by the First World War. I mean, he was ruined physically – he was terribly wounded. I don't mean ruined as a human being, but I think the war was terrible for people who had grown up in the great era of progress. To suddenly meet the *holocaust* of the trenches head-on like my father did – it was just terrible!'[8]

George Parkin Grant was born on November 13, 1918, two days after the armistice. He had three older sisters: Margaret, Charity, and Alison. He grew up in Toronto and attended Upper Canada College under the watchful eye of his father. He was not particularly good at sports and gravitated instead to the arts, particularly music, art, and literature. He became a pacifist after reading Graeme Watson's *Cry Havoc* in 1934, although he obviously had a predisposition toward a negative view of fighting and war from his father's horrific experience.

Grant graduated from Upper Canada College in 1936 near the top of his class. With family prodding and approval, he entered Queen's University, where he was seen, and saw himself, as the distinguished past principal's grandson. Again, George felt tied to family and tradition. He majored in history, in which he won a number of awards. His professors encouraged him to apply for a Rhodes Scholarship (established by his grandfather Parkin), which he won in 1939. Once again, he questioned whether he had won it on intellectual merit or family association. Grant attended Balliol College like his father before him, where A.D. Lindsay, a family friend, was master. He decided to major in law instead of taking a program that would have led to a teaching career. But a few years later, he decided to return to Oxford to do a DPhil in theology, not with the expectation of being a minister but rather a professor of theology and thus to carry on the family tradition of service to one's country by teaching its youth.

From the beginning, George never found law particularly stimulating. This alone was reason enough to escape and do something more meaningful. Pressing him in the same direction was the outbreak of the Second World War in the year he arrived at Oxford. The war was transforming Oxford, as the First World War had done, with men going off to fight and debates on the meaning of the war occurring on a regular basis. During this time of deliberation, Grant read William James, the American pragmatist philosopher, whose views delighted him. 'I was a sharp, ambitious little pragmatist,'[9] he recalled later in life at a time when many of his fellow students spoke the language of transcendence and the war was being presented as a moral struggle against fascism and brute force.

When Balliol College threatened to temporarily close its doors, George decided to train with the University Ambulance Unit. He experienced the gruesome conditions of wounded soldiers and civilians during the bombing of London. The scene caused him to write, 'This holocaust makes 1914 seem like a picnic and it is not lack of courage that made France fall but human endurance limits. Hate's most gruesome orgy.'[10] When the Ambulance Unit work seemed unfulfilling, he volunteered to work at Bermondsey Club, an Oxford settlement house in the poorer district of London. This was the most rewarding time in his life to this point, not least of all because for the first time he was his own man and not a family appendage. Most of all, however, he felt the reward of service to his fellow human beings.

It was during this time of uncertainty and angst that George experienced what he would call the most profound moment in his life, a religious conversion. Upon closing a gate during a countryside excursion on his bicycle, he suddenly felt that 'all was finally well, that God existed.' The moment would affect the direction of his life thereafter. In reflecting on the experience in later life, Grant claimed, 'I think it was a kind of affirmation that beyond time and space there is order. Now, all the psychologists and psychiatrists could just say that I was looking for order, but for me it was an affirmation about what is, an affirmation that ultimately there is order. And that is what one means by God, isn't it? That ultimately the world is not a maniacal chaos – I think that's what the affirmation was.'[11]

The war also affected his views of Western civilization. He recalled in later life that the war drove him 'out of the bourgeois altruistic liberalism'[12] because he came to associate it with all that was repugnant with modernity, destroying not only bodies but minds as well and personifying the worst of the industrial world. Amid the war, Grant wrote to his mother, 'The boys and girls I know in London whose work has no coordination with the rest of their lives, who seek, after a tiring day in their factory, escape in the movies, or alcohol or the less productive experiments of sex, are merely typical of the fact that there has been no kind of adjustment to keep the individual in line with the changes of the industrial world ... The whole world is in it now. We gave the Japanese western industrial civilization & heavens they are using it.'[13]

Grant was anxious to return to Canada. Ever since he had left Canadian shores for Oxford, he knew that his future lay in his native country; by tradition and temperament, he was Canadian. He returned in 1947, his degree in the philosophy of religion not yet in hand, to a position at Dalhousie

University. He became conscious of how materialistic North America was compared to Britain. While he was critical of that materialism, he was also aware of the comforts it provided. In a speech to the Community Chest in Halifax, Grant pointed out that 'there was a good side to the new mass world: surplus wealth made earning a living easier and thus generated leisure.'[14] 'In the pre-industrial age,' Grant pointed out, 'the mass of people was so tied to back breaking labour that only the very few had the possibility of any freedom from the continual task for twelve or fourteen hours a day of earning just enough to stay alive.'[15] He made a similar point from a personal perspective in a letter to his mother in 1954: 'My great weakness is that I care tremendously for the pleasant things of civilized existence – not so much the material things as the civilized way of life that you so wonderfully taught us and I have a fear of what my children will be like without them. Of course this is a failure of faith in God and yet that failure exists. Yet I am caught because I do not think I could give up philosophy for other aspects of life that bring more money.'[16] Grant believed that the virtues of materialism needed to be tempered, as did the philosophy behind it – that is, liberalism – by the virtues of moral uplift and tradition that the British philosophy of conservatism could provide.

He became aware of his need as a philosopher who was interested in linking philosophy to real life to come to terms with industrialism every time he made a trip from Halifax to Toronto. On one occasion, he wrote to his mother, 'If sometimes I seemed strange in Toronto, it was that it was all such a great impression to me. It is my job as a philosopher to try and see the industrial society as it is with all its virtues and its failures or else my philosophy will be a barren intellectual game and you cannot imagine how overwhelming an impression the expanding society like Toronto is to an outsider.'[17]

Grant addressed the issue of materialism versus spiritualism in the context of Canada's destiny in the postwar era in a little pamphlet entitled *The Empire, Yes or No?* He associated materialism with the United States and spiritualism with the British Empire, a perspective similar to that of his grandfathers Grant and Parkin, who were imperial federationists. Grant also saw materialism as a by-product of industrialism and the war the worst manifestation of that same industrialism. Together they were creating a new human being who was 'brutal and unreasonable, unethical and material, and who [was] ruthlessly dominated by his appetites,' a sort of cross between Marx's economic man and Freud's sexual beings. Such beings were the antithesis of Christian, ethical man. Grant believed that only Canada's membership in the British Commonwealth could prevent such debased, materialistic, power-hungry individuals from seizing control and destroying all that was fine and virtuous about his country. He summed up his views eloquently in his conclusion:

> We cannot judge the British Commonwealth from our petty interests alone (however well these are satisfied,) but on the highest criteria of political morality. For today in the modern world, with it more than with any other political institution, lies the hope of Christian man, of ethical man, of man the reasonable, moral being who stands before God and history ... If we believe in Christian man, the finest flower of all that Western civilization has produced, then there can be no doubt that our chief hope in the survival of such values is in the survival of the British Commonwealth.[18]

Grant was expounding a view of the British Empire that Sandford Fleming, Thomas Chandler Haliburton, and Thomas Keefer had developed nearly a century earlier, a view that saw technology, or materialism, and empire as compatible, because the spiritualism of the empire offset or balanced the materialism of North American society.

In his pamphlet, Grant dealt with another important aspect of his thinking on technology: the association of technology and materialism with the philosophy of liberalism. Liberalism, he noted, put the focus on the immediate pleasures of the greatest numbers, a sort of Benthamite viewpoint, rather than on eternal and transcendent values that came from the spirit through a fusion of philosophy and theology. The essence of liberalism was individual freedom, which Grant admired, but free individuals whose aspirations in life rose no higher than increasing their own personal wealth and greater material goods did nothing to improve the world socially, morally, or aesthetically. Industrialism could lead to a better world, but only if the values behind industrial advancement were ones that served the public good through equal treatment of all individuals within society. That required a philosophy of conservatism.

The problem was that social and political leaders in the modern industrial age had lost the sense of moral responsibility that a philosophy of conservatism had instilled in societal leaders a century earlier. That was why liberalism had to be tempered by conservatism. Grant believed that an association of liberalism and conservatism in North America could only come through intellectual contact with the roots of that conservatism in Europe. Since the United States had begun as a nation after the age of conservatism and when liberalism was in full flower and the liberal belief in human progress was most extenuated, that country lacked an innate philosophy of conservatism. Canada was also a North American nation, with all the virtues and vices of its southern neighbour. But it had a conservative tradition, however weak, as a result of its continued link with Britain. If that link should disappear, there would be nothing to prevent Canada from slipping into the arms of the United States. Therefore, Canada's hope remained in taking the best of its British and North American traditions while rejecting the worst. This meant keeping its association with the British Commonwealth to offset the powerful pull of the United States, especially in the post–Second World War era, when the United States emerged as a superpower in the world. The idea was commonplace at that time; the novel twist Grant gave to it was to emphasize the negative effects of the technological character of American society, which he believed Canadians were unaware of. That was as far as Grant would push his argument in 1945, but he knew that he was on to an issue that would be of major concern for the Western world in the mid-twentieth century.

What he wanted to do at this point was summarize the nature of North American philosophical thought that underlay its technological dynamism as a means of understanding the roots of Western thought out of which the most advanced expression of that technology evolved. The opportunity came in the fall of 1958, when the CBC asked him 'to inaugurate a new experiment in educational public broadcasting, a sort of university of the air.'[19] He had already put a great deal of thought into what he wanted to say and thus found it relatively easy to produce eight half-hour broadcasts on the topic 'Philosophy in the Mass Age.'

Grant decided to present the philosophy of the modern or mass age by contrasting it with that of the ancients. In doing so, he drew upon the ideas of Charles Cochrane, one of the great Canadian classicists and a professor of classics at the University of Toronto. Grant was attracted to Cochrane because of the classicist's ability to relate the ideas of the past to those of the present. Grant had first read Cochrane's seminal study *Christianity and Classical Culture* (1940) while completing his doctoral dissertation on John Oman, the Scottish liberal theologian, at Oxford in the late 1940s. He addressed the importance of Cochrane's study in a review of the book on CBC radio in October 1954 and praised Cochrane for showing the parallels between the decay of Roman civilization as the result of the rise of Christianity in the ancient world with what was happening in the modern world. 'It is surely true,' Grant emphasized, 'that our civilization – the civilization of the west – has reached [the] point where the signs of its intellectual, religious and moral disintegration are present at every point,' just as had occurred when Christianity overtook Roman civilization. He wrote, 'Internally we have only to look at the breakdown of our educational and religious traditions – our desperate trust that the problems of human spirit can be faced by psychological manipulation or political arrangement.'[20] Grant attributed such a decline to the current belief in the West that science in association with technology could alone make possible a 'fuller humanity' through the control of nature. As he saw it, it was 'a society which holds that the control of nature by technology is the chief purpose of human existence and so from that belief a community is built where all else is subordinated to that purpose.' He described this perspective as 'the religion of the manipulation of nature for short-term economic gains,'[21] a 'religion,' he noted in his address to the National Conference on Adult Education in May of 1954, that was very much Protestant oriented. In the address, Grant simply noted the association of science, technology, and Protestantism without pursuing it further.

He developed his ideas more fully in *Philosophy in the Mass Age*. He began by exploring what he saw about the modern 'mass age' that made it unique compared to previous ages. He reduced that uniqueness to two characteristics. 'First, it is scientific,' he wrote. 'It concentrates on the domination of man over nature through knowledge and its application.'[22] Second, it is a society dominated by economic, political, and military elites who have tremendous power to dominate individuals, including, paradoxically, the very individuals who make up that elite: 'All our institutions express the way in which one lot of men dedicated to certain ends impose their dominance over other men.'[23] In the Western world, this elite was made up of capitalists. Grant dismissed from the outset the liberal belief that people could choose to change this social dynamic because modern society was so novel from past societies as to defy judgment until intimately understood. He then quoted Hegel's observation that had been central in Harold Innis's thinking on technology and Western civilization, namely, that 'the owl of Minerva only takes its flight at twilight.'[24] Grant interpreted Hegel as meaning that 'we take thought about the meaning of our lives when an old system of meaning has disappeared with an old society, and when we recognize that the new society which is coming to be raises new questions which cannot be understood within the old system.'[25] Clearly Grant, like Innis and McLuhan, saw the modern age in Western civilization as radically different from any previous age. While Grant did not at this point explicitly identify technology as the defining novelty of the modern age, it was implicit in his analysis of modernity.[26]

Grant maintained that what shape this new age would take was difficult to know because institutions and beliefs perpetuate a present perspective or even hearken to the past rather than address the future. Even reason, he noted, is 'an instrument ... for the control of nature and the adjustment of the masses to what is required of them by the commercial society.' Despite his negative perspective of rationality, science, and technology, he reminded his audience of the benefits that technology provided. He wrote, 'Always before in history, the mass of men had to give most of their energy to sheer, hard work because of the fact of scarcity. The conquest of nature by man through technology means human energy is liberated to attain objectives beyond those practically necessary. As this becomes ever more realized, vast numbers of men are able to devote their time to the free play of their individual faculties.'[27] This freedom was not only from scarcity, Grant noted, but also from tradition, thus enabling young people to pursue their own individual goals rather than those imposed by society as in the past.

Having noted the positive attributes of modern technological society in the form of greater material goods, Grant then questioned the values underlying the technological age by contrasting them with those of ancient philosophy. In the past, humans believed in a divine order beyond historical or individualistic events into which everything fit and in which meaning in the form of wisdom occurred. In contrast, in the modern world, humans believed that all events were unique, new, and significant in and of themselves, because 'we [humans] and not God are the creators of history.' In the past, morality – the judgment of right and wrong – was set down as absolute in line with natural law, which humans did not make but were made to live within. The aim of education was to bring human reason in union with natural law or divine reason. In contrast, in the modern world, morality was relative to the views of each individual, who was alone responsible for his or her actions. In the modern world, Grant summed up, 'we create value, we do not participate in a value already given. We make what order there is; we are not made by it. In this sense we are our own; we are independent. We are not bound by any dependence on anything more powerful than ourselves. We are authentically free because what happens in the world depends on us, not on some providence beyond our control. The fate of man is in his own hands.'[28]

For Grant, what had caused such a change in perspective was the key question to answer. His answer was Christianity. Biblical Christianity taught that historical events were 'concrete expressions of the divine will,' with an ultimate purpose, 'the redemption of the Jewish people through the Messiah' in the future. Time thus became the fulfillment of God's will. During the Renaissance and Reformation periods, Grant explained, the religious perspective was replaced with the 'idea of history as progress,' culminating not in the Kingdom of God but in the kingdom of man. What tied the two visions together was the idea of freedom: 'Conscious of themselves as free, men came to believe that history could be shaped to their own ends.'[29] At first, freedom was only seen in religion, when Luther insisted on the right of each individual to find God for oneself. But it soon extended into every sphere. Today, Grant pointed out in 1958, 'time is still oriented to the future, but it is a future which will be dominated by man's activity. The idea of human freedom merges with Judeo-Christian hope and produces the idea of progress. This means an entirely new kind of humanism, ... [one] which put[s] science and technology at its centre, as the means of redemption.'[30]

Grant argued that it was in North America, especially in the United States, that the new humanist and secular religion of progress was most successful, thanks to a combination of Calvinist Protestantism in the form of Puritanism and a continent of abundant natural resources waiting to be 'conquered.' Calvinism was a practical religion, he pointed out, because of the belief in hard work in order to have assurance that as one of the elect the person was saved. 'This combined with the doctrine of the priesthood of all believers led to what Max Weber has brilliantly defined as "worldly ascetism" – the saints living practical lives in the world,' Grant wrote. 'Their worldly ascetism made the Calvinists an immense force in shaping history, as democratic reformers and as capitalists.'[31] He elaborated further: 'The Calvinist doctrine of the Hidden God meant that ... one could only contemplate God in Jesus Christ, and go out and act as best one could. It is this tradition of acting for the best in the world which has been of such influence in creating our modern North American practicality. When freed from all theological context, it becomes pragmatism.[32] In turn, pragmatism defined truth as that which works best to bring about change. What 'works best' was a value judgment that individuals alone could make.

Grant explored the implications of the American philosophy of pragmatism for an 'ethic of community' in an article by that title in *Social Purpose for Canada* (1961), a book spearheaded by members of the Co-operative Commonwealth Federation Party who wanted to explore ways of making the party more relevant for the mid-twentieth century. Grant noted that in North American society the philosophy of pragmatism, combined with capitalism and technology, upheld the rights of the individual over the good of society as a whole. While the United States held the possibility through technology to create a society in which all would benefit (equality), the country failed to do so because the ideology of capitalism worked against such an ideal. Thus, the only hope for a truly egalitarian society lay in a socialist state. Grant sketched out the moral values that should underlie such a state. First, there had to be a belief in the equality of all human beings qua human beings and not as some economic or political entity. This necessitated a Christian belief in equality, since all other forms of belief, such as Marxism, capitalism, or communism, saw individuals as economic and/or political entities rather than moral human beings. Second, such equality would also need to be based on the right of every individual to enjoy greater leisure, the fruit of his or her labour. Grant argued that this was possible thanks to technology. Greater leisure would in turn necessitate educating people on how to use in meaningful and constructive ways the leisure that technological advancement had given them. Grant summed up his argument as follows: 'North America is the first continent called to bring human excellence to birth throughout the whole range of the technological society. At the moment, the survival of its capitalist ethic, more than anything else, stands in the way of realizing that opportunity. The only basis on which it could be realized is a clearly defined ethic of community which understands the dignity of every person and is determined on ways of fulfilling that dignity in our new conditions.'[33] At this point in his thinking, Grant believed that it was capitalism, not technology, that hindered the creation of 'an ethic of community.'

Two important events occurred in Grant's life in the early 1960s that greatly advanced his thinking on technology. One was his move from Dalhousie University in Halifax, where he had taught since 1947, to McMaster University in Hamilton, Ontario, the industrial and technological heartland of Canada. The other was his introduction to the debate about 'the good society' between American philosopher Leo Strauss and French thinker Alexandre Kojève.

Grant discussed the differences between these two European analysts in his article 'Tyranny and Wisdom.' He noted by way of introduction that Strauss favoured the ancient view of tyranny as rule by a meritorious elite that had a sense of noblesse oblige over the modern version of the 'universal and homogeneous state' in which all were under the illusion that they were free and equal. Kojève believed that the universal and homogeneous state would be superior because it would be based on the freedom and equality of every individual, achievable through technology. When realized, it would be 'the ideal social order' and thus the end of history and philosophy. Strauss replied that, far from being a society of equality and freedom, it would be 'a tyranny ... destructive of humanity.'[34] A few would rule while the rest would allow the rulers unlimited power under the illusion that they were free because they would be liberated from work by technology. Grant clearly favoured Strauss's perspective because it placed moral judgment beyond human will and the good society beyond the materialism that technology could provide.

Strauss also reinforced an idea that Grant had introduced in *Philosophy in the Mass Age,* namely, that 'Biblical religion' (what Grant defined simply as Christianity) was the source of the modern belief in the universal and homogeneous state. The fundamental difference between the ancient and modern views of the world was not due to technology itself but rather to the religious-liberal philosophy that lay behind that technology; it was fundamentally a difference in philosophical thought. Grant acknowledged his debt to Strauss for widening and enriching his understanding of ancient philosophy; when *Philosophy in the Mass Age* was reprinted in 1966, Grant praised Strauss in his new Introduction: 'I count it a high blessing to have been acquainted with this man's thought.'[35]

That new Introduction revealed the contribution of another thinker to the evolution of Grant's views on modern technological society: Jacques Ellul. Grant claimed that, in Ellul's *The Technological Society,* 'the structure of modern society is made plain as in nothing else I have read.'[36] Later, he recalled what he specifically liked about Ellul. It was not so much his ideas or his theoretical understanding of what technology was 'but just his recognition of its presence at every point in modern society.'[37] Grant agreed with Ellul that the difficulty of understanding and critically assessing modern thought lay in the 'dominance of technique over all aspects of our lives.'[38] The modern faith in technique had become the new morality, the new religion. 'The pursuit of technological advance,' Grant uncategorically stated, 'is what constitutes human excellence in our age and therefore it is our morality.' In North America, 'the theology of technique goes by the name of liberalism. I mean by liberalism the belief that man's essence is his freedom.'[39]

More than anything else, his new Introduction revealed the maturing of his thought on technological society over the seven years since the first publication of *Philosophy in the Mass Age*. Now he claimed that the book reflected a stage in his thinking on the modern technological society that he had passed beyond. He therefore hoped, although he never stated it explicitly, that its reprinting might help others get beyond the dominant liberal-technological paradigm that had dominated Western thought for the past 400 years. Like Innis, Grant was a moralist and teacher who wanted to use his wisdom and insight into the technological society to warn others of its dangers and limitations. One cannot do better than to quote Grant himself as to his changed perspective on technology thanks in part to his reading of Strauss, Kojève, and Ellul.

> The book [*Philosophy in the Mass Age*] is permeated with the faith that human history for all its pain and ambiguities is somehow to be seen as the progressive incarnation of reason. What had been lost in the immediacy of the North American technological drive would be regained, and regained at a higher level because of the leisure made possible by technology ... When technology had reached a certain stage it would once again serve human purposes ...
>
> Since that day my mind has changed. In the practical realm, I am much less optimistic about the effects that a society dominated by technology has on the individuals which comprise it. I no longer believe that technology is simply a matter of means which men can use well or badly. As an end in itself, it inhibits the pursuit of other ends in the society it controls. Thus, its effect is debasing our conceptions of human excellence. So pervasive and deep-rooted is the faith that all human problems will be solved by unlimited technological development that it is a terrible moment for the individual when he crosses the rubicon and puts that faith into question.[40]

Grant had come to realize that technology was not something that could be accepted or rejected at will. It shaped our ends, and in that sense technique became our morality. This was Grant's awareness of technology as volition. He also became aware of the tension that underlay the relationship between the technological imperative and the moral imperative.

Grant wanted to write a book that encapsulated his new ideas of technological society within the context of current events. The defeat of John Diefenbaker's Conservative government on April 8, 1963, by Lester Pearson's

Liberals over what he saw as the key issue – the acquisition of nuclear weapons from the Americans to be used on Canadian missiles, which the Conservatives opposed and the Liberals supported – was the immediate stimulus; *Lament for a Nation,* Grant's most popular and controversial book, was the product. When asked in an interview in 1973 what had motivated him to write the book, he replied, 'My motive for writing that book was rage, just rage, that they'd [the Liberals] brought atomic weapons into Canada.'[41]

Lament for a Nation is a strange book precisely because it does what Grant intended it to do: blend philosophy and politics. On the philosophical level, it addressed the issue of the impact of the United States, as the modern universal and homogeneous state and the most advanced technological society, on Canada as the country most directly threatened by the American leviathan. On the political level, it dealt with the wrangling over nuclear warheads in the context of Canadian and American politics. At times, the interrelationship of philosophy and Canadian politics in the book is insightful, while at other times it seems trite and strained. Nevertheless, the book did address the issues of technology, imperialism, nationalism, and the meaning of a good society in a way that succeeded in generating controversy and thus thrust Grant – as McLuhan's *Understanding Media* did for him – into popular celebrity.

As David Cayley notes, *Lament for a Nation* is 'constructed in layers, or concentric rings, beginning with the particular circumstances of Diefenbaker's fall, and ending with a meditation on the idea that Canada's redundancy was only an instance of a more general tendency in technological civilization to extinguish all local differences.'[42] The book began with a punch: the malicious attacks on John George Diefenbaker. 'Never has such a torrent of abuse been poured on any Canadian figure as that during the years from 1960 to 1965,'[43] Grant wrote emphatically. What angered him, however, was not the sabotaging of this Canadian political figure by the establishment, since this was 'fair game' in politics, but rather what the attack on Diefenbaker had come to symbolize: the death of the nation of Canada as a conservative alternative on the North American continent to the pervasive liberal-technological society of the United States.

Here Grant first introduced Strauss's concept of technology as the offspring of the universal and homogeneous state that in turn was an offshoot of the liberal concept of progress. Grant argued that such a state restricted more local culture and alternative political ideology than liberalism in its relentless drive to dominance and used technology to achieve its mastery. Thus, it was impossible for Canada to exist as a conservative nation in a technological world based on the philosophy of liberalism and living next door to the 'universal and homogeneous state' of the United States. 'The impossibility of conservatism in our era,' Grant concluded, 'is the impossibility of Canada. As Canadians we attempted a ridiculous task in trying to build a conservative nation in the age of progress, on a continent we share with the most dynamic nation on earth. The current of modern history was against us.'[44]

However, Grant still continued to be ambivalent about the modern technological society. Despite Leo Strauss's warning about the tyranny of the modern universal and homogeneous state as a result of technological dominance, despite his adverse reaction to living in Hamilton, Ontario, the heartland of Canada's industrial and technological society, and despite his rage regarding the attacks on Diefenbaker who was attempting to prevent Canada from being absorbed into the imperial, technological, universal, and homogeneous American state, Grant concluded *Lament for a Nation* with a whimper. Rather than a clarion call for Canadians to rally behind their nation, Grant offered his readers only an open-ended choice, refusing even to indicate which of the alternatives he himself favoured. In fact, Grant presented such a good summary of the ideas of the continentalists in his concluding chapter that he almost negated the counter-argument he had made in the rest of the book. He reminded those who opposed progress and technology that what they were rejecting was good as well as bad:

> It can only be with an enormous sense of hesitation that one dares to question modern political philosophy. If its assumptions are false, the age of progress has been a tragic aberration in the history of the species. To assert such a proposition lightly would be the height of irresponsibility. Has it not been in the age of progress that disease and overwork, hunger and poverty, have been drastically reduced? Those who criticize our age must at the same time contemplate pain, infant mortality, crop failures in isolated areas, and the sixteen-hour day. As soon as that is said, facts about our age must also be remembered: the increasing outbreaks of impersonal ferocity, the banality of existence in technological societies, the pursuit of expansion as an end in itself.[45]

Thus, the choice Grant left his readers at the conclusion of the book was indeed open-ended and not a rhetorical choice. It revealed his own remaining uncertainty about the nature of technological society: 'The question as to whether it is good that Canada should disappear must be left unsettled. If the best social order is the universal and homogeneous state, then the disappearance of Canada can be understood as a step toward that order. If the universal and homogeneous state would be a tyranny, then the disappearance of even this indigenous culture can be seen as the removal of a minor barrier on the road to that tyranny. As the central issue is left undecided, the propriety of lamenting must also be left unsettled.'[46]

The writing of *Lament for a Nation* led to a new perspective on or at least a clarification of one aspect of technology for Grant: its imperialistic nature. He came to realize that the American universal and homogeneous state rooted in technology did not arise phoenix-like in North America but had its roots in its mother empire, Britain. 'Great Britain was the chief centre from which the progressive civilization spread around the world,' Grant noted in *Lament for a Nation*. 'Politically, it became the leading imperial power of the West.'[47] Britain therefore did not offer an alternative to the American universal and homogeneous state with its imperial technological dominance but was instead its source. How then could Canada's link to Great Britain be a source and inspiration for Canadian conservatism? Now he realized that it could not. Grant believed that Great Britain had offered that alternative perspective early on in its history before the age of progress and imperial dominance but not once Britain became an imperial power. At that point, the country accepted the liberal ethos of progress and faith in technology. Grant had come to realize a point that Plato had made and he paraphrased: that 'an imperialistic power cannot have a conservative society as its home base.'[48] Liberalism, progress, technology, *and* imperialism were all inextricably linked. Grant had a good understanding of the interconnection of the first three of these concepts; what he had not understood until he wrote *Lament for a Nation* was the interconnection of the first three concepts, and especially technology, with imperialism. Here was the title for his next study of technology, *Technology and Empire,* a collection of essays published in 1969.

If *Lament for a Nation* was a warning to Canadians to be aware of the implications of the absorption of their country into the universal and homogeneous state of the American technological leviathan, then *Technology and Empire* was a warning that Canada's fate was reflective of the fate of Western civilization as a whole as technology spread from its imperial centre in the United States to the rest of the world. Grant noted in one of the essays, 'Canadian Fate and Imperialism,' 'What lies behind the small practical question of Canadian nationalism is the larger context of the fate of western civilisation. By that fate I mean not merely the relations of our massive empire to the rest of the world, but even more the kind of existence which is becoming universal in advanced technological societies. What is worth doing in the midst of this barren twilight is the incredibly difficult question.'[49]

Technology and Empire was intended to be a monograph in which Grant further explored the intellectual roots of the American technological empire. It was also intended to offer the Western world an understanding of how he, as one Canadian philosopher on the fringe of that empire, had at least a glimmer of what it would be like when the universal and homogeneous technological society had achieved world domination. However, the book was instead a collection of divergent articles on a common theme, written at different times for different journals in the 1960s, except for the introductory essay, 'In Defence of North America,' and the concluding one, 'A

Platitude.' These two were written specifically for the book as an opening and ending to hold together the essays in between. As a result, *Technology and Empire* never reached the heights of popularity of the less intellectually sophisticated *Lament for a Nation*. Nevertheless, the collection of essays presented his ideas and, more importantly, his concerns about the new technological society, none better than 'In Defence of North America.'

It was a curious title for an article that criticized more than it defended North America. But the article succeeded in revealing the origins of American technological thought in its Calvinist roots, from the time of the early Puritan contact with the New World. In a sense, then, the United States was 'fated' to be the imperial centre of the modern technological society as a result of the country's historical roots in the same way that Canada was 'fated' to be in the vortex of imperial struggles and a part of the Western industrial and technological complex as a result of its historical evolution under the domination of the French, British, and American Empires. Given these 'inevitable' historical circumstances and the worldwide dominance of technology, Grant's 'defence' of North America could be seen as a justification and explanation for how the United States, possibly even against its own will, was destined to be the imperial centre of modern technology.

Grant explained that this American 'technological civilisation'[50] may have had its roots in European thought, but it was different from European civilization in a few important ways. First, North Americans broke away from Europe at the point when liberalism and its faith in progress had become the dominant paradigm in European thought. As a result, 'the U.S. is the only society which has no history (truly its own) from before the age of progress.' Thus, Americans saw everything from the perspective of progress alone. The will to conquer, or to progress, became the incentive to control the new and unexplored North American continent. Grant wrote, 'If the will to mastery is essential to the modern, our wills were burnished in that battle with the land. We were made ready to be leaders of the civilisation which was incubating in Europe.' Second, North Americans, unlike Europeans, had an inability to draw on Greek thought before the age when Christian thought became dominant. Finally, North Americans were unlike their European progenitor in that they were the offspring of a group of Europeans – Calvinist Protestants – who were themselves breaking from Europe, 'a turning away from the Greeks in the name of what was found in the Bible. We brought to the meeting with the land a particular non-Mediterranean Europeanness of the seventeenth century which was itself the beginning of something new.'[51]

That 'something new' was the new sciences, a Baconian account of science based on the desire to control nature and 'to free the minds of men from the formulations of mediaeval Aristotelianism,'[52] the same medieval thinking that the new Calvinist Protestants were reacting against. Contrary to popular belief, Grant noted, this new science was premised on Protestant Christian beliefs, not the denial of those beliefs. In an interview with David Cayley, Grant explained the connection. 'I think Christianity is essentially responsible for the demystifying of nature, because when you look at what comes between the Greeks and modern science during the long period in which modern science was being prepared in Western Europe, what comes between is Christianity. That is just a fact.'[53] The new science, with its faith in technology, became the means by which humans could work hard and advance so as to ensure their salvation, while the new Calvinist religion provided the 'worldly asceticism' necessary to use science and technology to achieve God's Kingdom on Earth.

As the 'primal' instinct to achieve the good Christian life ceased in the nineteenth and twentieth centuries to have a religious motive behind it, the ideal continued to shape North Americans in practical worldly ways. 'It shapes us above all,' Grant noted, 'as the omnipresence of that practicality which trusts in technology to create the rationalised kingdom of man.'[54] 'What makes the drive to technology so strong [in North America],' he emphasized, 'is that it is carried on by men who still identify what they are doing with the liberation of mankind.'[55] He explained:

> Indeed the technological society is not for most North Americans, at least at the level of consciousness, a 'terra incognita' into which we must move with hesitation, moderation and in wonder, but a comprehended promised land which we have discovered by the use of calculating reason and which we can ever more completely inherit by the continued use of calculation. Man has at last come of age in the evolutionary process, has taken his fate into his own hands and is freeing himself for happiness against the old necessities of hunger and disease and overwork, and the consequent oppressions and repressions.[56]

Grant argued that this faith in technology underlay all ideological differences from the Marxists on the left to the capitalists on the right; they differed only in *means,* not in the *end* or *vision*. They all believed in a perfect society achieved through technology.

Even the values North Americans might use to judge what was good or bad in the new technological society were shaped by technique. Grant noted, 'The moral discourse of "values" and "freedom" is not independent of the will to technology, but a language fashioned in the same forge together with the will to technology. To try to think [of] them separately is to move more deeply into their common origin.' The good society became for modern beings the universal and homogeneous state in which all were 'free and equal and increasingly able to realise their concrete individuality,' which, North Americans believed, could only be brought about by technology. North

Americans had come to implicitly and uncritically accept the very technological values they needed to judge and question. 'As moderns,' Grant stated emphatically, 'we have no standards by which to judge particular techniques, except standards welling up with our faith in technical expansion.'[57]

To Grant, this was simply another way of saying that North Americans lacked the ability to get outside of their technological mindset to be able to see the relative nature of their thinking on technology. This could only be done by experiencing the world before the Christian faith in progress. It required looking at nature not as an object to utilize for human ends, good or bad, but in terms of contemplation as the Greeks had done. Grant went on to explain what he meant by contemplation by comparing it to the modern pragmatic view of human nature and the natural world:

> It may perhaps be said negatively that what has been absent for us is the affirmation of a possible apprehension of the world beyond that as a field of objects considered as pragmata – an apprehension present not only in its height as 'theory' but as the undergirding of our loves and friendships, of our arts and reverences, and indeed as the setting for our dealing with the objects of the human and non-human world. Perhaps we are lacking the recognition that our response to the whole should not most deeply be that of doing, nor even that of terror and anguish, but that of wondering or marvelling at what is, being amazed or astonished by it, or perhaps best, in a discarded English usage, admiring it; and that such a stance, as beyond all bargains and conveniences, is the only source from which purposes may be manifest to us for our necessary calculating.[58]

What enabled Grant to see what others could not? It was being on the fringe of modern thought as a Christian philosopher who looked deeply for the meaning of theology beyond the surface expression of it in biblical religion. It was being on the fringe of the modern liberal-technological society as a Canadian intellectual who looked to his nation's heritage, his own family tradition of moral teaching, and distrust of the United States that had always kept the country distinct despite the tremendous pull of its southern neighbour. It meant abandoning the 'pervasive pragmatic liberalism' in which he was educated so as to find a meaning of life that went deeper than the dominant liberal-technological paradigm. It meant literally and figuratively 'going mad,' for only in madness could he understand the insanity of modern technological thought. It meant hanging on to the moral imperative when the whole thrust of the technological imperative brought into question the validity of the moral imperative. Grant ended his essay with a masterful summary of the dilemma faced by humans in the modern technological world. He wrote:

> We live then in the most realized technological society which has yet been; one which is, moreover, the chief imperial centre from which technique is spread around the world. It might seem then that because we are destined so to be, we might also be the people best able to comprehend what it is to be so. Because we are first and most fully there, the need might seem to press upon us to try to know where we are in this new found land which is so obviously a 'terra incognita.' Yet the very substance of our existing which has made us the leaders in technique, stands as a barrier to any thinking which might be able to comprehend technique from beyond its own dynamism.[59]

To many, his conclusion appeared pessimistic, as did the book in general. But Grant believed that he was simply providing a realistic and honest statement of where he saw the world heading if the liberal-technological paradigm was not challenged and ultimately replaced. He saw his task as holding forth that promise that there was an alternative, even if in discovering it people had to go mad. Was madness not better – and a greater sense of freedom – than simply acquiescing in a technological society in which humans were mere automata, the mechanized beings that technology produced under the illusion of perfection and freedom?

It is difficult to concisely describe what George Grant meant at this point in time by the term 'technology,' or 'technique,' the term he now preferred using because he felt the term had a wider meaning than 'technology.' Fortunately, he himself grappled with its meaning and in so doing came as close as possible to defining it. In an interview in 1969, when pressed to define 'technology,' he replied with a definition that was strikingly similar to that of Jacques Ellul, which Grant quoted often in his work. Ellul defined technology as technique and technique as follows: 'The term *technique,* as I use it, does not mean machines, technology, or this or that procedure for attaining an end. In our technological society *technique* is *the totality of methods rationally arrived at and having absolute efficiency* (for a given stage of development) in *every* field of human activity. Its characteristics are new; the technique of the present has no common measure with that of the past.'[60] Grant's definition may have been less eloquent but no less encompassing. When asked 'what do you mean when you describe our society as a technological society?' Grant replied:

> I mean that this is a society in which people think of the world around them as mere indifferent stuff that they are absolutely free to control any way they want through technology. I don't think of the technological society as something outside us, you know, like just a bunch of machines. It's a whole way of looking at the world, the basic way western men experience their own existence in the world. Out of it come large organisations,

> bureaucracy, machines, and the belief that all problems can be solved scientifically, in an immediate quantifiable way. The technological society is one in which men are bent on dominating and controlling human and non-human nature.[61]

If this was the positive and encompassing way of defining technology or technique, then the negative and empty way of seeing technology or technique was presented by Grant in 'A Platitude,' the concluding article in *Technology and Empire.* A platitude is a commonplace remark, especially solemnly delivered. On one level, such a title was pure irony, as what he declared in his platitude was as deep, personally agonizing, and profound as any moral statement; there was nothing commonplace in the depth, sincerity, and elegance of its message. On another level, however, Grant had come to realize that technology was so all-encompassing that it absorbed not only our thoughts but also our emotions, our beliefs, and our very faith in God. Humankind's belief in the freedom of words, emotions, nature, and faith was illusory; everything was simply an extension of the technological 'new God,' the technological imperative. The profundity of the message and its moral emptiness was so great as to make it simple and commonplace – hence a platitude. Grant wrote, 'It is difficult to think whether we are deprived of anything essential to our happiness, just because the coming to be of the technological society has stripped us above all of the very systems of meaning which disclosed the highest purposes of man, and in terms of which, therefore, we could judge whether an absence of something was in fact a deprival.'[62] Even if modern thinkers argued that what had been lost were only 'illusions, horizons, superstitions, taboos which bound men from taking their fate into their own hands,' that did not negate the fact that individuals today had lost something that gave meaning to their lives in the past. They had lost a sense of freedom as something greater than individual freedom, a freedom to be used for a higher and nobler purpose than simply the liberation of the individual. Grant wrote, 'Such may indeed be the true account of the human situation: an unlimited freedom to make the world as we want in a universe indifferent to what purposes we choose. But if our situation is such, then we do not have a system of meaning.' What then was the answer? 'If we cannot ... speak, then we can either only celebrate or stand in silence before that drive. Only in listening for the intimations of deprival could we live critically in the dynamo.' Maybe in so doing, in the language of silence, we will 'see the beautiful as the image, in the world, of the good.'[63]

George Grant's work stimulated a great deal of thinking on how to find meaning in a world dominated by technology. Dennis Lee, noted Canadian poet, felt the sense of silence that Grant articulated. Moved by *Lament for a Nation,* Lee wrote his epic poem *Civil Elegies,*[64] which reflected his personal search for meaning as a Canadian living in the colonial space of a North American technological society. In his essay 'Cadence, Country, Silence: Writing in Colonial Space,' Lee attributed his thoughts in *Civil Elegies* directly to the influence of George Grant: 'To find one's tongue-tied sense of civil loss and bafflement given words at last, to hear one's own most inarticulate hunches out loud, because most immediate in the bloodstream – and not prettied up, and in prose like a fastidious groundswell – was to stand erect in one's own space. I do not expect to spend the rest of my life in agreeing with George Grant. But in my experience at least, the somber Canadian has enabled us to say for the first time where we are, who we are – to become articulate.'[65]

Just as Grant found overwhelming silence to be the only response to technology, so too did Lee. As Lee sat in the centre of Nathan Phillips Square in Toronto, with the New City Hall, designed by Finnish architect Viljo Revell, behind him, and the Archer, the abstract sculpture by Henry Moore, in front of him, he wrote his poem about being an alienated being, a Canadian, in the universal and homogeneous world of technology. How could one be a responsible citizen in a world that discouraged civic responsibility and had no sense of 'the good' beyond the immediate goal of a more comfortable material life? The question of civic responsibility proved especially difficult for Canadians to answer because they had never owned their own space, their own country; they always lived in colonial space under imperial control. 'How empire permeates!'[66] Lee wrote.

> Many were born in Canada, and living unlived lives they died
> of course but died truncated, stunted, never at
> home in native space and not yet
> citizens of a human body of kind. And it is Canada
> that specialized in this deprivation.[67]

'We live on occupied soil,' Lee noted. 'For we are a conquered nation: sea to sea we bartered everything that counts, till we have nothing to lose but our forebears' will to lose.'[68]

From the beginning, Canadians had sold out to modernity; they had forfeited the opportunity to offer an alternative to America.

> And what can we do here now, for at last we have no notion
> of what we might have come to be in America, alternative,
> and how make public
> a presence which is not sold out utterly to the modern? utterly? to the
> savage inflictions of what is for real, it pays off, it is only
> accidentally less than human?[69]

Even God appeared to be absent, thus negating meaning and purpose beyond the present moment, denying the nation a mission beyond the practical one of surviving as a conquered people.

But this was not always so.

> Master and Lord, there was a
> measure once.
> There was a time when men could say
> my life, my job, my home
> and still feel clean.
> The poets spoke of earth and heaven. There were no symbols.[70]

Once, people lived by God's presence and judged their lives and found meaning in him. But this was no longer the case in the modern technological world. Now only a void existed, the absence and negation of God. In the name of a void – 'consenting citizens of a minor and docile colony ... cogs in a useful tool'[71] – Canadians had succumbed to American pressure to fight in the Vietnam War and to kill innocent people abroad.

> And this is void, to participate in an
> abomination larger than yourself. It is to fashion
> other men's napalm and know it, to be a
> Canadian safe in the square and watch the children dance and
> dance and smell the lissome burning
> bodies to be born in
> old necessity to breathe polluted air and
> come of age in Canada with lies and vertical on earth no man has drawn a
> breath that was not lethal to some brother it is
> yank and gook and hogtown linked in
> guilty genesis it is sorry mortal
> sellout burning kids by proxy acquiescent
> still though still denying it is merely to be human.[72]

For Lee, the answer was to move into and through the void, since it too would pass, like God and the soul. And Canadians would find new meaning in the written word.

> Freely out of its dignity the void must
> supplant itself. Like God like the soul it must
> surrender its ownness, like eternity it must
> re-instil itself in the texture of our being here.
> And though we have seen our most precious words
> withdraw, like smudges of wind from a widening water-calm,

> though they will not be charged with presence again
> in our lifetime that is
> well, for now we have access to new nouns –
> as water, copout, tower, body, land.[73]

Because of Grant's reputation, he was invited by the CBC to deliver the ninth radio broadcast of the Massey Lectures, the series so named in honour of his uncle Vincent. Grant chose to speak on the topic of 'Time as History' and focused on the ideas of Friedrich Nietzsche, the German thinker who had gone furthest in exploring the idea of time as history and especially its implication for modern thought. Although the words 'technology' and 'technique' were seldom used in the series of lectures, the meaning of these terms was implicit throughout in two ways. First, like Nietzsche, Grant assumed that the idea of time as history was a very 'modern' concept and thus intricately tied to an understanding of technology as seminal to modern Western liberal thought. As he stated at the outset of his lectures, 'history' is one of those words that 'express our civilization.'[74] It conveyed the belief unique to modern Western civilization that 'man is essentially an historical being and that therefore the riddle of what he is may be unfolded in those studies.'[75] Since an underlining characteristic of modernity in Western civilization was the dominance of technology, Grant assumed that a study of time as history would shed light on the nature of the Western liberal thought that underlay the dominance of technology in the West. Second, he argued that the idea of time as history was the chief means by which European intellectuals, especially Nietzsche, undermined the concept of time as eternal and absolute – and therefore above history – that contained moral values and standards that were true for all times. According to Grant, Nietzsche convincingly showed that the idea of time as history was Christian and held meaning only so long as the ideas and beliefs underlying Christianity – most importantly that God existed – were upheld. Now that God was dead, Nietzsche argued, so too was the belief in time as an absolute or in values as eternal. Grant pointed out that the idea of time as eternal and absolute lingered on in the Western belief in progress, even after Christianity waned. However, even that idea had outlived its 'time' as a historical concept, leaving humans with the need to face an uncertain and chaotic world and a future without any ultimate meaning. In the meantime, the idea of progress of the world to a perfect state – the Kingdom of God on Earth – had been transformed in modern times into faith in technology – the kingdom of man. In essence, technology had become the 'new God.' Thus, Grant showed how a study of time as history could lead to an understanding of how Western civilization had put such faith in technology that it could reach the point of being the central idea in the paradigm of Western liberal thought. In his concluding lecture, Grant set out the tenets of that liberal thought:

'The mastery of human and non-human nature in experimental science and technique, the primacy of the will, man as the creator of his own values, the finality of becoming, the assertion that potentiality is higher than actuality, that motion is nobler than rest, that dynamism rather than peace is the height.'[76] All he needed to add was the statement, implicit throughout his study, that these 'themes' or ideas were both essential underlying assumptions of and beliefs inherent in modern technological thought.

By the beginning of the 1970s, Grant was ready to write his magnum opus on the values and beliefs underlying 'technological civilisation.' He had a working title: *Technology and Justice*. Instead, two publications were produced: *English-Speaking Justice* (1974), the Josiah Wood Lectures given at Mount Allison University in Sackville, New Brunswick, in 1974, and *Technology and Justice,* a collection of essays published in 1986.

In *English-Speaking Justice,* Grant made a significant shift back to using the term 'technology' instead of 'technique' or 'technical' (which he had preferred in *Technology and Empire*) to describe the modern mindset. The change in terminology reflected his new perspective on the meaning of the word 'technology.' 'Technique,' he now realized, only referred to half of the process that was going on in shaping Western civilization into a 'technological civilisation.' As mentioned earlier, it was a term he had borrowed from Jacques Ellul and one that had been used frequently by Leo Strauss, both men specifically referring to the impact of technology on Western thought. But Grant realized that the relationship between technology and thought was a symbiotic relationship whereby Western thought contributed as much to the shaping and understanding of technology as technology did to the shaping of Western thought. That influence came through the influence of political liberalism. Grant best explained his shift in perspective in his article 'Thinking about Technology' in *Technology and Justice*. He pointed out that the word 'technology' combined the Greek words for 'art' – 'techne' (what today would be referred to as the 'practical arts') – and the systematic study of art – 'logos.' He explained the significance of this combination: 'What is given in the neologism – consciously or not – is the idea that modern civilisation is distinguished from all previous civilisations because our activities of knowing and making have been brought together in a way which does not allow the once-clear distinguishing of them. In fact, the coining of the word "technology" catches the novelty of that co-penetration of knowing and making. It also implies that we have brought the sciences and the arts into a new unity in our will to be masters of the earth and beyond.'[77]

Grant realized that for people like Ellul and Strauss the domination of technology in Western thought was certainly pervasive and subtle, but the hope remained that through an understanding of technology human beings could create a regime in which moral considerations would govern the use of technology. However, partly through re-reading Martin Heidegger's *The Question Concerning Technology,* Grant had come to a more complex and profound understanding, namely, that Western thought, especially in the guise of Calvinist Protestantism, shaped our modern understanding of technology just as technology shaped that thought, and therefore technology was our morality. The choice was not either technology or morality but rather both interacting together. Grant therefore began to focus on the *religious* beliefs underlying Western thought that put such unquestioning faith in technology as its 'new religion.'

In essence, Grant's perspective on technology was even more pessimistic than that of Ellul or Strauss. Knowledge of how Western thought shaped our perspective on technology was even more difficult to grasp because it required understanding and critically assessing the a priori beliefs and assumptions of modern thought within a paradigm that, due to the dominance of technological thought, did not allow for perspective, objectivity, or distance. Yet Grant denied such pessimism, often reacting angrily to those who accused him of being a pessimist. On one occasion, out of sheer frustration, he responded, 'I'm not being pessimistic at all. I think God will eventually destroy this technological civilization. I'm very optimistic about that.'[78] In that statement, Grant revealed both his deep religious conviction and his determination to understand the roots of technological thought.

English-Speaking Justice focused on the question of whether it was possible to define good or justice as absolute moral values in a modern liberal-technological society that did not believe in moral absolutes or the existence of God and no longer saw nature as ontology. Grant used the recently published book *A Theory of Justice* (1971) by John Rawls, the Harvard liberal political philosopher, as the departure point for his discussion of English-speaking justice and liberalism in relation to technology because he believed the book best expressed the modern liberal moral philosophy that he questioned and wanted to challenge. According to Grant, Rawls argued that humans were free to set their own moral laws based on what best served their individual interests because they contractually agreed to what freedoms they would surrender to the state in return for the protection of those freedoms by the state. The best political regimes were those that facilitated that freedom to choose for all individuals, thus ensuring a sense of equality. Grant noted the implication of such a perspective: 'The categorical imperative presents to us good without restriction. Moreover that justice which is our good depends upon our willing of it. We are the makers of our own laws; we are the cause of the growth of justice among our species.'[79]

But what happened if the rights of individuals conflicted with those of the state? Grant reminded his readers that, in the modern technological society, technology depended on efficiency, and efficiency required a disciplined workforce that cybernetics was geared to produce, as Heidegger, Ellul, and Wiener had already pointed out. The practical question then became 'whether a society in which technology must be oriented to cybernetics can maintain the institutions of free politics and the protection by law of the rights of the individual. Behind that lies the theoretical question about modern liberalism itself. What were the modern assumptions which at one and the same time exalted human freedom and encouraged that cybernetic mastery which now threatens freedom?'[80]

Grant believed that these questions lay at the heart of the Supreme Court case *Roe v. Wade*. The issue was whether the 'state has the right to pass legislation, which would prevent a citizen from receiving an abortion during the first six months of pregnancy.' Justice Harry Blackmun concluded that the state did not have such a right since individual rights or freedom preceded and took precedence over the rights of the state. In this case, the rights of the mother took precedence over the fetus because before six months fetuses in the womb 'are not persons, and as non-persons can have no status in the litigation.' Grant argued that in this one decision lay a number of premises of modern liberal thought concerning freedom, justice, and morality:

> The decision then speaks modern liberalism in its pure contractual form: right prior to good; a foundational contract protecting individual rights; the neutrality of the state concerning moral 'values'; social pluralism supported by and supporting this neutrality ... A quick name for this is 'technology.' I mean by that word the endeavour which summons forth everything (both human and non-human) to give its reasons, and through the summoning forth of those reasons turns the world into potential raw material, at the disposal of our 'creative' wills. The definition is circular in the sense that what is 'creatively' willed is further expansion of that union of knowing and making given in the linguistic union of 'techne,' and 'logos.'[81]

Thus, Grant concluded that the assumptions underlying liberalism and technology 'come from the same matrix of modern thought.'[82] It was therefore unfeasible to think of the one independent of the other and near impossible to get outside technological thought to realize its dangers.

Grant's final collection of essays, *Technology and Justice* (1986), started out as a full-scale book with the tentative title 'Technique(s) and Good.' Two of the articles, 'Thinking about Technology' and 'Faith and the Multiversity,' are seminal to understanding Grant's further thoughts on technology. In his Preface to the book, he found the old Spanish proverb 'take what you want, said God – take it and pay for it' particularly apt for understanding the dilemma facing modern Western civilization. Modern society was free to achieve unlimited power through technology, but there was a price to pay for that freedom and power. Grant's aim in the essays was for us to 'think what we have taken and how we have paid' for technology. He noted that in particular we have paid a price for our understanding of justice, and it was in terms of this issue that Grant explored the interconnection of technology and liberalism in Western thought.

'Thinking about Technology,' the first article in *Technology and Justice,* is reminiscent of Martin Heidegger's *The Question Concerning Technology,* and the article did reflect Grant's continued and, to a degree, renewed interest in this major twentieth-century German thinker on technology.[83] The article was written in the style of a Socratic lesson, reflective of Grant's approach as a teacher. He began by setting out the context of the problem he wanted to 'think about' – that is, technology. Next he raised for reflection, discussion, and debate a germane and controversial statement on the subject – in this case, the recent statement by a computer scientist: 'The computer does not impose on us the ways it should be used.' Then he proceeded to dissect what the statement might mean by posing a series of questions about it with the hope that, in doing so, the discussion would shed light on modern thought.

Grant pointed out that implied in the statement 'The computer does not impose on us the ways it should be used' were the following assumptions about these 'machines': 'They are instruments, made by human skill for the purpose of achieving certain human goals. They are neutral instruments in the sense that the morality of the goals for which they are used is determined outside them.'[84] In other words, they were merely 'machines,' in the old-fashioned (liberal) sense of objects external to ourselves that we could use at will (freely) according to rational human objectives.

But were they just 'machines' or 'neutral instruments'? For one thing, Grant noted, computers 'have been made within the new science and its mathematics. That science is a particular paradigm of knowledge and, as any paradigm of knowledge, is to be understood as the relation between an aspiration of human thought and the effective conditions for its realisation.'[85] That paradigm *did* 'impose' itself on us, and therefore, Grant pointed out, 'the computer *does* impose.'[86]

Next, Grant raised the question as to what was meant by the word 'ways' in the assertion that 'the computer does not impose the ways.' Clearly, the purpose of computers and what they were best used for was storing and transmitting information. The very concept of knowledge as 'information' was to see knowledge not as thought but as something that could be classified – that is, an object. And 'it is the very nature of any classifying to homogenise.' Thus, like all other forms of technology, computers contributed to a universal and homogeneous state. They were 'instruments of the imperialism of certain communities towards other communities.' Grant noted that this would be true if the political basis of the dominant community had as its ideology capitalist liberalism, communist Marxism, or national socialist historicism. All three of these modern political ideologies were premised on 'the same account of reason'[87] that produced the technology of the computer.

This led Grant to question the meaning of the third key word, 'should,' (besides 'impose' and 'ways') in the syllogism 'The computer does not impose on us the ways it should be used.' Implicit in the sentence was the concern 'with the just use of the machine as instrument. "Should" expresses that we ought to use it justly.'[88] The question was, Grant pointed out, who decides how the computer – or any other form of technology – should be used, or what is the just way that it should be used, or even for whom it 'should' be justly used? These, he noted, were 'value' judgments, and, going back to his discussion in *English-Speaking Justice,* 'value' was not a neutral-free word; it meant something quite different in our modern liberal-technological society than it did for the ancients. Today, Grant reminded his readers, we believe that human beings alone, based on what they conceive to be 'good' for them, decide what is of value or good. In contrast, the ancients believed 'the good' or truth of justice existed in an absolute realm beyond human 'manipulation' or historicity. Thus, 'the coming to be of technology has required changes in what we think is good, what we think good is, how we conceive sanity and madness, justice and injustice, rationality and irrationality, beauty and ugliness.'[89] To understand what was involved in accepting technological thought, Grant compared it to accepting a 'package deal' over individual items:

> To put the matter crudely: when we represent technology to ourselves through its own common sense we think of ourselves as picking and choosing in a supermarket, rather than within the analogy of the package deal. We have bought a package deal of far more fundamental novelness than simply a set of instruments under our control. It is a destiny which enfolds us in its own conceptions of instrumentality, neutrality and purposiveness. It is in this sense that it has been truthfully said: technology is the ontology of the age ... Unless we comprehend the package deal we obscure from ourselves the central difficulty in our present destiny: we apprehend our destiny by forms of thought which are themselves the very core of that destiny.[90]

'Faith and the Multiversity,' the second article written for *Technology and Justice,* is a unique and very important article because, of all of Grant's writings on technology, it is the one in which he attempted to go beyond an analysis or critique of technological civilization to offer a possible response to that civilization. In an interview with David Cayley, Grant commented on his purpose in writing the article: 'I want to think less about what is wrong with the modern and more about the truth of what is not present in the modern.'[91] This was not easy for Grant, who had spent all his life as a philosopher and writer showing how comprehensive and all-encompassing technology was – that in essence technology was 'us,' our very modern ways of acting, thinking, and living. How could one possibly get outside of such a self-contained and closed world? Grant had one simple but profound answer: faith.

The format of the article followed that of 'Thinking about Technology': an introductory section setting out the issue to be discussed – in this case, 'the relation between faith and modern science.' He followed this by a statement raised for discussion, which was Simone Weil's definition of faith: 'Faith is the experience that the intelligence is enlightened by love.'[92] Then he debated the points raised in that quotation through a series of questions and answers in the Socratic mode.

Grant began by discussing the modern paradigm of knowledge, which he claimed to be 'the project of reason to gain objective knowledge.' He argued that this perspective looked at all things as 'objects' (even human beings) to be judged in terms of their usefulness or reason for existence.[93] It was humans who decided the value of such objects, thus placing them over and against the objective world.

What, Grant asked, was the relation of this paradigm of knowledge to faith? According to Weil, faith placed 'love' over and against human intelligence, a belief that a sense of love did not come from human perception alone but from a realm beyond. 'The key difficulty in receiving the beauty of the world [as love] these days,' Grant wrote, 'is that such teaching is rooted in the act of looking at the world as it is, while the dominant science is rooted in the desire to change it.' He noted that an indication of this was our inability today to see nature as more than just a resource. Even people became 'resources,' as in the statement 'Canada's greatest resource is its people.'[94]

In turn, justice had become defined as 'something human beings make and impose for human convenience' rather than 'something in which we participate as we come to understand the nature of things through love and knowledge.' Knowledge of justice was based then on 'calculation.' 'This disjunction of beauty and truth,' Grant went on to argue, 'is the very heart of what has made technological civilisation.' To get beyond the paradigm of technological civilization to be able to see 'the truth that the world proceeds from goodness itself'[95] was near impossible, except through faith in something greater than ourselves as human beings. By way of conclusion, Grant wrote:

> Modern scientists, by placing before us their seamless web of necessity and chance, which excludes the lovable, may help to reteach us the truth about

> the distance which separates the orders of good and necessity. One of Nietzsche's superb accounts of modern history was that Christianity had produced its own gravediggers. Christianity had prepared the soil of rationalism from which modern science came, and its discoveries showed that the Christian God was dead. That formula gets close to the truth of western history, but is nevertheless not true. The web of necessity which the modern paradigm of knowledge lays before us does not tell us that God is dead, but reminds us of what western Christianity seemed to forget in its moment of pride: how powerful is the necessity which love must cross. Christianity did not produce its own gravedigger, but the means to its own purification.[96]

But how could one write about an antiquated subject like faith in the modern, secular, liberal, and technological society of the West? In essence, one could not, Grant came to realize. One could only live it, and that was only possible for a saint.

Grant's editor for *Technology and Justice* at Anansi Press perceptively noted the gap between the expectation of Grant's readers and admirers for the definitive work on technology and the inability of Grant to produce it. 'I am always puzzled,' Jim Polk wrote, 'at how some critics expect a "big" definitive tome from you ... It would be nice of course, but it seems very Germanic to ask for a *Summa Theologica* or a *Critique of Pure Reason* or any huge system building structure from a philosopher in the late 20th century.'[97] The best a philosopher could do was reflect on the means by which that faith could be found, that path that needed to be taken, a path made all the more difficult in the modern technological world. In so reflecting, George Grant was one of the most profound modern thinkers on technology. However, his failure to find an answer to the riddle of lack of faith in a technological world that most needed that faith made him appear as the prophet of technological doom rather than the saviour.

Conclusion

If 'technology is the metaphysics of our age,' as George Grant noted in the quotation that opened this book, then who have been the metaphysicians? In the Anglo Canadian intellectual tradition, they have been the thinkers who believed that technology had become the new intellectual imperative, what I have identified in this book as the technological imperative. Initially, in the first flowering of thought on technology, Anglo Canadian theorists of technology believed that this technological imperative was compatible with and even complementary to the traditional moral imperative that had guided Western civilization for centuries. However, the more deeply Anglo Canadian thinkers delved into the complex nature of technology and its pervasive influence on society and especially on the very thought process of the modern age, the more they became aware of the magnitude and power of the technological imperative. Ultimately, like Grant, some came to believe that the technological imperative had consumed the moral imperative. We became technological beings.

In looking at the technological imperative juxtaposed against a moral imperative, these Canadian thinkers created a tension that has been part of Canadian thought from the mid-nineteenth century to the present. This tension kept both imperatives in check and prevented either one from becoming dominant or absolute. However, it also led to a one-dimensional perspective: technology or the technological imperative invariably came to be looked at only from the vantage point of its association with morality or the moral imperative rather than as a subject in and of itself.

Thomas Keefer, T.C. Haliburton, and Sandford Fleming, writing in the age of the first fine careless rapture of technology, appealed to their fellow Canadians to accept the new technology of railways as a means of instilling moral values that would advance the good of society. They argued that railways would enable Canadians to rise above their parochial and isolated existence on the northern half of the North American continent to become citizens of the world. In their minds, railways were lifelines to other regions of the continent and, in association with steamships, to the British Empire, the centre of Western civilization. Thomas Keefer made a clarion call for Canadians to 'awake' from their 'Sleepy Hollows' through the building of railways to a world that offered so much more in terms of intellectual stimulation and moral regeneration. T.C. Haliburton appealed to his fellow Nova Scotians to take the progressive nature of the Americans – their technological 'know-how' – and to combine it with the moral imperative of their British tradition – their 'know-what' – to become the most enlightened people on the face of the Earth. He believed that railways were the elixir that would bring about this magical transformation. Sandford Fleming envisioned railways in conjunction with inter-oceanic telegraph cables as the means to partake in and contribute to the intellectual currents radiating out from the mother country to her distant colonial possessions. Situated midway between the British Isles and Asia, Canada could, through the new communication technology, assist in the moral well-being of the empire and the world. Alexander Graham Bell saw the wonders of technology, particularly the telephone, creating a closer world in which unity, peace, and moral goodness would prevail. For this generation of Canadian theorists, technology was objects or machines; the greatest were those associated with communication because they annihilated time and space and drew the British Empire and ultimately the world together. They accepted without question that faster and more far-reaching communication meant better communication. The world would become one in peace and harmony. In their minds, the moral and the technological imperatives interacted to the benefit of the moral well-being of society, because technology was the revelation of God's mind.

Hearing the call from these earlier exponents to accept technology, advocates of technical education in the late nineteenth and early twentieth centuries argued that technical education was the best means to create a morally good and just society. Advocates of manual training and domestic science in elementary and secondary schools pointed out that a technical education combined the use of a child's or young adult's mental and physical faculties; both faculties, they believed, were important for cultivating the human imagination, which in turn was essential for creativity. Supporters of faculties of applied science and later faculties of engineering in universities noted the value of technical education not only for the economic well-being of society but also for its moral well-being by cultivating ethical standards deemed important for the advancement of civilization. Thus, technology set the standards by which education was judged to be a success. Those standards were the values of productivity, industriousness, and efficiency, values deemed essential for progress in the modern world. What would later be criticized as a regimented system of education that inhibited rather than enhanced creativity became the new pedagogy for generations of students in the late nineteenth and early twentieth centuries.

Technology's rampage across the battlefields of Europe during the First World War brought to a halt its unscathed march across humankind's belief that technology would help forge a perfect world. Technology took on a new complexity as a process of industrialism and mechanization. Still clinging to the belief in the compatibility of the moral and technological imperatives, Canadian theorists of technology during the First World War and the interwar years looked to regenerated individuals or a reformed society as the means to bring about reconciliation. Political economy and the social sciences in conjunction with moral philosophy and social reform were the new cathartic disciplines of learning that would usher in a utopia. But beneath such optimism lay nagging doubts about technology in observations of its power to bring out the demonic side of human nature and to create societies devoid of moral values. Dreams of utopias turned into nightmarish dystopias: stories of Frankensteinian monsters creating world wars and worldwide industrial unrest, of humans becoming the servants to technological masters, of a world gone mad in a technological frenzy, of moral values giving way to the relentless demands of technology. Such themes became the subjects of poems and fiction, while for the first time technology became a defined and ever-present subject in the mental landscape of the Canadian imagination.

The debate over the relationship between the moral imperative and the technological imperative in the pre–Second World War era occurred within the secure belief that technology was something external to the human psyche to which humans reacted and could control once aware of technology's power. These pre–Second World War theorists believed that humans were free to accept or reject the lure of technology. However, the nature of the discussion changed in the post–Second World War period. Canadian theorists of technology came to realize that technology was not something to resist from the outside; it was something to recognize as being within our psyche – indeed, as the essence of our being. In recognizing technology as a state of mind that was itself technologically induced, Canadian theorists fed the technological imperative by warning of its unbound dominance and pervasiveness. Technology as volition took on a whole new meaning when one realized that the very values by which to judge technology as baneful or beneficial were themselves technologically driven. The ability to be free in a world in which technology was omnipresent became questionable; the need to be free of technology's powerful grip became all the more imperative. Each theorist searched for an avenue of escape from the technological leviathan.

What avenues of escape did Canadian theorists of technology in the post–Second World War period offer? For Harold Innis, it was a return to the moral values of the oral tradition found in societies on the margins. Marshall McLuhan put faith in the 'resiliency and adaptability' of humans to use technology, especially electronic technology, to build a more interactive and ethically enriched world. Northrop Frye saw the power of the indomitable human imagination to visualize an ideal world and then use the power of words, or myths, to bring it about. Even George Grant, considered the most pessimistic of Canadian theorists, offered a way out of the 'technological hell' that he believed entrapped us; it was through faith, which he defined as a belief in a higher power.

What perspective on technology have Canadian theorists of technology since taken, and what solutions have they offered? They too have looked at technology as an imperative over and against the moral imperative. However, rather than attempt to reconcile the two, they have looked for ways to negotiate with the technological imperative. They have also been less inclined to look at technology from a philosophical perspective and have avoided raising open-ended questions like 'What is technology?' which preoccupied Innis, McLuhan, Frye, and Grant. Instead, they have seen technology as 'a given' and have looked at the practical ways in which it has had an impact on the individual and on society, in hopes of salvaging something of the moral being in the process. Thus, they approach the technological imperative from the vantage point of how individuals can negotiate a modicum of freedom in a world that daily tends toward conformity. Equally important, for the first time in Canadian thought, the theorists of technology include women who look at the issue of technology from a feminist perspective. As a result, the edges of the debate are challenged and extended beyond the confines of the male prerogative to enrich our understanding of the relationship between the technological imperative and the moral imperative with regard to women.

Ursula Franklin, a professor of metallurgy and archaeometry (the latter being an analysis of technological materials applied to archaeology) at the University of Toronto, explores the nature of our modern world of technology in her CBC Massey Lectures, *The Real World of Technology*. She takes technology 'as a given,' 'the *ways of doing something*,'[1] and therefore as part of the modern mindset in Western thought. She then shows the many ways that this technological mindset 'as *practice*'[2] influences our everyday activities and ways of thinking. Part of that mindset, she argues, is to divide technology into two different types, holistic and prescriptive. The former are 'normally associated with the notion of crafts,' where the work performed is of a creative nature usually done by one person, a craftsman. The latter, prescriptive technologies, are ones where the division of labour is broken down into identifiable steps, each step carried out by a different worker, or group of workers, who are knowledgeable only about that component of the overall creation and who have limited skills beyond their immediate task. Franklin notes that the latter has become the norm today. While prescriptive technologies 'are often exceedingly effective and efficient, they come with an enormous social mortgage': a culture of compliance and conformity. The individual in essence is in danger of becoming a slave to the productive process.

Franklin argues that the prescriptive approach has come to be applied as well to the natural environment, an area of growing concern among current Canadian theorists of technology. Even the use of the term 'environment' instead of 'nature,' she notes, is designed to break the natural world down into component parts. Then it is looked at only from one perspective – that of human beings – on the assumption that they alone are the only important component to consider when dealing with issues related to the natural world. Franklin argues, 'Such a mindset makes nature into a construct rather than seeing nature as a force or entity with its own dynamics.'[3] It also sees nature as merely 'natural resources,' as industrial materials to be utilized for human control and consumption. As well, it is the same mindset that judges the value of things or actions simply on the basis of 'efficiency' or 'productivity.'

Franklin offers several solutions to the issue. One is to begin a public discourse that will 'break away from the technological mindset to focus on justice, fairness, and equality' as worthwhile values of judgment.[4] Such an approach would mean putting the needs of people above those of technology – in other words, putting values associated with the moral imperative above those of the technological imperative.

She also suggests the need for more female input into the dialogue on technology. She is the first Canadian theorist of technology to note the absence of a feminist perspective on technology and to suggest that feminist understanding of the nature of technology offers a different approach from that of the dominant male theorists. She argues that rather than a perspective that is 'fragmented, specified and prescribed,' and 'fully scheduled and carried out without reference to context,' as is the case of male theorists of technology, a female perspective would 'arise in contexts out of specific needs.' Those 'needs' would be '"copability," the ability to deal and cope adequately with a variety of circumstances,' 'minimizing disaster,'[5] and an emphasis on quality over quantity.

Franklin suggests that one way in which individual women may go about implementing these values is through 'imagining' – that is, placing themselves 'into a technical or social setting in the future and "project[ing] back" the resources and constraints required to get from the present to that future ordering.' As well, women need to find ways of overcoming or circumventing the fragmentation of work that technology fosters and demands so as to bring decision making and responsibility back to the individual. Finally, Franklin appeals to women to fight for the reintroduction of 'the skills of listening, of developing compromise and fostering co-operation and improvisation,'[6] all essential female qualities for the modern world of technology.

The result would be the fostering of different technologies, particularly those that put moral values above economic interests.

Heather Menzies, adjunct professor at Carleton University in Ottawa, joins Franklin in fighting for a feminist critique of technology. Menzies claims that the realization of the need for a feminist critique came to her in 1980, when she was asked to do a study on 'the effects of the new technology of informatics on women employed in the service-producing sector' for the Institute for Research on Public Policy. Her report, *Women and the Chip: Case Studies of the Effects of Informatics on Employment in Canada,* argued that women were particularly adversely affected by modern electronic technologies, such as computers and telecommunications, since these are areas of employment in which women traditionally dominate.

This study prompted her to examine the roots and values of the technological age beginning with the Industrial Revolution. In *Computers on the Job: Surviving Canada's Microcomputer Revolution,* Menzies notes that women and children have consistently been unemployed or underemployed since the time of the Industrial Revolution because they have been seen as 'expendable commodities.' She found the roots of such thinking in the religious and scientific values of the sixteenth and seventeenth centuries, which she claims are still with us today. Those values, which have become associated with the technological imperative, are enshrined in the ideologies of individualism and capitalism. 'Once people were reduced to marketable units of labour, and land and resources were similarly given fixed value terms,' she writes, 'these factors could then be bought and sold, and thereby controlled.'[7] Menzies likens the computer revolution to a 'second industrial revolution, because, like mass production machinery before them, computers are transforming society, not just adding something to its traditional constitution.'[8]

In the process of writing *Computers on the Job,* Menzies realized that in drawing an analogy between the machine age and the computer age she blurred the profound differences between the two ages. She notes that the machine age was one in which technology was synonymous with tools and machinery, whereas the computer age is a highly complex technological system or process. It is a system that measures everything in a one-dimensional way: by output and by economic scale. It is also a system that someone controls and manipulates. That someone may be an individual or a group of people, but for certain, Menzies claims, it will be of the male gender (hence the title of her article 'In His Image: Science and Technology,' in which she explores the idea of technology as gendered). Menzies sees Francis Bacon, the 'father of modern science,' as the first person to see the potential of dominating nature by controlling the technology that could manipulate nature. Menzies points out that Bacon's denigration of nature 'as putty in the hands of technological man' to be 'forced out of her natural state and squeezed and moulded ... to establish and extend the power and dominion of the human race itself over the universe,'[9] coincided with the age of the witch hunt, which saw 'the denigration of women and women's way of healing.'[10] This was not coincidental, she argues; both attitudes marginalized women. This accounts in part for the continued marginalization of women today in the modern world of technology.

In subsequent works, Menzies takes a broader perspective on technology beyond its impact on women only. She has become concerned about the one-dimensional nature of the 'language of technology.' To be aware of the linear thinking of technology requires being more cognizant of the language technocrats use to define, describe, and anaesthetize the technological process, so that they can implement profound changes while claiming to make only minor 'adjustments' to the system. More importantly, people need to be aware of the fact that there are other, equally legitimate ways of knowing beyond the technological, such as 'the embodied, personal way grounded in felt values and shared tradition.' She spells out what this more humane perspective on technology might entail:

> Reclaiming the personal power of naming. Recovering the voices from the margins, and the sense of the whole social context. Raising the personal and collective consciousness of what is at stake, what is being lost, in the current economic and technological restructuring. It is all vital to reclaiming a democratic middle ground of shared values on which a new social contract can be defined and negotiated, globally and locally; a common ground of indivisible social goals which can yield diverse participation in a de-massified society without the perils and the injustice of polarization.[11]

Mark Kingwell, a philosophy professor at the University of Toronto who has taken on the persona of a public intellectual by dealing with complex philosophical issues in a way that makes them understandable and meaningful to the general public, addresses the issue of technology in a number of his writings. In *Dreams of Millennium: Report from a Culture on the Brink,* he claims that he first became aware of the impact of technology upon seeing two science-fiction films: *2001: A Space Odyssey* and *Forbidden Planet.* Both films, and especially the latter, show how technology brings out the worst in human nature, our inner violence, 'to make our dreams dangerously real. We have the power to destroy ourselves, the film suggests, only when our own unconscious wishes and fears are sent into the world.'[12] Such a perspective reminds one of the theme of Mary Shelley's *Frankenstein* and of William Lyon Mackenzie King's concern about technology unleashing the worst in human nature. Kingwell notes that the dream of using technology to transform human nature into a super form of being 'evolving beyond the limitations of our corporeal bodies into a sort of half-human/half-machine creature'[13] was particularly prominent during the approach of the millennium in the twenty-first century.

Kingwell notes that, for many people, 'the messiah of the Second Coming is not a man; the messiah is the Net itself.'[14] He cautions against accepting such hyperboles as 'gospel truth.' What the Internet revolution has brought about is not an emancipation of the individual, not a higher form of human consciousness, not a new level of spirituality, but the exact opposite: the enslavement of individuals, the trivializing of knowledge, and the loss of the soul. 'The new medium of electronic communication has become ... candy floss for the mind – quick, sweet, lacking in nutrients,' Kingwell writes. 'More deeply, we are not moving closer to a single planetary soul. Instead, as the Net grows, it begins more and more to take on a mind of its own ... The dark inner logic of technology, the desire for mastery that nestles within the hopes of emancipation, comes back to haunt us, aided by the cheerleading of those the Unabomber so vividly calls "the technophiles."'[15]

Kingwell is especially concerned about how to be an effective civic citizen in the modern world of technology that militates against rational, effective, and meaningful action: 'Technology no longer arises, as it did for first-generation Marxists, as an issue of the mode of production: the developmental level of the machines used to make goods,' he notes. 'Now technology is both more important and less discernable: it has become part of us on a deeper level.'[16] Technology, by numbing or desensitizing us, causes us to turn inward into our private lives and private (domestic) space and to relinquish responsibility for what occurs beyond our own immediate sphere. To overcome these temptations requires awareness on the part of responsible citizens as to the debilitating effect of technology so as to resist the urge to turn inward and to accept the need to become citizens of the nation-state and the world.

Derrick de Kerckhove, the current director of the McLuhan Program in Culture and Technology at the University of Toronto, has followed in the footsteps of his mentor, Marshall McLuhan. His foray into the study of technology came as a result of an examination of the Greek alphabet as a technological tool of communication that shaped Western thought, an idea explored by Innis, Havelock, and McLuhan. De Kerckhove maintains that the Greek orientation of writing and reading from left to right favoured the left hemisphere of the brain, the sphere of the brain that processes temporal sequences, as opposed to the right sphere, which processes visiospatial relationships. In turn, this orientation favoured rationalism. It was also a 'mathematical model,' de Kerckhove observes, thus making it especially applicable to the computer age: 'With full phonetization, writing seems to have acquired a precision, flexibility, and a paradoxical meaninglessness that is comparable to computer programming codes. I do not mean by this that alphabetic writing has turned people into computerized automatons, but that it made language available for a kind of information processing which is, technically, and especially in scientific investigations, very close to a mathematical model.'[17]

Like McLuhan, de Kerckhove is interested in the impact of the electronic era on Western thought. 'Electricity is the core of technology,' he writes emphatically and with the certainty of a McLuhanite. 'It has dethroned the domination of the mechanical principle and reversed many of the explosive and fragmenting tendencies of the alphabet.'[18] Like McLuhan, he sees electricity as a tactile medium. 'Electricity puts everything in touch just as the alphabet put everything in perspective.'[19] And as McLuhan also said, being 'in touch' requires the utilization of all the senses.

De Kerckhove is particularly interested in the impact of electronic technology on the individual. Here he notes a paradox: electronic technology both enhances and diminishes the individual. It enhances the individual because electronic information is implosive, or brought to the individual, rather than explosive, or projected out from the individual, thus making the individual the focus and the receptor of information. As well, the individual is able to acquire this information on his or her own, without having to be dependent on anyone else and even without having to physically interact with others.

Also, the individual can be connected to a worldwide community and can partake of a 'collective mentality.' De Kerckhove states, 'The Web ... allows and encourages the input of individuals within a "collective" medium' where both the individual and the worldwide virtual community are 'connected and individual at the same time.' Connectedness, he argues, 'is one of mankind's most powerful resources. It is a condition for the accelerated growth of human intellectual production.'[20]

On the negative side, the individual can feel dwarfed and insignificant in the virtual community where she or he has no sense of identity, no opportunity for physical interaction, and no purpose. It can result in a crisis of identity, since the distinction between the 'natural' self and the 'virtual' self is blurred. Indeed, de Kerckhove reverses the idea put forward by McLuhan, that technologies are extensions of the human body, to argue that the human body is an extension of technology: 'It used to be so comfortable to say technology is an extension of the body: it is less comfortable to say the body has become an extension of technology. Nevertheless, that is what is happening, because there is more and more of it (the body) out there, so much in fact that the balance between what is "out there" and what is "in here" has changed completely. It is quite obvious that we are becoming the organic core or organic extension of our own brilliantly sophisticated machines.'[21]

By the end of his life, George Grant had come to realize that 'technology is being.' What he had not foreseen was the possibility that 'being' could be a virtual being, not a human being, a self created by technology, for technology, and in the image of technology. Such a perspective raises new questions and concerns about what it means to be 'human' in a moral sense in the modern world of technology. Thus, once again, the issue of the tension between the technological imperative and the moral imperative surfaces.

Arthur Kroker, holder of the Canadian Research Chair in Technology, Culture, and Theory at the University of Victoria, explores the modern world of the virtual being. His foray into the world of technology began with a brief study of the ideas of Innis, McLuhan, and Grant on technology. He concluded that this generation 'brought us to the edge of the technological dynamo, but it's our fate now to experience the designed environments of technology as *the* most pervasive and basic fact of human existence.'[22]

Kroker maintains that this new technological world is ambiguous and paradoxical with 'opposing tendencies towards domination and freedom, radical pessimism and wild optimism ... Indeed,' he writes, 'central to the human situation in the twentieth-century is the profound *paradox* of modern technology as simultaneously a prison-house and a pleasure-palace.'[23] Such extremities have necessitated a re-examination of moral and ethical values for the new age at a time when the world lacks 'a language by which to rethink technology in late twentieth-century experience.'[24] What is required is a revolution in thinking about technology comparable to the Copernican Revolution that replaced the old medieval view of the world.

In *Data Trash: The Theory of the Virtual Class,* Kroker and Michael A. Weinstein further explore the new cyber-world. In it, the human body disappears, they claim, as does history, both victims of technology's 'will to virtuality.'[25] This virtuality 'becomes the primal impulse of pan-capitalism (virtual political economy), the mediascape (virtual culture), and post-history (virtual history). In understanding how and why this happens, McLuhan's 'outerized' nervous system and Nietzsche's 'last man' serve as starting points. However, like de Kerckhove, Kroker challenges McLuhan's view of technologies as 'extensions of man.' De Kerckhove saw technology as the replacement of human beings by the virtual being; Kroker and Weinstein see technologies as 'humiliations of the flesh, which remains [sic] as an embarrassment after "man" dies.'[26] If man is dead, they reason, then so is history, since humankind is the subject of history. All that is left is 'the endless exchange of data,' devoid of meaning and purpose, a by-product of cybernetics. Such a world is not a utopia, as the technophiles predicted it would be, but rather a dystopia, a world gone mad, destroying itself on the very technology the world believes is its saviour.

As someone who has spent considerable time analyzing the ideas of Canadian theorists of technology from the perspective of the relationship between the technological imperative and the moral imperative, and in putting forward their solutions, what perspective and solutions do I offer? As an intellectual historian, I feel more comfortable analyzing the ideas of those in the past than offering my own thoughts. Intellectual history is a sobering subject because in examining the ideas of others who have thought deeply and extensively on their subject – in this case, technology – the challenge is to come up with a 'better perspective.' The advantage the intellectual historian has is her or his ability to build on the ideas of others, to have a perspective that incorporates all that has gone before.

The tension between the moral imperative and the technological imperative that has dominated Anglo Canadian thought on technology pits the past against the future. It is based on the assumption that to achieve something new, and ipso facto better, as technophiles invariably promise, requires the obliteration of the past. The tension is also based on the assumption by technophobes that whatever technology creates in the future will compromise, or possibly obliterate, the moral roots of the past. Such opposing perspectives are premised on the belief that ultimately one of the imperatives will triumph over the other. It is the case of either/or. Yet the history of Canadian thought concerning technology proves otherwise. Neither imperative has disappeared from the Canadian intellectual landscape since the debate over the two imperatives began in the mid-nineteenth century. The moral imperative continues to reassert itself, despite the dire predictions and concerns of some twentieth-century Canadian theorists of technology, while the technological imperative certainly shows no sign of weakening, despite the fear on the part of late-nineteenth-century technophiles that this would be the case. The realization of the strength and endurance of both imperatives – and the inability of either imperative to dominate and become the absolute imperative – is in itself a way out of the intellectual conundrum that these Canadian intellectuals created. The choice is not either/or but and/and; the tension of opposites becomes the calming effect of compromise.

Both imperatives have been powerful forces in Canadian thought since the mid-nineteenth century. The moral imperative can never again enjoy the dominance it held at that time, but this is not to negate its continuous importance in Canadian thought. Equally, the technological imperative has been a very powerful force in Canadian thought throughout the twentieth century – *the* most powerful imperative, I would argue – but that force has always been tempered by the need to take the moral imperative into consideration. In essence, the issue that Canadian theorists of technology have identified since the advent of technological thought in Canada in the mid-nineteenth century to the present – namely, the tension between the moral imperative and the technological imperative – is age-old. It is the Canadian version of the conflict between freedom and power within the context of technology that was first identified in Aeschylus's play *Prometheus Bound*. While the technological will to power has always threatened to suppress the moral will to freedom, neither emerges triumphant from the intellectual battle.

Notes

Introduction

1 Gad Horowitz and George Grant, 'A Conversation on Technology and Man,' *Journal of Canadian Studies* 4, 3 (1969): 3.
2 A.B. McKillop, *A Disciplined Intelligence: Critical Inquiry and Canadian Thought in the Victorian Era* [1979] (Montreal and Kingston: McGill-Queen's University Press, 2001).
3 Ramsay Cook, *The Regenerators: Social Criticism in Late Victorian English Canada* (Toronto: University of Toronto Press, 1985).
4 Carl Mitcham, 'Philosophy of Technology,' in *A Guide to the Culture of Science, Technology, and Medicine,* ed. Paul T. Durbin (New York: Free Press, 1980), 306.
5 Ibid., 314.
6 Ibid., 308.
7 Ibid., 316.

Chapter 1: Perspectives on Technology

1 For a summary and analysis of Lewis Mumford's views on technology, see William Kuhns, *The Post-Industrial Prophets: Interpretations of Technology* (New York: Weybright and Talley, 1971), 32-64.
2 Lewis Mumford, *Technics and Civilization* (New York: Harcourt, Brace, and Company, 1934), 12.
3 Ibid., 156.
4 For example, in *Technics and Civilization,* he wrote, 'Behind all the great material inventions of the last century and a half was not merely a long internal development of technics: there was also a change of mind. Before the new industrial processes could take hold on a great scale, a reorientation of wishes, habits, ideas, goals was necessary' (3).
5 Ibid., 51.
6 Lewis Mumford, *The Myth of the Machine: Technics and Human Development* (New York: Harcourt, 1966), 51.
7 Quoted in Kuhns, 53.
8 See Benjamin Farrington, *Francis Bacon: Philosopher of Industrial Science* (New York: Octagon Books, 1979).
9 William Leiss, *Under Technology's Thumb* (Montreal and Kingston: McGill-Queen's University Press, 1990), 13. An explanation is required as to why William Leiss, a Canadian analyst of technology, is included among the international analysts of technology in this chapter rather than in my subsequent chapters in which I deal with Canadian thinkers on technology. William Leiss has become a recognized international authority in his analysis of the ideas of individuals who have looked at technology as knowledge. But in analyzing the ideas of others, he has not formulated his own theory of technology. In other words, Leiss is an analyst of technology or an intellectual historian of the ideas of international thinkers

on technology as knowledge as opposed to being a theorist of technology like the Canadian thinkers discussed in the remainder of this book.

10 Ibid., 17.

11 For an analysis of John Kenneth Galbraith's ideas, see Langdon Winner, *Autonomous Technology: Technics-out-of-Control as a Theme in Political Thought* (Cambridge, MA: MIT Press, 1977), 162ff.

12 John Kenneth Galbraith, *The New Industrial State,* 3rd ed. (Boston: Houghton Mifflin Company, 1978). In Galbraith's earlier study, *The Affluent Society* (Boston: Houghton Mifflin, 1960), he emphasized the increasing importance of education and an educated and diverse workforce in the modern, highly complex world of technology. He also noted that many of the inventions of the Industrial Revolution, such as the spinning jenny, the spinning frame, and the steam engine, were made by inventors with little education but 'with a mechanical turn of mind' (271). In the mid-to-late twentieth century, technological advancement would only occur with a well-educated populace. 'Innovation has become a highly organized enterprise,' Galbraith wrote, one that requires a heavy investment in education. 'Investment in human resources needs today to be in step with investment in material capital' (272).

13 Ibid., 7, 8.

14 Quoted in Leiss, 20.

15 Ibid., 22.

16 Ibid., 148.

17 Karl Marx, *Capital: A Critique of Political Economy,* Vol. 1 [1867] (New York: Random House, 1976), 495.

18 Ibid., 503, 504.

19 Ibid., 531-32.

20 For an analysis of Giedion's ideas on technology, see Kuhns, 65-81.

21 Siegfried Giedion, *Mechanization Takes Command: A Contribution to Anonymous History* (New York: Oxford University Press, 1948), 34.

22 Ibid., v.

23 Ibid., 77.

24 Ibid., 43.

25 Ibid., 5.

26 Ibid., 14.

27 Ibid., 34.

28 Kuhns, 73.

29 Giedion, 720.

30 Ibid., 714.

31 Ibid.

32 Ibid., 715.

33 Carl Mitcham, 'Philosophy of Technology,' in *A Guide to the Culture of Science, Technology, and Medicine,* ed. Paul T. Durbin (New York: Free Press, 1980), 316.

34 For a reflective discussion of Heidegger's concept of technology, and the extent to which it accords with the record of American cultural history, see Leo Marx, 'On Heidegger's Conception of "Technology" and Its Historical Validity,' *The Massachusetts Review* 25, 4 (1984): 638-79.

35 Martin Heidegger, *The Question Concerning Technology and Other Essays,* trans. William Lovitt (New York: Harper and Row Publishers, 1977), 4.

36 Ibid., 5.

37 Mitcham, 319.

38 William Lovitt, Introduction, *The Question Concerning Technology and Other Essays,* by Martin Heidegger, xxvi.

39 Heidegger, 26-27. Original italics.

40 Gregory Bruce Smith, "Heidegger, Technology, and Postmodernity," *Social Science Journal* 28, 3 (1991): 377.

41 Heidegger, 30. Original italics.

42 Ibid., 34.
43 Ibid., 35.
44 For a discussion of Ellul's ideas, see the essays in Clifford G. Christians and Jay M. Van Hook, eds., *Jacques Ellul: Interpretative Essays* (Urbana: University of Illinois Press, 1981). See also Kuhns, 82-111.
45 Jacques Ellul, *The Technological Society* [1964] (New York: Alfred A. Knopf, 1967), xxv. Original italics.
46 Ibid., 6. Original italics.
47 Ibid., 78-79.
48 Ibid., 79-80. Original italics.
49 Ibid., 94.
50 Ibid., 128. Original italics.
51 Ibid., 325.
52 Ibid., 429.
53 Kuhns, 83.
54 For a discussion of Norbert Wiener's ideas, see Kuhns, 205-19.
55 Norbert Wiener, *Human Use of Human Beings: Cybernetics and Society* (New York: Avon Books, 1950), 25.
56 Ibid., 253-54.
57 Ibid., 254.

Chapter 2: T.C. Keefer, T.C. Haliburton, Sandford Fleming, and Alexander Graham Bell

1 Quoted in Leo Marx, *The Machine in the Garden: Technology and the Pastoral Ideal in America* (London: Oxford University Press, 1964), 170.
2 Ibid., 171.
3 Ibid., 174.
4 Wolfgang Schivelbusch, *The Railway Journey: The Industrialization of Time and Space in the 19th Century* (Berkeley, CA: University of California Press, 1986), 1.
5 Quoted in Walter E. Houghton, *The Victorian Frame of Mind, 1830-1870* (New Haven, CT: Yale University Press, 1957), 3. Original italics.
6 Ibid., 4.
7 Quoted in Leo Marx, 'The Impact of the Railroad on the "American Imagination," as a Possible Comparison for the Space Impact,' in *The Railroad and the Space Program: An Exploration in Historical Analogy,* ed. Bruce Mazlish (Cambridge, MA: MIT Press, 1965), 202.
8 Quoted in Lawrence Surtees, 'Alexander Graham Bell,' *Dictionary of Canadian Biography,* Vol. 15, 1921-1930 (Toronto: University of Toronto Press, 2005), 84.
9 T.C. Keefer, *Philosophy of Railroads and Other Essays* (Toronto: University of Toronto Press, 1972), 10. All subsequent quotations are from this edition.
10 Lewis Mumford, *The Myth of the Machine: Technics and Human Development* (New York: Harcourt, 1966), 282.
11 Quoted in Marx, *The Machine in the Garden,* 234.
12 Keefer, *Philosophy of Railroads,* 37-38. Original italics.
13 For a biographical sketch of Keefer, see H.V. Nelles, 'Thomas Coltrin Keefer,' *Dictionary of Canadian Biography,* Vol. 14, 1911-1920 (Toronto: University of Toronto Press, 1998), 552-55.
14 Keefer, *Philosophy of Railroads,* 3.
15 Ibid., 6.
16 Ibid., 7.
17 Ibid., 7-8.
18 Ibid., 9.
19 Ibid., 10.
20 Ibid., 10-11.
21 Ibid., 11.
22 Ibid., 32.
23 Ibid., 38.

24 Keefer, 'Montreal,' reprinted in *Philosophy of Railroads*, 67.
25 Ibid., 71.
26 Ibid., 86.
27 Ibid., 87.
28 Ibid., 88.
29 Ibid.
30 Ibid., 89.
31 See the section 'A Sequel to the Philosophy of Railroads' with lectures on civil engineering, letters back and forth, and editorials from the newspaper all focused on the issue of morality and railways; *Philosophy of Railroads*, 92-126.
32 Keefer, 'Extracts from Lectures on Civil Engineering, Delivered at McGill University, 1855-56,' *Philosophy of Railroads*, 93.
33 T.C. Keefer, 'Travel and Transportation,' in *Eighty Years' Progress of British North America, Showing the Wonderful Development of Its Natural Resources, by the Unbounded Energy and Enterprise of Its Inhabitants*, ed. H.Y. Hind et al. (Toronto, 1864), 192.
34 Fred Cogswell, 'Thomas Chandler Haliburton,' *Dictionary of Canadian Biography*, Vol. 9, 1861-1870 (Toronto: University of Toronto Press, 1976), 352.
35 Quoted in A.H. O'Brien, *Haliburton: A Sketch and Bibliography*, 2nd ed. (Montreal, 1909), 6.
36 For a discussion of Haliburton's non-fictional writings on Nova Scotia, see M. Brook Taylor, 'Haliburton as a Historian,' in *The Thomas Chandler Haliburton Symposium*, ed. Frank M. Tierney (Ottawa: University of Ottawa Press, 1985), 103-22.
37 Sam Slick, *The Clockmaker, Series One* [1836]. *The Clockmaker, Series One, Two, and Three*, ed. George L. Parker (Ottawa: Carleton University Press, 1995), 18, 19-20. All subsequent quotations from 'The Clockmaker Series' are taken from this edition.
38 George L. Parker, Introduction, *The Clockmaker, Series One, Two, and Three*, by Thomas Chandler Haliburton (Ottawa: Carleton University Press, 1995), xxiv.
39 Slick, *The Clockmaker, Series One*, 98.
40 Ibid., 33. Not all Nova Scotians shared Haliburton's railroad building schemes. In June 1836, in a letter to the *Novascotian*, the writer, signing himself as 'Us, or U. & S.,' argued that '"prudent going folks never think of making such things as Railroads and Canals" until they have the money and the population to support them ... He used the Clockmaker to attack what he called "the present Railroad excitement": "You ask what it is besides time? Why, it requires that (in the language of the Clockmaker) we should not be asleep and lose our time; and it requires (in my own language) that we should not be crazy, and waste our time – both of which we have been very much in the habit of doing."' Quoted in Parker, Introduction, xxvii.
41 Slick, *The Clockmaker, Series One*, 35.
42 Robert L. McDougall notes that at the heart of Haliburton's serious and humorous writings was a 'preoccupation with the mechanics of linking together the parts of a world community which he believed was about to be born – his preoccupation, for example, with what he called "the responsibility of steam," with the building of canals and roads and bridges, with river and lake systems, with steamship and railway routes.' Robert L. McDougall, 'Thomas Chandler Haliburton,' in *Our Living Tradition: Seven Canadians*, ed. Claude T. Bissell (Toronto: University of Toronto Press, 1959), 24.
43 Thomas C. Haliburton, *The Season Ticket* [1860] (Toronto: University of Toronto Press, 1973), 21.
44 Ibid., 44.
45 Haliburton, *The Clockmaker, or, The Sayings and Doings of Samuel Slick, Series Three* [1840], 443.
46 On Fleming's role in the development of the technology of communication, see Graham M. Thompson, 'Sandford Fleming and the Pacific Cable: The Institutional Politics of Nineteenth-Century of [sic] Imperial Communications,' *Canadian Journal of Communication* 15, 2 (May 1990): 64-75; and Robert S. Fortner, 'The Canadian Search for Identity, 1846-1914: Communication in an Imperial Context,' *Canadian Journal of Communication* 6 (Summer 1979): 24-31. This article is the first of a four-part series by Fortner in the *Canadian Journal of Communication*. The other three parts are 'Communication and Canadian National Destiny,' 6 (Fall 1979): 43-57; 'Communication and Regional/Provincial Imperatives,' 6

(Winter 1979): 32-46; and 'Communications and Canadian-American Relations,' 7 (Summer 1980): 37-52.

47 Quoted in Clark Blaise, *Time Lord: The Remarkable Canadian Who Missed His Train and Changed the World* (Toronto: Knopf, 2000), 125.

48 Biographical information on Sandford Fleming is taken from Blaise; Lawrence Burpee, *Sandford Fleming: Empire Builder* (London: Oxford University Press, 1915); Hugh MacLean, *Man of Steel: The Story of 'Sir Sandford Fleming'* (Toronto: Ryerson, 1969); Lorne Edmond Green, *Sandford Fleming* (Toronto: Fitzhenry and Whiteside, 1980); as well as his *Chief Engineer: Life of a Nation-Builder – Sandford Fleming* (Toronto: Dundurn Press, 1993); and Mario Creet, 'Sir Sandford Fleming,' *Dictionary of Canadian Biography,* Vol. 14, 1911-1920 (Toronto: University of Toronto Press, 1998), 359-62.

49 In his biography of Sandford Fleming, Clark Blaise notes how the stereotype has 'carried forth to our own fantasies in which "Scotty" worked his mechanical magic on the starship *Enterprise*' (x).

50 Sandford Fleming, *Canada and Its Vast Undeveloped Interior* (n.p., 1878), 248.

51 Sandford Fleming, *Terrestrial Time: A Memoir* (London, 1876), 2.

52 Quoted in Blaise, 35.

53 On the importance of standard time, see Stephen Kern, *The Culture of Time and Space, 1880-1918* (Cambridge, MA: Harvard University Press, 1983).

54 Library and Archives Canada (LAC), Sandford Fleming Papers [Fleming Papers], vol. 2, file 9: Americans: 'Memorandum Issued by the Special Committee of the American Society of Civil Engineers on Uniform Standard Time,' 10-11.

55 Fleming, *Canada and Its Vast Undeveloped Interior,* 253.

56 LAC, Fleming Papers, vol. 102, file 32: Standard Time: 'Time Reckoning for the Twentieth Century,' paper given to the Canadian Institute, January 1879, 345.

57 Quoted in Fortner, 26.

58 LAC, Fleming Papers, vol. 93, file 2: Calendar Reform: 'Reform of the Almanac,' 3.

59 Biographical information on Bell is taken from Charlotte Gray, *Reluctant Genius: The Passionate Life and Inventive Mind of Alexander Graham Bell* (Toronto: HarperCollins, 2006); Lawrence Surtees, 'Alexander Graham Bell,' *Dictionary of Canadian Biography,* Vol. 15, 1921-1930 (Toronto: University of Toronto Press, 2005), 78-88; and Robert V. Bruce, 'Alexander Graham Bell and the Conquest of Solitude,' in *Technology in America: A History of Individuals and Ideas,* 2nd ed., ed. Carroll W. Pursell Jr. (Cambridge, MA: MIT Press, 1990), 105-16.

60 Bruce, 106.

61 Quoted in Gray, 372.

62 Bruce, 106.

63 Ibid., 209, 211-12.

64 Alexander Graham Bell National Historic Site [AGBNHS], Alexander Graham Bell Papers [Bell Papers], Bell to his wife Mabel, May 25, 1899.

65 Bell was always conscious of acknowledging the help of others in the inventions he made. In the case of the telephone, he wrote, 'Right in the beginning I want to state that while I invented the telephone, credit for much of its development belongs to a number of able and conscientious co-workers, and men who came into the work later.' AGBNHS, Bell Papers, 'Contributors to Knowledge: The Telephone,' vol. 2, no. 38 [binder 22], 86. He was also practical, realizing the difficulty of acknowledging when different inventors shared knowledge of their subject. He noted:

> I do not think there can be effective cooperation between men of active independent minds, excepting upon an altruistic basis, and the subordination of self in the interests of the subject. I do not think that the true authorship of suggestive thoughts can be preserved unless all communications are made in writing, nor can the true history of the development of an idea be unravelled unless the different steps in the evolution of the idea exist in written in tangible form. This thought lies at the basis of the scribbling-books I use in my laboratory and elsewhere.

AGBNHS, Bell Papers, 'Contributions to Knowledge: Graphophone,' vol. 2, no. 33 [binder 12], Bell to David F. Fairchild, November 6, 1907, 85.

66 Alexander Graham Bell, 'The Substance of My Latest Research,' Empire Club of Canada, *Addresses,* 15-16 (1917-18, 1918), 2.

67 Quoted in Surtees, 83. On another occasion, Bell wrote, 'The great advantage it [the telephone] possesses over every other form of electrical apparatus consists in the fact that it requires no skill to operate the instrument. All other telegraphic machines produce signals which require to be translated by experts and such instruments are therefore extremely limited in their applications, but the Telephone actually *speaks* and for this reason it can be utilized for nearly every purpose for which speech is employed.' AGBNHS, Bell Papers, 'Contributions to Knowledge: The Telephone,' vol. 2, no. 33 [binder 22], Bell to the Capitalists of the Electric Telephone Company, March 25, 1878, 41.

68 'Contributions to Knowledge: The Telephone,' 89.

69 Quoted in Sean Dennis Cashman, *America in the Gilded Age: From the Death of Lincoln to the Rise of Theodore Roosevelt,* 3rd ed. (New York: New York University Press, 1984), 22.

70 Bruce, 111.

71 Asa Briggs, 'The Pleasure Telephone: A Chapter in the Prehistory of the Media,' in *The Social Impact of the Telephone,* ed. Ithiel de Sola Pool (Cambridge, MA: MIT Press, 1977), 40.

72 Bell, 'The Substance of My Latest Research,' 1.

73 The fact that Bell's research and writing were often highly technical did not negate the fact that he believed such research should be used for the benefit of society. When a debate arose in the late nineteenth century between the 'pure' and 'applied' scientists as to the ultimate value of each, Bell clearly sided with the 'applied scientists': 'Research is none the less genuine, investigation none the less worthy,' he wrote, 'because the truth it discovers is utilizable for the benefit of mankind.' Quoted in James Rodger Fleming, 'Science and Technology in the Second Half of the Nineteenth Century,' in *The Gilded Age: Essays on the Origins of Modern America,* ed. Charles W. Calhoun (Washington: Scholarly Resources, 1996), 30.

74 Bell, 'The Substance of My Latest Research,' 4.

75 Sandford Fleming, 'The Pacific Cable,' *Queen's Quarterly* 5, 3 (January 1898): 226. On the importance of the telegraph and the underwater cable as technological inventions, see James W. Carey, 'Technology and Ideology: The Case of the Telegraph,' *Prospects: The Annual of American Cultural Studies* 8 (1983): 303-25; and James Carey and John J. Quirk, 'The Mythos of the Electronic Revolution,' *The American Scholar* Part 1, 39, 2 (Spring 1970): 219-41; Part 2, 39, 3 (Summer 1970): 395-425.

76 Fleming, 'The Pacific Cable,' 231. English Canadian imperialist George Parkin was of the same opinion. In 1905 he observed that 'for a people scattered as is our British race in all quarters of the globe, and yet aspiring to closer commercial intercourse, to complete political unity and effective mutual support, rapidity, ease, and cheapness of communication are of the very essence of our needs.' Quoted in Fortner, 26.

77 LAC, Fleming Papers, vol. 99, file 21, 'Pacific Cable – Addenda – Completion of the Trans-Pacific Cable with the Views of Sir Sandford Fleming' (n.d.), 1.

78 Sandford Fleming, *Canada and Ocean Highways,* Address at the Eighth Ordinary General Meeting of the Royal Colonial Institute, London, Tuesday, June 9, 1896, 419-20.

Fleming was not alone in extolling the virtues of technology for the advancement of civilization. On the eve of the First World War, W.F. King reflected on 'The Value of Science' in his presidential address to the Royal Society of Canada. He began by quoting from Francis Bacon: 'The real and legitimate goal of the sciences is the endowment of human life with new inventions and riches.' Three centuries later, King noted, Western civilization was benefiting from scientific ingenuity. 'The conditions of of [sic] life have been completely transformed: travel, by the steamship and the railway train; communication, by the telegraph and the telephone; hygiene, medicine and surgery have made immense advances; by means of the steam engine and the dynamo the forces of nature are compelled to the service of man in all kinds of labour saving machinery; innumerable conveniences and comforts of our daily life we owe to applications of science.' W.F. King, 'The Value of Science,' Presidential Address, *Proceedings and Transactions of the Royal Society of Canada* 6, 3 (1912): Appendix A, xxxix.

79 See M.O. Scott, 'Marconi in Canada,' *Canadian Magazine* 18 (1902): 338-40; and L.W. Gill, 'Wireless Telegraphy,' *Queen's Quarterly* 10, 3 (January 1903): 268-73.

80 On the nature of the dispute, see Thompson, 64-75.

81 In 1889, in a paper given to the Royal Society of Canada on yet another of Fleming's many 'causes' (political reform), Fleming maintained that in its constant evolution, the world was moving beyond the stage of warfare and belligerency:

> This much will be conceded: the chronic feuds between tribes and races which characterized the history of the human family in a less advanced stage of civilization no longer exist. War is manifestly not the normal condition of society in our time. Is it not therefore an anachronism to perpetuate hostility in the internal affairs of a nation [through party politics]? Is it not in the highest interests of the state that each member of the community, in every matter which concerns him as a citizen, should have the fullest opportunity of acting up to the injunction, 'Live peaceably with all men.' If the age of belligerency has passed away, is it not eminently fit and proper that we should seek for the removal of the last vestiges of a belligerent age which still remain in our political system?

Sandford Fleming, 'A Problem in Political Science,' *Transactions of the Royal Society of Canada* 7, 3 (1889): 39-40.

82 Henry David Thoreau, *Walden* [1854], ed. J. Lyndon Shanley (Princeton: Princeton University Press, 1971), 52.

Chapter 3: Advocates of Technical Education

1 William Leiss, *Under Technology's Thumb* (Montreal and Kingston: McGill-Queen's University Press, 1990), 5-6.

2 Biographical information on Hoodless is taken from Terry Crowley, 'Adelaide Sophia (Hoodless) Hunter,' *Dictionary of Canadian Biography,* Vol. 13, 1901-1910 (Toronto: University of Toronto Press, 1994), 488-93. For a good discussion of Adelaide Hoodless's efforts to introduce domestic science courses in the context of urbanization and industrialization, see Robert M. Stamp, 'Teaching Girls Their "God Given Place in Life,"' *Atlantis* 2, 2 (1977): 18-34.

3 University of Guelph Archives [UGA], Adelaide Hoodless Papers [Hoodless Papers], box 3, file: Newspaper Clippings (1899), 'The Address by Mrs. Hoodless,' 3-4.

4 UGA, Hoodless Papers, box 3, file: Newspaper Clippings (1902), Adelaide Hoodless, 'A New Education for Women,' 2.

5 UGA, Hoodless Papers, box 3, file: Newspaper Clippings (1899), 'Technical Education and Domestic Science: An Address by M.S. Hoodless,' 3.

6 UGA, Hoodless Papers, box 1, file: Correspondence, 1876-1909, Adelaide Hoodless, 'Domestic Science Instruction in the Schools of Canada,' (n.d.), 3.

7 Quoted in Stamp, 33.

8 James Cappon, 'In Memoriam' [N.F. Dupuis], *Queen's Quarterly* 25, 2 (1917): 126.

9 Ibid., 128.

10 Nathan Fellowes Dupuis, *An Address Delivered at the Opening of the Thirty-First Session of Queen's College, October 2, 1872* (Kingston, 1872), 9.

11 Ibid., 11. In *Inventing Secondary Education,* R.D. Gidney and W.P.J. Millar note that, 'until the early 1870s, science education and technical education were often treated as close to synonymous.' Then arose a chorus of enthusiasm for technical education: 'for training in art and design, agriculture, engineering, metallurgy, industrial chemistry, and other similar applied sciences.' But advocates disagreed as to where technical education should fit in the curriculum, whether in separate schools of technology or in the regular school and university system. This issue is reflected in the following discussion about technical education, which often centred on this exact issue. R.D. Gidney and W.P.J. Miller, *Inventing Secondary Education: The Rise of the High School in Nineteenth-Century Ontario* (Montreal and Kingston: McGill-Queen's University Press, 1990), 286ff.

12 Dupuis, 11. Original italics.

13 Ibid., 13.

14 Ibid., 14. Original italics.

15 N.F.D. [Dupuis], 'Some of the Factors of Modern Civilization,' *Queen's Quarterly* 4, 1 (1896): 49, 50, 51, 52.

16 Ibid., 53, 54.
17 For background on Galbraith, see C.R. Young, *Early Engineering Education at Toronto, 1851-1919* (Toronto: University of Toronto Press, 1958), 71-139 passim; Catherine Moriarty, *John Galbraith, 1846-1914: Engineer and Educator, a Portrait* (Toronto: Faculty of Applied Science and Engineering, University of Toronto, 1989); and Richard White, *The Skule Story: The University of Toronto Faculty of Applied Science and Engineering, 1873-2000* (Toronto: University of Toronto Press, 2000). For an attempt to link John Galbraith, first principal of the School of Practical Sciences and later the first dean of the faculty of applied science and engineering at the University of Toronto, with the John Galbraith who is author of *In the New Capital; or, The City of Ottawa in 1999* (Toronto: Toronto News Company, 1897), a utopian and reformist novel, see John Galbraith, *In the New Capital; or, The City of Ottawa in 1999*, reprint with an introduction by R. Douglas Francis (Ottawa: Penumbra Press, 2000).
18 Quoted in Young, 65.
19 University of Toronto Archives [UTA], John Galbraith Papers [Galbraith Papers], John Galbraith, *Technical Education: Address Delivered by Professor Galbraith at the Opening of the Engineering Laboratory of the School of Practical Science, Toronto, February 24, 1892* (Toronto, 1892), 3.
20 Ibid., 6.
21 Ibid., 7-8.
22 John Galbraith, 'The Function of the School of Applied Science in the Education of the Engineer,' *University of Toronto Monthly* 1, 5 (1901), 153.
23 Ibid., 157.
24 Suzanne Zeller, '"Merchants of Light": The Culture of Science in Daniel Wilson's Ontario, 1853-1892,' in *Thinking with Both Hands: Sir Daniel Wilson in the Old World and the New*, ed. Elizabeth Hulse (Toronto: University of Toronto Press, 1999), 116. See also Carl Berger, 'Sir Daniel Wilson,' *Dictionary of Canadian Biography*, Vol. 12, 1891-1900 (Toronto: University of Toronto Press, 1990), 1109-14.
25 For a summary of Wilson's views on evolutionary theory, see Carl Berger, *Science, God, and Nature in Victorian Canada* (Toronto: University of Toronto Press, 1983).
26 UTA, Daniel Wilson Papers [Wilson Papers], box: Pamphlets, Daniel Wilson, *Address at the Convocation of University College, Oct. 16, 1885*, 15.
27 Ibid., Daniel Wilson, *On the Practical Uses of Science in the Daily Business of Life* (1881), 6.
28 UTA, James Loudon Papers [Loudon Papers], box 15, file 53, Daniel Wilson, *Convocation Address, University of Toronto, Oct. 1, 1890*, 19.
29 Ibid., 14.
30 UTA, Wilson Papers, box: Pamphlets, Sir Daniel Wilson, *Convocation Address at the University of Toronto, Oct. 5, 1891*, 15.
31 Wilson, *On the Practical Use of Science*, 16.
32 Ibid., 17.
33 For Loudon's views on the university as research institution, see President Loudon, 'The Universities in Relation to Research,' *University of Toronto Monthly* 2, 9 (1902): 234-44; originally given as his presidential address to the Royal Society of Canada, 1902.
34 UTA, Loudon Papers, box 15, file 25, Technical Education, James Loudon, 'Technical Education,' Address given at Frederickton [sic], New Brunswick, 1894, 1, 2.
35 Ibid., 2.
36 Ibid., 5.
37 Ibid., 8.
38 UTA, Loudon Papers, box 15, file 24, James Loudon, *Convocation Address, October 2, 1899* (Toronto, 1899), 7.
39 Ibid., 14.
40 Henry T. Bovey, 'The Fundamental Conceptions which Enter into Technology,' *McGill University Magazine* 4, 1 (1905): 35-51.
41 Biographical information is taken from an obituary in McGill University Archives, 'Henry Taylor Bovey,' in *The Storied Province of Quebec*, 393.
42 Henry T. Bovey, 'Presidential Address,' *Transactions of the Royal Society of Canada* 2, 3 (1896): 3.

43 Bovey, 'Fundamental Conceptions,' 35.
44 Ibid., 37, 38.
45 Ibid., 37. Original italics.
46 Ibid., 39.
47 Ibid., 40.
48 Ibid., 41.
49 Ibid., 43. Original italics.
50 Ibid., 50.
51 Ibid., 50-51. Original italics.
52 Ibid., 51. Original italics.
53 See Rodney Millard, *The Master Spirit of the Age: Canadian Engineers and the Politics of Professionalism* (Toronto: University of Toronto Press, 1988).

Chapter 4: George Sidney Brett and the Debate on Technology as War

1 George S. Brett, 'The Revolt against Reason: A Contribution to the History of Thought,' *Transactions of the Royal Society of Canada* 13, 2 (1919): 9-17.
2 John A. Irving, 'The Achievement of George Sidney Brett (1879-1944),' *University of Toronto Quarterly* 14, 4 (1945): 329-65.
3 Michael Gauvreau, 'Philosophy, Psychology, and History: George Sidney Brett and the Quest for a Social Science at the University of Toronto, 1910-1940,' Canadian Historical Association, *Historical Papers* (1988): 225.
4 R.S. Peters, ed., *Brett's History of Psychology,* abrid. (London: George Allen and Unwin, 1953), 32.
5 Ibid.
6 Ibid., 716.
7 Ibid., 356.
8 Brett, 'Revolt against Reason,' 9.
9 Ibid., 9-10.
10 Ibid., 10. Dr. J.W. McCullough, Chief Officer of Health for Ontario, drew the classic analogy of humans to machines in an article entitled 'Industrial Hygiene': 'Machinery has worked miracles in labour, but with machinery has come increased danger to life. Of all the labour-saving machines invented, however, there is none so vital and indispensable in the production of wealth as the human machine – and at the same time there is none so sensitive and delicate. Both from the standpoint of humanity and the standpoint of economy the human machine deserves greater care and consideration than any other mechanism engaged in the production of wealth. Yet it is a fact that while the ordinary machine is oiled, cleaned and cooled and kept in the most careful repair, the human machine is expected to run day in and day out with no particular care on the part of the employer that this delicate mechanism shall be kept in constant repair. It has been the habit to work the human machine for all it is worth, which when worn out and dies, is forgotten the next day.' J.W.S. McCullough, 'Industrial Hygiene,' *Public Health Journal* 11, 6 (1920): 245-56.
11 Brett, 'Revolt against Reason,' 11.
12 Ibid., 14.
13 Ibid., 17.
14 George S. Brett, 'The History of Science as a Factor in Modern Education,' *Transactions of the Royal Society of Canada* 29, 3 (1925): 39.
15 Ibid., 44.
16 Ibid., 39.
17 Ibid., 45.
18 George S. Brett, 'Makers of Science,' *University of Toronto Quarterly* 5, 4 (1936): 611.
19 Lindsay Crawford, 'Current Events,' *Canadian Magazine* 45, 4 (1915): 350.
20 J. Squair, 'Germany's Megalomania,' *University of Toronto Monthly* 15 (1915): 136.
21 Robert Falconer, 'The President's Opening Address,' *University of Toronto Monthly* 15 (1915): 21.
22 H.H.L., Review of R.A. Falconer's *The German Tragedy and Its Meaning for Canada, University of Toronto Monthly* 15 (1915): 394.

23 T. Brailsford Robertson, 'Science and the War,' *Canadian Magazine* 51, 6 (1918): 447, 448.
24 C. Lintern Sibley, 'Britain's Intellectual Empire,' *Canadian Magazine* 44, 6 (1915): 486.
25 Francis Mills Turner Jr., 'Our Great National Waste,' *Canadian Magazine* 46, 1 (1915): 4.
26 Ibid., 6.
27 A.B. MacCallum, 'The Old Knowledge and the New,' Presidential Address, *Proceedings and Transactions of the Royal Society of Canada* 2, 3 (1917): Appendix A, 62.
28 A. Stanley Mackenzie, 'The War and Science,' *Transactions of the Royal Society of Canada* 12, 3 (1918): 1.
29 Clarence M. Warner, 'The Growth of Canadian National Feeling,' *Canadian Magazine* 45, 4 (1915): 281.
30 C. Lintern Sibley, 'Canada's Mighty Gains from the War,' *Canadian Magazine* 46, 2 (1915): 163-69.
31 Harold Garnet Black, 'Literature and Life,' *Canadian Magazine* 45, 6 (1915): 466.
32 H.T.J. Coleman, 'The Teacher and the New Age,' *Queen's Quarterly* 25, 4 (1918): 396.
33 Ibid.
34 Ibid., 401.
35 J.K. Robertson, 'Pure Science and the Humanities,' *Queen's Quarterly* 26, 1 (1918): 54.
36 Ibid., 65.
37 Ira A. MacKay, 'Educational Preparedness,' *Canadian Magazine* 52, 4 (1919): 810.
38 Ibid., 817.
39 Ibid., 818.
40 C.W. Mitchell, 'The Future of Applied Science,' *University of Toronto Monthly* 20, 2 (1919): 58.
41 'The Need of Men for Industrial Research,' *University of Toronto Monthly* 20 (March 1919): 179.
42 R.F. Ruttan, 'International Co-Operation in Science,' Presidential Address, *Proceedings and Transactions of the Royal Society of Canada* 14, 3 (1920): Appendix A, 38.
43 E.F. Scott, 'The Effects of War on Literature and Learning,' *Queen's Quarterly* 27, 2 (1919): 147.
44 Ibid., 153.
45 James A. Lindsay, 'On Thinking Biologically,' *Dalhousie Review* 1, 1 (1921): 8.
46 'The President's Opening Address,' *University of Toronto Monthly* 21, 2 (1920): 68.
47 E.H. Blake, 'The New Year,' *Canadian Forum* 1, 4 (1921): 103.
48 Harry Elmer Barnes, 'Dynamic History and Social Reform,' *Canadian Forum* 4, 47 (1924): 331, 332.
49 Richard de Brisay, 'The Conquest of War,' *Canadian Forum* 7, 83 (1927): 329.
50 Ibid.
51 UTA, C.R. Young Papers, box 6, file 3, R.W. Leonard, 'Presidential Address,' Annual Meeting of the Engineering Institute of Canada, *Transactions of the Royal Society of Canada* 34, 1 (January 27, 1920): 5.
52 A.R. MacDougall, 'Cultural Values in Old Countries and New,' *Dalhousie Review* 5 (1925-26): 352-53.
53 For an excellent discussion of Western civilization as a neurotic society, see George Frankl, *The Social History of the Unconscious* (London: Open Gate Press, 1989).
54 Newton MacTavish, 'The Tide Now Running,' *Canadian Magazine* 52, 5 (1919): 942.
55 J. Clark Murray, 'Pragmatism,' *University Magazine* 14, 1 (1915): 103.
56 Ibid., 111.
57 D. Fraser Harris, 'Science and Faith-Healing,' *Dalhousie Review* 2, 1 (1922): 59, 60. Original italics.
58 George S. Brett, 'The Modern Mind and Modernism,' *Canadian Journal of Religious Thought* 5, 2 (1928): 91-104.
59 Ibid., 92. Brett commented further on man's aggressive nature in a review of *Civilization, War, and Death: Selections from Three Works by Sigmund Freud, Queen's Quarterly* 46, 2 (1939): 247: 'What of War? The whole problem is illuminated if, as Freud claims, there lies, deep down in the unconscious of each of us, a tremendous wish to destroy. For them the task must not be to eliminate human aggression. That cannot be done; one might as well propose to eliminate human arms because they may be used to wield bayonets, or human eyes,

because they are necessary in order to lay guns. The task is rather to recognize aggression as an elemental urge and use it for the ends of society.'

60 Brett, 'Modern Mind and Modernism,' 92.
61 Ibid., 94.
62 Ibid.
63 Ibid., 95.
64 Ibid., 96.
65 Ibid.

Chapter 5: William Lyon Mackenzie King and Frederick Philip Grove

1 William Lyon Mackenzie King, *Industry and Humanity: A Study in the Principles Underlying Industrial Reconstruction* [1918] (Toronto: University of Toronto Press, 1973), 38. All subsequent quotations are taken from this edition. For a good analysis of King's ideas as outlined in *Industry and Humanity,* see Reginald Whitaker, 'The Liberal Corporatist Ideas of Mackenzie King,' *Labour/Le Travail* 2 (1977): 137-69. See also Henry Ferns and Bernard Ostry, *The Age of Mackenzie King* (Toronto: James Lorimer, 1976), 243-82.
2 King, *Industry and Humanity,* 40.
3 Ibid., 34.
4 Ibid., 40.
5 Ibid., 18.
6 Ibid., 12.
7 Ibid., 10. Original italics.
8 Marshall Berman, *All That Is Solid Melts into Air: The Experience of Modernity* (New York: Penguin Books, 1988).
9 C.P. Stacey, *A Very Double Life: The Private Life of Mackenzie King* (Toronto: Macmillan, 1976). It is interesting to note that King's friend and confidant, Henry Albert Harper, about whom King would write a tribute in the form of a 'novelette' in later life, saw King on first acquaintance as 'one that could be a saint or a devil, a "Dr. Jekyll and Mr. Hyde."' King went on to record in his diary: 'He [Harper] has learned to see beyond the external nature which I do not wonder presents that picture to many.' Library and Archives Canada [LAC], W.L.M. King Papers [King Papers], *Diaries,* September 7, 1895.
10 LAC, King Papers, *Diaries,* March 3, 1915.
11 King, *Industry and Humanity,* 14, 15.
12 Ibid.
13 Robert H. Wiebe, *The Search for Order 1877-1920* (New York: Hill and Wang, 1967), viii.
14 LAC, King Papers, *Diaries,* October 15, 1893.
15 Ibid., November 3, 1897.
16 King, *Industry and Humanity,* 107.
17 Ibid., 17.
18 Ibid., 18.
19 Ibid., 19.
20 Ibid., 15. Italics added.
21 Ibid., 22. Italics added.
22 Ibid., 23.
23 Quoted in King, *Industry and Humanity,* 20.
24 Frederick Philip Grove, *The Master of the Mill* [1944] (Toronto: McClelland and Stewart, 1967), 21. All subsequent quotations are from this edition.
25 For a discussion of this theme in literature, see William Leiss, *Under Technology's Thumb* (Montreal and Kingston: McGill-Queen's University Press, 1990), 40ff.
26 Grove, *The Master of the Mill,* 192.
27 Ibid., 193.
28 Ibid., 193-94.
29 Ibid., 39.
30 Ibid., 287.
31 Ibid., 327-28.
32 LAC, King Papers, *Diaries,* June 22, 1895.

33 Ibid., November 29, 1894.
34 Ibid., April 27, 1895.
35 Quoted in Whitaker, 140.
36 William Lyon Mackenzie King, *The Secret of Heroism: A Memoir of Henry Albert Harper* [1906] (Toronto: Hunter-Ross, 1919), 114.
37 J.G. Hume, *Political Economy and Ethics* (Toronto, 1892), 39.
38 Quoted in Craufurd D.W. Goodwin, *Canadian Economic Thought: The Political Economy of a Developing Nation, 1814-1914* (Durham, NC: Duke University Press, 1961), 159n29.
39 Quoted in ibid., 177.
40 Quoted in Paul Craven, *'An Impartial Umpire': Industrial Relations and the Canadian State 1900-1911* (Toronto: University of Toronto Press, 1980), 42. Original italics.
41 Quoted in ibid., 43.
42 See S.E.D Shortt's chapter 'James Mavor: The Empirical Ideal,' in *The Search for an Ideal: Six Canadian Intellectuals and Their Convictions in an Age of Transition, 1890-1930* (Toronto: University of Toronto Press, 1976), 119-36.
43 Quoted in Goodwin, 160.
44 Quoted in Shortt, 131.
45 LAC, King Papers, *Diaries,* July 11, 1894.
46 Arnold Toynbee, *Lectures on the Industrial Revolution of the Eighteenth Century in England* [1884] (London: Longmans, Green, and Co., 1928), 151.
47 An example of how King saw reform and technology working hand in hand was his recording of a sermon by one of his favourite ministers, Reverend Jordan of St. James Square Church. King attended a lot of sermons; in fact, at one point he noted in his diary, 'There is nothing in the world I like better than a good sermon[,] nothing I hate more than a bad one'; July 8, 1894. Most sermons went unrecorded in his diary, but the one by Reverend Jordon on January 21, 1894, particularly intrigued him because of an unusual analogy the minister made. King recalled that Rev. Jordon showed how in the present day there were greater possibilities for good works than in Christ's day, thanks to modern inventions. He 'showed how limited was Christ's field, compared to ours, the steam engine etc. are used for good ends & brings his message to the uttermost parts of the earth, and how graves are being opened, graves of iniquity, eyes opened, in that the spirituality [sic] blind now see etc.'
48 Toynbee, 6-7.
49 LAC, King Papers, *Diaries,* September 29, 1894.
50 Ibid., August 30, 1897.
51 Ibid., June 27, 1916.
52 Toynbee, 6.
53 Quoted in Craven, 35.
54 Ibid., 36.
55 LAC, King Papers, *Diaries,* July 20 and July 21, 1895.
56 Ibid., July 28, 1895.
57 For an interesting discussion of changed perspectives on social reform in Toronto at the turn of the twentieth century, see Cathy James, 'Reforming Reform: Toronto's Settlement House Movement, 1900-20,' *Canadian Historical Review* 82, 1 (2001): 55-90.
58 Joseph Dorfman, *The Economic Mind in American Civilization: Vol. 3, 1865-1918* (New York: Viking, 1949), 438.
59 John P. Diggins, *The Bard of Savagery: Thorstein Veblen and Modern Social Theory* (New York: Seabury Press, 1978), 78-79.
60 LAC, King Papers, *Diaries,* August 12, 1897.
61 Ibid., May 21 and May 28, 1897.
62 Ibid., August 13, 1897.
63 For an interesting comparison of the ideas of King and Marx, see Barry Cooper, 'On Reading *Industry and Humanity:* A Study in the Rhetoric Underlying Liberal Management,' *Journal of Canadian Studies* 13, 4 (1978-79): 28-39.
64 Karl Marx and Frederick Engels, *The Communist Manifesto* (New York: Pathfinder Press, 1987), 17.

65 Ibid., 19.
66 Ibid., 32.
67 Dorfman, 265.
68 LAC, King Papers, *Diaries,* January 7, 1898.
69 Ibid., January 9, 1898.
70 Ibid., January 27, 1898.
71 Quoted in Dorfman, 266.
72 See W. Cunningham, *The Growth of English Industry and Commerce in Modern Times* (Cambridge: Cambridge University Press, 1882); *Outlines of English Industrial History* (Cambridge: Cambridge University Press, 1895); and for his Christian approach, *Christianity and Social Questions* (New York, 1910).
73 LAC, King Papers, *Diaries,* October 6, 1897; March 17, 1899.
74 Quoted in Craven, 57.
75 LAC, King Papers, *Diaries,* December 6-11, 1896.
76 Gustave Le Bon, *The Psychology of Peoples: Its Influence on Their Evolution* (London: T. Fisher Unwin, 1899), 145.
77 Ibid., 201, 204.
78 LAC, King Papers, *Diaries,* April 19, 1899.
79 For Le Play's ideas, see Frédéric Le Play, *On Family, Work, and Social Change,* ed. and trans. Catherine Bodard Silver (Chicago: University of Chicago Press, 1982). On the life and ideas of Le Play, see Dorothy Herbertson, *The Life of Frédéric Le Play* (Ledbury, Herefordshire: Le Play House Press, 1950); and Michael Z. Brooke, *Le Play: Engineer and Social Scientist* (London: Longman, 1970).
80 Brooke, 2.
81 Herbertson, 131.
82 Quoted in ibid., 115.
83 LAC, King Papers, *Diaries,* January 10, 1900.
84 Ibid., January 27, 1900.
85 King, *Secret of Heroism,* 122-23.
86 Quoted in J. Dover Wilson, Introduction, in *Culture and Anarchy,* by Matthew Arnold (Cambridge: Cambridge University Press, 1960), xviii.
87 LAC, King Papers, *Diaries,* January 10, 1901.
88 King, *Industry and Humanity,* 10.
89 Ibid., 88, 89.
90 Frederick Philip Grove, 'Apologia Pro Vita Et Opere Suo,' *Canadian Forum* 11, 131 (1931): 420-21.
91 Frederick Philip Grove, *In Search of Myself* [1946] (Toronto: McClelland and Stewart, 1974), 369. Original italics. It has been discovered that *In Search of Myself* is not an accurate autobiography but a fictitious account; nevertheless, the book does accurately depict Grove's attitudes.
92 Ibid., 448.
93 Frederick Philip Grove, *It Needs to Be Said* [1929] (Ottawa: Tecumseh Press, 1982), 18.
94 Ibid., 5-6.
95 Mikael Härd and Andrew Jamison, eds., *The Intellectual Appropriation of Technology: Discourses on Modernity* (Boston: MIT Press, 1998), 1.

Chapter 6: Stephen Leacock and Archibald Lampman

1 Stephen Leacock, *The Unsolved Riddle of Social Justice* [1920], reprinted in *The Social Criticism of Stephen Leacock,* ed. Alan Bowker (Toronto: University of Toronto Press, 1973), 76. All subsequent quotations are from this edition.
2 Archibald Lampman, 'At the Ferry,' in *The Poems of Archibald Lampman (Including 'At the Long Sault'),* ed. Margaret Coulby Whitridge (Toronto: University of Toronto Press, 1974), 151-52.
3 Archibald Lampman, 'The Railway Station,' in ibid., 116.
4 Archibald Lampman, 'The City,' in ibid., 215.

5 Ibid., 215, 217.
6 Archibald Lampman, 'To a Millionaire,' in *The Poems of Archibald Lampman,* 3rd ed., ed. Duncan Campbell Scott (Toronto: Morang and Co., 1905), 276-77.
7 Archibald Lampman, 'The Land of Pallas,' in ibid., 201, 202-3.
8 Ibid., 204.
9 Ibid., 209.
10 Archibald Lampman, 'The City of the End of Things,' in ibid., 180-82.
11 W.E. Collins, 'Archibald Lampman,' in *Archibald Lampman,* ed. Michael Gnarowski (Toronto: Ryerson Press, 1970), 135.
12 Archibald Lampman, 'Socialism,' in *The Essays and Reviews of Archibald Lampman,* ed. D.M.R. Bentley (London, ON: Canadian Poetry Press, 1996), 186.
13 Ibid., 187.
14 Ibid.
15 Ibid., 188.
16 Ibid., 189-90.
17 Leacock, *Unsolved Riddle of Social Justice,* 79, 78.
18 Ibid., 82.
19 Ibid., 75.
20 Ibid.
21 Ibid., 77. Leacock talked about the radical transformation brought about by the Industrial Revolution in one of his lectures:

> Then came into the world a great light like the dawning of a new day over the hills. This was around the middle of the eighteenth century, in other words, in time's long record only yesterday. At this point began the two great moving forces, the doctrine of individual liberty and the rise of the machine industry that have since transformed the world. The advance of science and the progress of invention began to accelerate man's power of production beyond all previous dreams. The purely abstract science of the previous century, the work of Galileo and Newton here bears fruit in the practical works of Watt and Arkwright and Stephenson, the use of steam power, the invention of mechanical spinning and weaving, the smelting of iron with coal, all the new glory of the machine, – drab and duty and clangorous but announcing a new world. Such was the Industrial Revolution that originated in England and within a century spread over western civilization, multiplying its power with every decade.

Library and Archives Canada [LAC], Stephen Leacock Papers [Leacock Papers], vol. 4, file: While There Is Time, Stephen Leacock, 'While There Is Time,' 19-20.
22 Leacock, *Unsolved Riddle of Social Justice,* 79.
23 Ibid., 78.
24 Ibid., 79-80. Leacock noted the paradox as late as 1942. In *Our Heritage of Liberty; Its Origins, Its Achievement, Its Crisis: A Book for War Time* (London: Bodley Head, 1942), he wrote, 'The paradox is still there, the inconsistency ever greater. Each new saving of labour seems to mean new wants. Each new mechanism of life turns to an added instrument of death. We are still seeking, collectively, our means of salvation' (46).
25 Leacock, *Unsolved Riddle of Social Justice,* 107.
26 Ibid., 80.
27 Ibid., 76.
28 Ibid., 86, 88.
29 Ibid., 91.
30 No one has analyzed Leacock's views on technology, but there are a number of good studies on Leacock as a social reformer. See Bowker, Introduction to *Social Criticism of Stephen Leacock;* Ramsay Cook, 'Stephen Leacock and the Age of Plutocracy, 1913-1921,' in *Character and Circumstance: Essays in Honour of Donald Grant Creighton,* ed. John S. Moir (Toronto: Macmillan, 1970), 163-81; Carl Berger, 'The Other Mr. Leacock,' *Canadian Literature* 55 (Winter 1973): 23-40; and F.W. Watt, 'Critic or Entertainer? Stephen Leacock and the Growth of Materialism,' *Canadian Literature* 5 (Summer 1960): 33-42.

31 The biographical sketch of Leacock is taken from Bowker, Introduction to *Social Criticism of Stephen Leacock;* Ralph Curry, *Stephen Leacock and His Works* (Toronto: ECW Press, n.d.); and David M. Legate, *Stephen Leacock: A Biography* (Toronto: Doubleday, 1970).

32 Quoted in Carl Spadoni, Introduction, *My Recollection of Chicago* and *The Doctrine of Laissez Faire,* by Stephen Leacock (Toronto: University of Toronto Press, 1998), xiii.

33 For a criticism of North American education as mechanical, specialized, and anti-intellectual, see Leacock's 'Literature and Education in America' [1909], in Bowker, ed., *Social Criticism of Stephen Leacock,* 13-26.

34 Stephen Leacock, *My Discovery of the West: A Discussion of East and West in Canada* (Toronto: Thomas Allen, 1937), 137.

35 For a discussion of the intellectual association of Leacock and Veblen, see Myron J. Frankman, 'Stephen Leacock, Economist: An Owl among the Parrots,' in *Stephen Leacock: A Reappraisal,* ed. David Staines (Ottawa: University of Ottawa Press, 1986), 51-58; Berger, 23-40; Claude T. Bissell, 'Haliburton, Leacock, and the American Humorous Tradition,' *Canadian Literature* 39 (Winter 1969): 5-18; Bowker, Introduction to *Social Criticism of Stephen Leacock,* ix-xliii; and Cook. On Veblen and his ideas, see Joseph Dorfman, *Thorstein Veblen and His America* (New York: Viking, 1961), and *The Economic Mind in American Civilization,* Vol. 3, 1865-1918 (New York: Viking, 1949), 434-55; and John P. Diggins, *The Bard of Savagery: Thorstein Veblen and Modern Social Theory* (New York: Seabury Press, 1978).

36 Stephen Leacock, 'Democracy and Social Progress,' in *The New Era in Canada: Essays Dealing with the Building of the Canadian Commonwealth,* ed. J.O. Miller (Toronto: J.M. Dent, 1917), 17.

37 Leacock, *My Discovery of the West,* 137.

38 Leacock, 'The Apology of a Professor: An Essay on Modern Learning' [1910], in Bowker, ed., *Social Criticism of Stephen Leacock,* 38.

39 Leacock, *My Discovery of the West,* 140.

40 In an address to the Empire Club in 1933 called 'The Riddle of the Depression,' Leacock stated, 'There is a man abroad, called the Technocrat, many of whose speeches and sayings are foolish in the extreme. He has made what he calls "an energy survey of the world" – mostly seen from Greenwich villages and the speakeasies. The Technocrat has told us all kinds of silly things about abandoning our customary standard of values and substituting the scientific standard which is usually measured by an "erg" or an "umph."' Empire Club of Canada, *Addresses* 31 (1933-34): 72.

41 For a good summary of Leacock's criticisms of laissez-faire and socialism in the context of the Great Depression, see his article 'What Is Left of Adam Smith?' in *Canadian Journal of Economics and Political Science* 1, 1 (1935): 41-51.

42 Quoted in Bowker, Introduction to *Social Criticism of Stephen Leacock,* ix.

43 Leacock believed that the role of the political economist was to offer judgments on the 'equity and inequity of the social system.' In one of his lecture notes entitled 'The Production of Wealth,' he notes, 'Political economy, it must be repeated, is not a moral nor a didactic science. But it presents here the basis for our judgment as to the equity or inequity of the social system, and its principles govern our view of the effectiveness of possible remedies for admitted social wrongs.' LAC, Leacock Papers, vol. 6, file: Teaching Notes: Practical Political Economy, Stephen Leacock, 'The Production of Wealth' (n.d.), 13.

44 Bissell, 14. For a further comparison of Veblen's *Theory of the Leisure Class* and Leacock's *Arcadian Adventures with the Idle Rich,* see J. Kusher and R.D. MacDonald, 'Leacock: Economist/Satirist in *Arcadian Adventures* and *Sunshine Sketches,*' *Dalhousie Review* 56, 3 (1976): 493-509.

45 Stephen Leacock, *Arcadian Adventures with the Idle Rich* [1914] (Toronto: McClelland and Stewart, 1959), 8. Leacock also noted the contrast of the rich in the city and the poor in the town, using the Mausoleum Club as his point of reference, in *Sunshine Sketches of a Little Town* (1912). Note in particular the last chapter, in which Leacock uses the train trip from the city to the little town of Mariposa as the occasion to reflect on the changed values in society as a result of industrialization and urbanization.

46 Leacock, *Arcadian Adventures,* 55.

47 Leacock, 'Democracy and Social Progress,' 15-16.

48 Ibid., 17-18.
49 Stephen Leacock, 'Our National Organization for the War,' in Miller, *The New Era in Canada,* 417.
50 Leacock noted the altered perspective on the laissez-faire theory and state intervention in the 1921 edition of his popular political science text, *Elements of Political Science* (Boston: Houghton Mifflin, 1921):

> The economists and political philosophers of the present time are prepared to defend a degree of state interference quite at variance with the doctrines of their predecessors. The reason for this remarkable alteration both in theory and practice is found in the altered circumstances of our industrial environment. We have seen in a previous chapter that the new rapid expansion of industry under the stimulus of the new mechanical processes of the industrial revolution seemed to demand its liberation from all forms of government restraint, and that the consequent removal of the standing impediments to the free movement of capital and labor was accompanied, at any rate as far as the total volume of production was concerned, with marked success. But it has been seen also that in reference to the welfare of the laboring class the system of free competition, particularly in regard to the work of women and children, was open to serious objection. (381-82)

51 Leacock, *Unsolved Riddle of Social Justice,* 135.
52 Ibid.
53 Ibid., 139. Leacock's concern for social equality for all children, and his faith in children as the promise of the morrow, remained with him throughout his life. In a text on Canada written during the Second World War, Leacock noted in his final chapter, 'Canada as a Future World Power': 'The war effort seems all the more worthwhile if we can see the vision possible beyond it. We need immigrants – not thousands, millions – not gradually, but in a mass. Above all we have to realize that the best immigrants – in fact the best of all general imports – are children. We need them, imported and homegrown, in cradlefuls. That way lies security. In no time they grow up; see them there in the air above us, the children of yesterday.' Stephen Leacock, *Canada: The Foundations of Its Future* (Montreal: privately printed, 1941), 246. Even more poignant is his article 'To Every Child' in *Last Leaves* (Toronto: McClelland and Stewart, 1945), in which he wrote, 'Each of us must stand appalled at the further existence, after the war of misery and poverty, of lives frustrated by want, of children underfed, of people sunk from their birth below a chance to live. We must decide that that must not be, just as we decided that savage conquest and brutality should not be ... Especially with the children lies our chief chance. Older people are battered out of shape, or were never battered into it. Faces all wrinkled and furrowed with care cannot be altered now. But to every child we must give the chance to live, to learn, to love' (106-7).
54 Quotations are taken from a draft of the article in LAC, Leacock Papers, vol. 3, file: Disarmament and Common Sense, Stephen Leacock, 'Disarmament and Common Sense,' 1.
55 Ibid., 2.
56 Ibid.
57 Note the reversal of the theme of the master/servant relationship with regard to technology that was popular in the interwar years.
58 Stephen Leacock, 'The Man in Asbestos: An Allegory of the Future,' in *Nonsense Novels* (New York: John Lane, 1921), 165. From this point on, Leacock began to make fun of new technology. With regard to the automobile, see, for example, 'Motor Signs and Morals: An Allegorical Fancy,' *Metropolitan Newspaper Service* (December 1, 1929); and LAC, Leacock Papers, vol. 3, file: The Gasoline Goodbye, Stephen Leacock, 'The Gasoline Goodbye: The Awkward Moment When the Car Refuses to Start'; vol. 2, file: Motor Signs and Morals, Stephen Leacock, 'Motor Signs and Morals.' With regard to the radio and the cinema, see vol. 3, file: Lecture Notes, Stephen Leacock, 'Short Circuits by Radio and Cinema,' and 'If We Only Had the Radio Sooner.' See also Stephen Leacock, 'Isn't Just Wonderful,' in *The Iron Man and the Iron Woman with Other Such Futurities: A Book of Little Sketches of To-Day and To-Morrow* (New York, 1929), 18-23. Concerning the airplane, see Stephen Leacock, 'Once to Every Man,' n.d., in LAC, Leacock Papers, vol. 3, file: Miscellaneous, 1910-33; and Stephen Leacock,

'Once to Every Man,' *The Sportsman Pilot* (December 1930): 17, 46. On robots, see Stephen Leacock, 'The Iron Man and the Tin Woman,' in *The Iron Man,* 1-6.

59 Leacock, 'The Man in Asbestos,' 170-71.

60 Ibid., 159.

61 Leacock's disillusionment with technology lasted till his dying days. In *Our Heritage of Liberty,* written during the Second World War, he observed:

> To-day there are no new lands, and the machine in a certain sense has become the master, mankind the slave. Most of the habitable world has been explored and appropriated. Invention still goes on, but finds its readiest application in the means of death. Nor can even the industry of peace follow its perpetual changes. Nor is there left any longer the escape from civilization, the new start in the wilderness. The last frontier is vanishing. From our narrowed world there is no getting away ... We cannot wonder that this imprisoned feeling, this loss of one's own control, breeds in many people something like despair, a wistful longing for the "good old times." (12)

Chapter 7: Harold A. Innis and Eric Havelock

1 Anti-modernism is here associated with a critique of modernity, particularly the tendency in modern thought to a pathological kind of present-mindedness, an uncritical acceptance of rationality, an acceptance of materialism, and a severing of moral and ethical values from a traditional moral imperative that had guided Western civilization for centuries.

2 For a study of Innis's ideas on modernity and technology, see R. Douglas Francis, 'Modernity and Canadian Civilization: The Ideas of Harold Innis,' in *Globality and Multiple Modernities: Comparative North American and Latin American Perspectives,* ed. Luis Roninger and Carlos H. Waisman (Brighton: Sussex Academic Press, 2002), 213-29. Judith Stamps examines the ideas of Harold Innis and Marshall McLuhan in the context of the ideas of the members of the Frankfurt School in *Unthinking Modernity: Innis, McLuhan, and the Frankfurt School* (Montreal and Kingston: McGill-Queen's University Press, 1995). See also Philip Massolin, *Canadian Intellectuals, the Tory Tradition, and the Challenge of Modernity, 1939-1970* (Toronto: University of Toronto Press, 2001).

3 This theme of technology and power in Innis's thought is developed further in R. Douglas Francis, 'The Anatomy of Power: A Theme in the Writings of Harold Innis,' in *Nation, Ideas, Identities: Essays in Honour of Ramsay Cook,* ed. Michael D. Behiels and Marcel Martel (Toronto: Oxford University Press, 2000), 26-40.

4 This biographical sketch of Innis is taken from Alexander John Watson, *Marginal Man: The Dark Vision of Harold Innis* (Toronto: University of Toronto Press, 2006); Donald Creighton, *Harold Adams Innis: Portrait of a Scholar* [1957] (Toronto: University of Toronto Press, 1978); Robert E. Babe, 'The Communication Thought of Harold Adams Innis (1894-1952),' in *Canadian Communication Thought: Ten Foundational Writers,* ed. Robert E. Babe (Toronto: University of Toronto Press, 2000), 51-88; and Daniel J. Czitrom, *Media and the American Mind: From Morse to McLuhan* (Chapel Hill: University of North Carolina Press, 1982), 147-82.

5 Carl Berger, *The Writing of Canadian History: Aspects of English-Canadian Historical Writing Since 1900,* 2nd ed. (Toronto: University of Toronto Press, 1986), 85.

6 Quoted in ibid.

7 Watson, 60.

8 For a discussion of the influence of Innis's Baptist faith on his social science perspective, see Michael Gauvreau, 'Baptist Religion and the Social Science of Harold Innis,' *Canadian Historical Review* 76, 2 (1995): 161-204. Gauvreau also draws a direct link between George Sidney Brett and Harold Innis in terms of their common belief, as colleagues at the University of Toronto, that the social science discipline should fuse humanistic and scientific knowledge by applying moral and ethical values to scientific studies. In his article 'Philosophy, Psychology, and History: George Sidney Brett and the Quest for a Social Science at the University of Toronto, 1910-1940,' Gauvreau states, 'The work of both Brett and Innis ... reflected the view that the social sciences involved simply the application of value-free techniques of description and measurement to social problems; any attempt to analyze

human behaviour returned the social thinker to the question of human will and values, where only the guidance of philosophy and history assured the possibility of rational action' (236).

9 Judith Stamps, 'Innis in the Canadian Dialectical Tradition,' in *Harold Innis in the New Century: Reflections and Refractions,* ed. Charles R. Acland and William J. Buxton (Montreal and Kingston: McGill-Queen's University Press, 1999), 56.

10 Harold A. Innis, *The Bias of Communication* [1951] (Toronto: University of Toronto Press, 1991), xxvii. All subsequent quotations are from this edition unless noted otherwise.

11 Quoted in William Christian, *Harold Innis as Economist and Moralist* (Guelph, ON: Department of Political Science, University of Guelph, 1981), 2.

12 Quoted in Berger, 86-87.

13 Harold A. Innis, 'The Returned Soldier,' MA thesis, McMaster University Library, 1918, 7.

14 Watson, 102.

15 Berger, 87.

16 On the Chicago School and its influence on Innis, see James Carey, 'Space, Time, and Communications: A Tribute to Harold Innis,' in *Communication as Culture: Essays on Media and Society* (Boston: Unwin Hyman, 1989), 142-72; and 'Innis "in" Chicago: Hope as the Sire of Discovery' in Acland and Buxton, eds., *Harold Innis in the New Century,* 81-104.

17 Marshall McLuhan, Introduction to *The Bias of Communication,* by Harold Innis (Toronto: University of Toronto Press, 1964), xv.

18 Ibid.

19 For a study of Veblen's influence on Innis, see Fletcher Baragar, 'Influence of Veblen on Harold Innis,' *Journal of Economic Issues* 30 (September 1996): 667-83.

20 Harold Innis, 'The Work of Thorstein Veblen,' in *Essays in Canadian Economic History,* ed. Mary Q. Innis (Toronto: University of Toronto Press, 1956), 23, 24.

21 Ibid., 26.

22 Quoted in Watson, 162.

23 Quoted in ibid., 161.

24 Harold A. Innis, *A History of the Canadian Pacific Railway* [1923] (Toronto: University of Toronto Press, 1971), 287, 294.

25 Harold A. Innis, 'A Trip through the Mackenzie River Basin,' *University of Toronto Monthly* 25, 4 (1925): 152.

26 Harold A. Innis, *The Fur Trade in Canada: An Introduction to Canadian Economic History* [1930], rev. ed. (Toronto: University of Toronto Press, 1979), 383.

27 Innis did make reference to a 'Canadian civilization' in *A History of the Canadian Pacific Railway* but only in a very tangential way. He noted in his conclusion to his chapter on 'Fulfilment of the Contract':

> The fulfilment of the contract in the completion of the main line of the road was a significant landmark in the spread of civilization throughout Canada. It was significant of the strength and character of the growth of civilization within the boundaries of three distinct areas which served as buttresses for the transcontinental bridge. With this addition to technological equipment, the civilization of these areas changed in its character, and its extent, and became more closely a part of a civilization narrowly described as Canadian, and typically, western. These changes are recorded to some extent in the history of the Canadian Pacific Railroad and the history of Canada. (128)

28 Innis, *The Fur Trade in Canada,* 388.

29 Ibid., 385-86.

30 Harold A. Innis, *The Cod Fisheries: The History of an International Economy* [1940], rev. ed. (Toronto: University of Toronto Press, 1954), 1.

31 Ibid., 508.

32 Marshall McLuhan, 'The Later Innis,' *Queen's Quarterly* 60, 3 (1953): 385.

33 Robert W. Cox, 'Civilizations: Encounters and Transformations,' *Studies in Political Economy* 47 (Summer 1995): 17.

34 The idea is presented in Robin Neill, *A New Theory of Value: The Canadian Economics of H.A. Innis* (Toronto: University of Toronto Press, 1972), 17.
35 Harold A. Innis, *Political Economy in the Modern State* (Toronto: Ryerson Press, 1946), xv-xvi.
36 Quoted in R. Douglas Francis, *Frank H. Underhill: Intellectual Provocateur* (Toronto: University of Toronto Press, 1986), 123.
37 Harold Innis, 'This Has Killed That: An Unpublished Paper,' *Journal of Canadian Studies* 12, 5 (1977): 3.
38 Ibid.
39 Ibid., 4.
40 Harold A. Innis, 'The Newspaper in Economic Development,' in *Political Economy in the Modern State* (Toronto: Ryerson Press, 1946), 32.
41 Harold A. Innis, 'Minerva's Owl,' in *The Bias of Communication,* 3.
42 The best analysis of Innis's theories on communication media and their social and cultural influence is still James W. Carey, 'Harold Adams Innis and Marshall McLuhan,' *Antioch Review* 27 (Spring 1967): 5-39.
43 See Harold A. Innis, *Empire and Communications* (Toronto: University of Toronto Press, 1972), and *The Bias of Communication,* especially 'Minerva's Owl.'
44 Harold A. Innis, 'A Critical Review,' in *The Bias of Communication,* 190.
45 Ibid., 194.
46 Harold A. Innis, 'The Bias of Communication,' in *The Bias of Communication,* 34.
47 Harold A. Innis, 'Technology and Public Opinion in the United States,' in *The Bias of Communication,* 187.
48 Ibid., 189.
49 Ibid., 156.
50 Harold A. Innis, 'A Plea for Time,' in *The Bias of Communication,* 62.
51 Innis, *Empire and Communications,* 3.
52 Ibid.
53 Ibid., 3-4.
54 Ibid., 159ff.
55 Ibid., 161.
56 Harold A. Innis, 'Technology and Public Opinion,' in *The Bias of Communication,* 160.
57 Harold A. Innis, 'Great Britain, the United States, and Canada' [1948], reprinted in *Staples, Markets, and Cultural Change: Selected Essays [by] Harold A. Innis,* ed. Daniel Drache (Montreal and Kingston: McGill-Queen's University Press, 1995), 287-88 .
58 Harold A. Innis, 'The Strategy of Culture with Special Reference to Canadian Literature: A Footnote to the Massey Report,' in *The Strategy of Culture* (Toronto: University of Toronto Press, 1952), 19-20.
59 Innis, *Empire and Communications,* 169-70.
60 McLuhan, 'The Later Innis,' 391-92.
61 Innis, Preface, in *The Strategy of Culture.*
62 E.A. Havelock, *The Crucifixion of Intellectual Man,* Introduction to a reprint of *Prometheus* [1951] (Seattle: University of Washington Press, 1968), 14. All subsequent quotations are from this source.
63 Ibid.
64 Ibid., 15.
65 Ibid., 21.
66 Ibid., 24.
67 Ibid., 31.
68 Ibid., 44.
69 Ibid., 58.
70 Ibid., 78.
71 Ibid., 91.
72 Ibid., 16.
73 Ibid., 93.
74 Ibid.

75 Ibid., 109.
76 Eric A. Havelock, *The Liberal Temper in Greek Politics* (New Haven: Yale University Press, 1957), 55.
77 Ibid., 119.
78 Eric A. Havelock, *The Literate Revolution in Greece and Its Cultural Consequences* (Princeton: Princeton University Press, 1982), 106.
79 Ibid.
80 Eric A. Havelock, *The Muse Learns to Write: Reflections on Orality and Literacy from Antiquity to the Present* (New Haven: Yale University Press, 1986), 61-62.
81 Ibid., 62.
82 Ibid., 120.
83 Alexander John Watson, 'Harold Innis and Classical Scholarship,' *Journal of Canadian Studies* 12, 5 (1977): 57.
84 Eric Havelock, *Harold A. Innis: A Memoir* (Toronto: The Harold Innis Foundation, University of Toronto, 1982), 42.
85 Ibid., 33-34.

Chapter 8: Marshall McLuhan

1 Marshall McLuhan, 'The Future of Man in the Electric Age' [1965], in *Understanding Me: Lectures and Interviews,* ed. Stephanie McLuhan and David Staines (Toronto: McClelland and Stewart, 2003), 57.
2 James W. Carey, 'McLuhan and Mumford: The Roots of Modern Media Analysis,' *Journal of Communication* 31, 3 (1981): 166.
3 Marshall McLuhan, *The Gutenberg Galaxy: The Making of Typographic Man* (Toronto: University of Toronto Press, 1962), 40.
4 Marshall McLuhan, *Understanding Media: The Extensions of Man* (New York: McGraw-Hill, 1965), 21.
5 Ibid., 9.
6 Marshall McLuhan with Quentin Fiore, *The Medium Is the Massage: An Invention of Effects* (New York: Bantam Books, 1967), 26.
7 McLuhan delighted in wordplay and the nuances of meaning in the slight variation of a word.
8 Marshall McLuhan, 'Playboy Interview: A Candid Conversation with the High Priest and Metaphysician of Media,' in *Essential McLuhan,* ed. Eric McLuhan and Frank Zingrone (Toronto: Anansi, 1995), 239.
9 Ibid., 243.
10 For an interesting discussion of how McLuhan's ideas on the impact of electronic technology on the future of Western civilization related to those of Lewis Mumford, see Carey, 162-78.
11 Biographical information on McLuhan has been taken from Philip Marchand, *Marshall McLuhan: The Medium and the Messenger* (Toronto: Random House, 1989); W. Terrence Gordon, *Marshall McLuhan: Escape into Understanding, a Biography* (Toronto: Stoddart, 1997); William Kuhns, 'The Sage of Aquarius: Marshall McLuhan,' in *The Post-Industrial Prophets: Interpretations of Technology* (New York: Weybright and Talley, 1971), 169-201; and Robert E. Babe, 'The Communication Thought of Marshall McLuhan (1911-1980),' in *Canadian Communication Thought: Ten Foundational Writers* (Toronto: University of Toronto Press, 2000), 266-306.
12 Marchand, 5.
13 Ibid.
14 Ibid., 23.
15 Ibid., 27.
16 McLuhan and Zingrone, 1.
17 Quoted in Marchand, 31.
18 Ibid., 33.
19 Ibid., 34.
20 Ibid.

21 Ibid., 35.
22 Ibid., 48.
23 Ibid., 49.
24 Ibid., 59.
25 Ibid., 69.
26 Ibid., 73.
27 Marshall McLuhan, *The Mechanical Bride: The Folklore of Industrial Man* (New York: Vanguard Press, 1951).
28 McLuhan was forced to reflect on what he was doing in *The Mechanical Bride*. In a letter to Archie Malloch, he wrote:

> I was told yesterday by a lad in the know that *The Mechanical Bride* type of intellectual activity is regarded officially as destructive of real social values. The Manichean view of this life as a hell built precariously of rotten sticks and mud forbids serious exercise of reason except for purposes of escape. Idea of applying reason to serious criticism or fashioning of this life is for Manicheans a blasphemy.
>
> However, Bergson in his *Two Sources of Religion and Morality* points out that technology really makes sense only in Oriental terms. Although a by-product of Western culture it really is the lot of our culture. The machine belongs to the East because we refuse to deify it.
>
> Naturally, had I know[n] these things I would not have bothered to write *The Bride*. All the writings of W. Lewis I now see are directed against the pagan cults and their systematic buggering of mankind and the Arts.

Library and Archives Canada [LAC], Archibald Edward Malloch Papers [Malloch Papers], vol. 1, file: Correspondence, 1953-55, letter from Marshall McLuhan, dated only 'Tuesday.'
29 McLuhan, *The Mechanical Bride,* vi.
30 McLuhan and Zingrone, 25-26.
31 Ibid., 27.
32 Quoted in Gordon, 149.
33 McLuhan, *The Gutenberg Galaxy,* 50.
34 Ibid., 216. In a letter to his student Archibald Edward Malloch, McLuhan wrote, 'Thanks for doing the job at MLA for me. I expect you noticed how I was trying out the Innis concept. I think much can be done with it.' LAC, Malloch Papers, vol. 1, file: Correspondence, 1952, letter dated only 'Wednesday.'
35 Marshall McLuhan, Introduction, *The Bias of Communication,* by Harold Innis (Toronto: University of Toronto Press, 1964), ix.
36 Quoted in Marchand, 113. Was this a self-reflective comment? McLuhan was raised a Methodist and might have wondered how a 'hick Methodist' arrived at such brilliant insights. Innis abandoned his Baptist faith during the First World War, preferring not to adhere to any religious denomination. McLuhan also later abandoned his Methodist faith, in his case to embrace Roman Catholicism.
37 Marshall McLuhan, 'Media and Cultural Change,' in McLuhan and Zingrone, 94. In his Introduction to a reprint of Innis's *The Bias of Communication,* McLuhan elaborated further on Innis's 'technological blindness ... in regarding radio and electric technology as a further extension of the patterns of mechanical technology ... After many historical demonstrations of the space-binding power of the eye and the time-binding power of the ear, Innis refrains from applying these structural principles to the action of radio. Suddenly, he shifts the ear world of radio into the visual orbit, attributing to radio all the centralizing powers of the eye and of visual culture ... Electric light and power, like all electric media, are profoundly decentralizing and separatist in their psychic and social consequences' (xii).
38 McLuhan, *The Gutenberg Galaxy,* Preface, 1.
39 Ibid., 3.
40 Ibid., 272.
41 McLuhan, 'Playboy Interview,' 237.
42 Ibid., 240.

43 Dennis Duffy, *Marshall McLuhan* (Toronto: McClelland and Stewart, 1969), 29.
44 McLuhan, 'Playboy Interview,' 244.
45 Ibid.
46 Ibid., 243.
47 Ibid.
48 McLuhan, *The Gutenberg Galaxy,* 44.
49 McLuhan, 'Playboy Interview,' 264.
50 McLuhan, *The Gutenberg Galaxy,* 141.
51 Ibid., 141-42.
52 Ibid., 141.
53 Lewis H. Lapham, Introduction, *Understanding Media: The Extensions of Man,* by Marshall McLuhan (Cambridge, MA: MIT Press, 1994), xi.
54 Marchand, 167.
55 McLuhan, *Understanding Media,* 3-4.
56 McLuhan, 'Playboy Interview,' 239.
57 McLuhan, *Understanding Media,* 4-5.
58 McLuhan, 'Playboy Interview,' 249.
59 McLuhan, *Understanding Media,* 25.
60 Ibid., 23.
61 McLuhan, 'Playboy Interview,' 245.
62 McLuhan, *Understanding Media,* 313-14. Original italics. Philip Marchand comments that 'McLuhan insisted on this point so strongly that his colleagues began to joke that McLuhan owned a poor television set' (123).
63 McLuhan, *Understanding Media,* 334.
64 Ibid., 335.
65 McLuhan, 'Playboy Interview,' 240.
66 Ibid., 262.
67 McLuhan, *Understanding Media,* 80.
68 McLuhan, 'Playboy Interview,' 263.
69 McLuhan, *Understanding Media,* 248.
70 Claude T. Bissell, 'Herbert Marshall McLuhan,' in *Marshall McLuhan: The Man and His Message,* ed. George Sanderson and Frank Macdonald (Golden, CO: Fulcrum, 1989), 8.
71 In light of McLuhan's popularity as a public speaker, it is interesting to note that he had little interest in the effect of his message, in using communication technology to communicate more effectively. He confessed to Archibald Edward Malloch, 'It's only when I actually get up on my feet that I realize that I ought to be at home working and learning. I hate the sound of my own voice and I don't care enough about [the] *effect* of my remarks to give them any shape or value.' LAC, Malloch Papers, vol. 1, file: Correspondence, 1953-55, letter dated Friday, January 23, 1953. Original italics.
72 Marchand, 193.
73 Quoted in ibid., 209.
74 Marshall McLuhan, 'A McLuhan Mosaic,' in Sanderson and Macdonald, 3.
75 Marshall McLuhan, *Counterblast* (Toronto: McClelland and Stewart, 1969), 5.
76 Ibid., 14.
77 Ibid., 16-17.
78 Marshall McLuhan and Eric McLuhan, *Laws of Media: The New Science* (Toronto: University of Toronto Press, 1988), 3.
79 Ibid., 33.
80 McLuhan, 'Playboy Interview,' 267.
81 Ibid., 258-59.
82 Ibid., 259.
83 McLuhan, *Understanding Media,* 252.
84 McLuhan, 'Playboy Interview,' 267. Original italics.
85 Ibid., 268.
86 Ibid., 239.

87 William Kuhns, *The Post-Industrial Prophets: Interpretations of Technology* (New York: Weybright and Talley, 1971), 123.
88 This idea that Innis emphasized the message senders and McLuhan the message recipients is developed in Robert E. Babe, *Canadian Communication Thought: Ten Foundational Writers* (Toronto: University of Toronto Press, 2000), 299-300.

Chapter 9: Northrop Frye and E.J. Pratt

1 David Cayley, *Northrop Frye in Conversation* (Toronto: House of Anansi, 1992), 123-24.
2 Ibid., 124.
3 Ibid., 159.
4 Ibid., 159-60.
5 Ibid., 160.
6 Ibid., 164.
7 Ibid.
8 Ibid., 161.
9 Northrop Frye, *The Critical Path: An Essay on the Social Context of Literary Criticism* (Toronto: The Harvester Press, 1983), 150.
10 Ibid.
11 Ibid., 150-51.
12 Ibid., 152.
13 The only attempt to date to look at Frye's views on technology in the context of his social criticism is David Cook, *Northrop Frye: A Vision of the New World* (Montreal: New World Perspectives, 1985).
14 Biographical material on Frye is taken from John Ayre, *Northrop Frye: A Biography* (Toronto: Random House, 1989), unless otherwise noted.
15 Ibid., 30.
16 Quoted in ibid., 34.
17 Quoted in ibid., 44.
18 Quoted in ibid., 45.
19 Ibid., 57.
20 Ibid., 74.
21 Quoted in ibid., 65.
22 Cayley, 48.
23 Quoted in Ayre, 103-4. Robert E. Babe notes that Frye missed the opportunity of exploring Blake's critique of the political-economic dimensions in his works: 'It is worth noting that, although in his later writings Frye explored some political-economic dimensions of myth and literature, he missed the opportunity Blake presented him. Blake, of course, is still renowned as the social critic who attacked the dark satanic mills of the industrial revolution, an autocratic and unfeeling church, an inhumane state, disease-infested streets, child labour, and exploitation of all kinds.' *Canadian Communication Thought: Ten Foundational Writers* (Toronto: University of Toronto Press, 2000), 234.
24 Northrop Frye, 'The Narrative Tradition in English-Canadian Poetry' [1946], reprinted in *The Bush Garden: Essays on the Canadian Imagination* (Toronto: Anansi, 1971), 145.
25 Northrop Frye, 'Preface to An Uncollected Anthology,' in *The Bush Garden,* 164.
26 Jonathan Hart, *Northrop Frye: The Theoretical Imagination* (London: Routledge, 1994), 97-98.
27 Frye, *The Critical Path,* 89.
28 Frye, 'The Narrative Tradition,' in *The Bush Garden,* 147.
29 Ibid., 154.
30 Ibid., 155.
31 Frye, 'Preface to An Uncollected Anthology,' in *The Bush Garden,* 166.
32 Ibid.
33 Louis Dudek, 'Poet of the Machine Age (1958),' in *E.J. Pratt,* ed. David G. Pitt (Toronto: Ryerson Press, 1969), 88.
34 Susan Gingell, ed., *E.J. Pratt on His Life and Poetry* (Toronto: University of Toronto Press, 1983), 136-37.

35 Sandra Djwa and R.G. Moyles, eds., Introduction, *E.J. Pratt: Complete Poems, Part I* (Toronto: University of Toronto Press, 1989), xxiv. *Clay* has been printed as Appendix B, ed. by Susan Gingell, in ibid., 305-57.

36 Sandra Djwa, *E.J. Pratt: The Evolutionary Vision* (Toronto: Copp Clark, 1974), 112. Pratt's views on the role of technology appeared to be influenced by personal developments as well. In 1914, he took part in an English honours reading group at the University of Toronto in which he read works of the Romantics, among them Shelley's *Prometheus Unbound*. Sandra Djwa and R.G. Moyles state that 'Prometheus, that primary romantic rebel and saviour (a Christ figure in turn of the century theology), recurs throughout Pratt's poems.' Prometheus becomes for Pratt not only the firegiver of Greek mythology and thus the originator of technology but is also representative of an atomistic and mechanical worldview. Such a view of the world went against Pratt's religious sensibilities but was in keeping with the Darwinian view of the world so prominent during his young and impressionable years. Equally influential in leading Pratt to the belief that 'the universe was little more than a great uncaring machine from which God, the artificer, had withdrawn' was Newfoundland, the place of his heritage, where 'the vagaries of nature were a daily observance and death a constant.'

37 Gingell, 96-97.

38 James Reaney, '*Towards the Last Spike:* The Treatment of a Western Subject' (1955), in Pitt, ed., *E.J. Pratt,* 75.

39 For a discussion of this technological nationalism, see Maurice Charland, 'Technological Nationalism,' *Canadian Journal of Political and Social Theory* 10, 1-2 (1986): 196-220.

40 Northrop Frye, 'Conclusion to a *Literary History of Canada,*' reprinted in *The Bush Garden,* 223.

41 Frye best explained his meaning of myth and especially the rise of both a myth of concern and a myth of freedom in *The Critical Path:*

> As a culture develops, its mythology tends to become encyclopaedic, expanding into a total myth covering a society's view of its past, present and future, its relation to its gods and its neighbours, its traditions, its social and religious duties, and its ultimate destiny ... A fully developed or encyclopaedic myth comprises everything that it most concerns its society to know, and I shall therefore speak of it as a mythology of concern, or more briefly as a myth of concern.
>
> The myth of concern exists to hold society together, so far as words can help to do this ... [Ultimately] a myth of concern develops different social, political, legal, and literary branches, and at this stage religion becomes more exclusively the myth of what [Paul] Tillich calls ultimate concern, the myth of man's relation to other worlds, other beings, other lives, other dimensions of time and space. For a long time this 'ultimate' aspect of religion remains in the centre of the total myth of concern. The myth of concern which European and American culture has inherited is, of course, the Judaeo-Christian myth as set out in the Bible, and as taught in the form of doctrine by the Christian Church. (36-37)

> The normal tendency of the truth of correspondence is non-mythical, appealing not directly to concern but to more self-validating criteria, such as logicality of argument or (usually a later stage) impersonal evidence and verification. The mental attitudes it develops, however, which include objectivity, suspension of judgment, tolerance, and respect for the individual, become social attitudes as well, and consolidate around a central relationship to society. The verbal expression of concern for these attitudes I shall call the myth of freedom. The myth of freedom is part of the myth of concern, but is a part that stresses the importance of the non-mythical elements in culture, of the truth and realities that are studied rather than created, provided by nature rather than by a social vision. It thus extends to the safeguarding of certain social values not directly connected with the myth of concern, such as the tolerance of opinion which dissents from it ... The myth of freedom thus constitutes the 'liberal' element in society, as the myth of concern constitutes the conservative one. (44-45)

42 Frye explained the difference between an open and a closed mythology as follows:

> A liberal or "open" mythology ... [is] a structure, but it is often so fluid that the solid metaphor of structure hardly applies to it at all. Each man has his own version of it, conditioned by what he knows best, and in fact he will probably adopt several differing versions in the course of his life. Myths are seldom if ever actual hypotheses that can be verified or refuted; that is not their function: they are co-ordinating or integrating ideas ... No idea is anything more than a half-truth unless it contains its own opposite, and is expanded by its own denial or qualifications.
>
> An open mythology of this kind is very different from a closed one, which is a structure of belief ... A closed mythology ... requires the statement of theoretical belief from everyone, and imposes a discipline that will make practice consistent with it. Thus the closed mythology is a statement both of what is believed to be true and of what is going to be made true by a certain course of action ...
>
> A closed mythology forms a body of major premises which is superior in authority to scholarship and art. A closed myth already contains all the answers, at least potentially: whatever scholarship or art produce has to be treated deductively, as reconcilable with the mythology, or, if irreconcilable, suppressed.

Northrop Frye, *The Modern Century* (Toronto: Oxford University Press, 1967), 115-17 passim.

43 Northrop Frye, *Fables of Identity: Studies in Poetic Mythology* (New York: Harcourt, 1963), 162. In *The Critical Path,* Frye again showed the link between the scientific-technological view and literature, especially poetry:

> The tradition associating the poet with an existential protest against the increased mechanizing of life goes back in English literature to Blake, with his 'dark Satanic mills' and his vision of society as a machine
> ... wheel without wheel, with cogs tyrannic
> Moving by compulsion on each other
> Blake links this growth of tyranny with Lockian philosophy and Newtonian science. For an attitude which begins in the detached study of nature or an objective world is very apt to end in a conviction that moral law is, or ought to be, as predictable as natural law. One can trace similar attitudes in Milton, what with the technological interests he give[s] to Satan, and his very ambiguous treatment of Galileo in *Paradise Lost* ... And just as Matthew Arnold is the last of the great humanists, so the turning point toward the present anti-technological attitude of the artist is marked by Ruskin, who also shows a profound interest in the cultural importance of mythology. (91-92)

44 Northrop Frye, *The Educated Imagination,* Massey Lectures: Second Series (Toronto: CBC, 1963), 67-68.

45 Northrop Frye, *The Modern Century,* Whidden Lectures (Toronto: Oxford University Press, 1967), 9.

46 Ibid., 20.

47 Ibid., 31, 24, 30.

48 Ibid., 30-31.

49 Ibid., 38.

50 Ibid., 59-60, 65, 66.

51 Ibid., 106.

52 Ibid., 120-21.

53 Ibid., 122-23.

54 Northrop Frye, 'Conclusion to a *Literary History of Canada,*' in *The Bush Garden,* 217.

55 Northrop Frye, 'Conclusion to *Literary History of Canada,* Second Edition,' in *Divisions on a Ground: Essays on Canadian Culture* (Toronto: Anansi, 1982), 76.

56 Ibid.

57 Ibid., 77.

58 Ibid., 77-78.

59 Northrop Frye, 'Culture as Interpenetration,' in *Divisions on a Ground,* 20.
60 Ibid.
61 Both of these examples are mentioned by Frye in the film *Journey without Arrival.*
62 Northrop Frye, 'Humanities in a New World,' in *Divisions on a Ground,* 105.
63 Cayley, 183-84.
64 Northrop Frye, *Words with Power: Being a Second Study of 'The Bible and Literature'* (Toronto: Viking, 1990), 295.
65 In *The Modern Century,* Frye notes, 'Man is not very good at the creating business; he is much better at destroying, for most of him, like an iceberg, is submerged in a destructive element' (121).
66 Frye, *Words with Power,* 309.
67 Ibid., 312-13.

Chapter 10: George Grant and Dennis Lee

1 George Grant, 'A Platitude,' in *Technology and Empire: Perspectives on North America* (Toronto: Anansi, 1969), 137.
2 Gad Horowitz and George Grant, 'A Conversation on Technology and Man,' *Journal of Canadian Studies* 4, 3 (1969): 3.
3 Grant, 'A Platitude,' 137.
4 Most of the biographical information is from William Christian's fine book, *George Grant: A Biography* (Toronto: University of Toronto Press, 1993), unless noted otherwise.
5 David Cayley, *George Grant in Conversation* (Toronto: Anansi, 1995), 46.
6 Christian, 8.
7 Ibid., 7.
8 Cayley, 45-46. Original italics.
9 Christian, 62.
10 Quoted in ibid., 64.
11 Cayley, 49.
12 Quoted in Christian, 103.
13 Quoted in ibid., 103-4.
14 Ibid., 166.
15 Quoted in ibid.
16 Quoted in ibid., 168.
17 Quoted in ibid., 166.
18 George Grant, 'The Empire: Yes or No,' in *The George Grant Reader,* ed. William Christian and Sheila Grant (Toronto: University of Toronto Press, 1998), 49-50.
19 Christian, 187.
20 Quoted in ibid., 169.
21 Quoted in ibid., 170.
22 George Grant, *Philosophy in the Mass Age* (Toronto: Copp Clark, 1966), 4.
23 Ibid.
24 Ibid., 5.
25 Ibid., 5-6.
26 For a comparison of Harold Innis's and George Grant's views on empires and technology, see R. Douglas Francis, 'Technology and Empire: The Ideas of Harold Innis and George P. Grant,' in *Canada and the End of Empire,* ed. Phillip Buckner (Vancouver: UBC Press, 2004), 409-29.
27 Grant, *Philosophy in the Mass Age,* 9, 10-11.
28 Ibid., 40.
29 Ibid., 45, 49.
30 Ibid., 51.
31 Ibid., 83. Later, in a series of interviews with David Cayley for CBC, Grant stated more emphatically the importance of Weber's book *Protestantism and the Spirit of Capitalism:* 'It really shows you why there is something frenetic in Protestantism – to get things done and to control the world.' Cayley, 138.
32 Ibid., 84.

33 George Grant, 'An Ethic of Community,' in *Social Purpose for Canada,* ed. Michael Oliver (Toronto: University of Toronto Press, 1961), 26.
34 George Grant, 'Tyranny and Wisdom,' in *Technology and Empire,* 92.
35 Grant, *Philosophy in the Mass Age,* ix.
36 Ibid.
37 Cayley, 133.
38 Grant, *Philosophy in the Mass Age,* iii.
39 Ibid., iv.
40 Ibid., vi, vii-viii.
41 Quoted in Christian, 245.
42 Cayley, 4.
43 George Grant, *Lament for a Nation: The Defeat of Canadian Nationalism* (Toronto: McClelland and Stewart, 1965), 1.
44 Ibid., 68.
45 Ibid., 94.
46 Ibid., 96.
47 Ibid., 73.
48 Ibid.
49 George Grant, 'Canadian Fate and Imperialism,' in *Technology and Empire,* 78.
50 George Grant, 'In Defence of North America,' in *Technology and Empire,* 16. It is interesting to note that, when Grant uses the word 'civilization' in the context of Western civilization he spells it with a 'z,' but when he uses it in the context of American civilisation he spells it with an 's.'
51 Ibid., 17, 17-18, 19.
52 Ibid., 20.
53 Cayley, 141.
54 Grant, 'In Defence of North America,' 25.
55 Ibid., 27.
56 Ibid., 28.
57 Ibid., 32, 33, 34.
58 Ibid., 35.
59 Ibid., 40.
60 Jacques Ellul, *The Technological Society* [1954] (New York: Knopf, 1964), xxv. Original italics.
61 Horowitz and Grant, 3.
62 Grant, 'A Platitude,' 137.
63 Ibid., 138, 141, 143. In his reference to 'the dynamo,' Grant may have been harkening back to an earlier critic of the technological age in the context of the moral values lost as machines took over, namely, Henry Adams, *The Education of Henry Adams: An Autobiography,* particularly Chapter 25, 'The Dynamo and the Virgin' [1900] (Boston: Houghton Mifflin, 1918), 379-90.
64 Dennis Lee, *Civil Elegies and Other Poems* (Toronto: Anansi, 1972). All quotations are from this edition.
65 Quoted in R.D. MacDonald, 'Lee's "Civil Elegies" in Relation to Grant's "Lament for a Nation,"' *Canadian Literature* 98 (Autumn 1983): 11.
66 Lee, 36.
67 Ibid., 33.
68 Ibid., 40, 56.
69 Ibid., 34.
70 Ibid., 38.
71 Ibid., 47.
72 Ibid., 48.
73 Ibid., 57.
74 George Grant, *Time as History,* Massey Lectures (Toronto: CBC, 1969), 2.
75 Ibid., 4.

76 Ibid., 44.
77 George Grant, *Technology and Justice* (Toronto: Anansi, 1986), 12.
78 Christian, 301.
79 George Grant, *English-Speaking Justice* [1974] (Toronto: Anansi, 1985), 26. All subsequent quotations are from this edition.
80 Ibid., 10.
81 Ibid., 70, 82.
82 Ibid., 86.
83 Grant had an ambivalent attitude toward Heidegger. This is evident in an interesting episode that David Cayley recounts in his Introduction in *George Grant in Conversation:* 'At the time I visited him he had a small photograph of Heidegger on the mantelpiece, and I noticed one morning that it was facing the wall rather than the room, as it had been the previous day. When I asked him about this, he said that he changed the orientation depending on how he was disposed to "the old bastard" that day' (25-26).
84 Grant, *Technology and Justice,* 20.
85 Ibid., 21.
86 Ibid., 23. Original italics.
87 Ibid., 23, 25, 26.
88 Ibid., 29.
89 Ibid., 32.
90 Ibid.
91 Cayley, 187.
92 Grant, 'Faith and the Multiversity,' in *Technology and Justice,* 38.
93 Ibid., 36.
94 Ibid., 50, 51.
95 Ibid., 60, 61, 66, 70.
96 Ibid., 'Appendix,' 76-77.
97 Christian, 361.

Conclusion

1 Ursula Franklin, *The Real World of Technology* [1992] (Toronto: Anansi, 1999), 6. Original italics. All subsequent quotations from this work are from the 1999 edition.
2 Ibid., 25. Original italics.
3 Ibid., 85.
4 Ibid., 125.
5 Ursula Martius Franklin, "Will Women Change Technology or Will Technology Change Women?" in *Knowledge Reconsidered: A Feminist Overview,* ed. Ursula Martius Franklin et al. (Toronto: Canadian Research Institute for the Advancement of Women, 1984), 84.
6 Ibid., 88, 89.
7 Heather Menzies, *Computers on the Job: Surviving Canada's Microcomputer Revolution* (Toronto: James Lorimer, 1982), x.
8 Ibid.
9 Heather Menzies, 'In His Image: Science and Technology,' in *Twist and Shout: A Decade of Feminist Writing in* This Magazine, ed. Susan Crean (Toronto: Second Story Press, 1992), 164-65.
10 Ibid., 164.
11 Heather Menzies, *Fast Forward and Out of Control: How Technology Is Changing Your Life* (Toronto: Macmillan, 1989), 243.
12 Mark Kingwell, *Dreams of Millennium: Report from a Culture on the Brink* (Boston: Faber and Faber, 1996), 14.
13 Ibid., 137.
14 Ibid., 139.
15 Ibid., 153, 154-55.
16 Mark Kingwell, *The World We Want: Virtue, Vice, and the Good Citizen* (Toronto: Penguin Books, 2000), 186.

17 Derrick de Kerckhove, 'Critical Brain Processes Involved in Deciphering the Greek Alphabet,' in *The Alphabet and the Brain: The Lateralization of Writing,* ed. Derrick de Kerckhove and Charles J. Lumsden (Berlin: Springer-Verlag, 1988), 417.
18 Derrick de Kerckhove, *The Architecture of Intelligence* (Basil: Birkhauser Publisher for Architecture, 2001), 22.
19 Ibid., 23.
20 Derrick de Kerckhove, *Connected Intelligence: The Arrival of the Web Society,* ed. Wade Rowland (Toronto: Somerville House, 1997), xxx-xxxi.
21 Ibid., 54-55.
22 Arthur Kroker, *Technology and the Canadian Mind: Innis/McLuhan/Grant* (Montreal: New World Perspectives, 1984), 128. Original italics.
23 Ibid., 125. Original italics.
24 Ibid., 127.
25 Arthur Kroker and Michael A. Weinstein, *Data Trash: The Theory of the Virtual Class* (New York: St. Martin's Press, 1994), 2.
26 Ibid., 143.

Selected Bibliography

Primary Sources

Archival Sources

Alexander Graham Bell National Historic Site (Baddeck, Nova Scotia)
Alexander Graham Bell Papers

Library and Archives Canada (Ottawa)
Archibald Edward Malloch Papers
Donald G. Creighton Papers
Elspeth Chisholm Papers
George Parkin Grant Papers
Lawren Harris Papers
Marshall McLuhan Papers
Mrs. W.L. Grant Papers
Stephen Leacock Papers
Sandford Fleming Papers
T.C. Keefer Papers
William Lyon Mackenzie King Papers

McGill University Archives (Montreal)
Henry Taylor Bovey Papers

University of Guelph Archival and Special Collections (Guelph)
Adelaide Hoodless Papers

University of Toronto Archives (Toronto)
C.R. Young Papers
Daniel Wilson Papers
Harold A. Innis Papers
James Loudon Papers
John Galbraith Papers
R.A. Falconer Papers

Books and Articles

Adams, Henry. *The Education of Henry Adams: An Autobiography*. Boston: Houghton Mifflin Company, 1918.

Bell, Alexander Graham. 'The Substance of My Latest Research.' Empire Club of Canada. *Addresses* (1917-18): 15-16.

Bentley, D.M.R., ed. *The Essays and Reviews of Archibald Lampman*. London, ON: Canadian Poetry Press, 1996.
Bovey, Henry T. 'The Fundamental Conceptions Which Enter into Technology.' *McGill University Magazine* 4, 1 (1905): 35-51.
–. 'Presidential Address.' *Transactions of the Royal Society of Canada* 2, 3 (1896): 3-24.
Brett, George Sidney. *A History of Psychology*. 3 vols. London: G. Allen, 1912-21.
–. 'The History of Science as a Factor in Modern Education.' *Transactions of the Royal Society of Canada* 29, 2 (1925): 39-46.
–. 'Makers of Science.' *University of Toronto Quarterly* 5, 4 (1936): 605-11.
–. 'The Modern Mind and Modernism.' *Canadian Journal of Religious Thought* 5, 2 (1928): 91-104.
–. 'The Revolt against Reason: A Contribution to the History of Thought.' *Transactions of the Royal Society of Canada* 13, 2 (1919): 9-17.
Butler, Samuel. *Erewhon*. [1872]. New York: Penguin Books, 1970.
Cappon, James. 'In Memoriam: Nathaniel Fellowes Dupuis.' *Queen's Quarterly* 25, 2 (1917): 125-40.
Cayley, David, ed. *George Grant in Conversation*. Toronto: Anansi, 1995.
–, ed. *Northrop Frye in Conversation*. Toronto: Anansi, 1992.
Christian, William, ed. *George Grant: Selected Letters*. Toronto: University of Toronto Press, 1996.
–. *The Idea File of Harold Adams Innis*. Toronto: University of Toronto Press, 1980.
–. *Innis on Russia: The Russian Diary and Other Writings*. Toronto: Harold Innis Foundation, Innis College, University of Toronto, 1981.
Christian, William, and Sheila Grant, eds. *George Grant Reader*. Toronto: University of Toronto Press, 1998.
Cunningham, W. *Christianity and Social Questions*. New York, 1910.
–. *The Growth of English Industry and Commerce in Modern Times*. Cambridge: Cambridge University Press, 1982.
–. *Outlines of English Industrial History*. Cambridge: Cambridge University Press, 1895.
De Kerckhove, Derrick. *The Architecture of Intelligence*. Basil, Switzerland: Birkhauser–Publisher for Architecture, 2001.
–. *Connected Intelligence: The Arrival of the Web Society*. Ed. Wade Rowland. Toronto: Somerville House, 1997.
–. 'Critical Brain Processes Involved in Deciphering the Greek Alphabet.' In *The Alphabet and the Brain: The Lateralization of Writing*. Ed. Derrick de Kerckhove and Charles J. Lumsden, 401-21. Berlin: Springer-Verlag, 1988.
–. 'Logical Principles Underlying the Layout of Greek Orthography.' In *The Alphabet and the Brain: The Lateralization of Writing*. Ed. Derrick de Kerckhove and Charles J. Lumsden, 153-72. Berlin: Springer-Verlag, 1988.
–. *The Skin of Culture: Investigating the New Electronic Reality*. Ed. Christopher Dewdney. Toronto: Somerville House, 1995.
Djwa, Sandra, and R.G. Moyles, eds. *E.J. Pratt: Complete Poems, Part I*. Toronto: University of Toronto Press, 1989.
Drache, Daniel, ed. *Staples, Markets, and Cultural Change: Selected Essays [by] Harold A. Innis*. Montreal and Kingston: McGill-Queen's University Press, 1995.
Dupuis, Nathan Fellowes. *An Address Delivered at the Opening of the Thirty-First Session of Queen's College, October 2, 1872*. Kingston, 1872.
–. 'Some of the Factors of Modern Civilization.' *Queen's Quarterly* 4, 1 (1896): 45-55.
Ellul, Jacques. *Technological Society*. Trans. John Wilkinson. New York: Vintage Books, 1964.
Fleming, Sandford. *Canada and Its Vast Undeveloped Interior*. N.p., 1878.
–. *Canada and Ocean Highways*. [London?], 1896.
–. 'Expeditions to the Pacific.' *Transactions of the Royal Society of Canada* 8, 2 (1889): 89-141.
–. 'Nomenclature in Time-Reckoning.' *Transactions of the Royal Society of Canada* 9, 3 (1891): 19-25.
–. 'Our Empire Cables.' *Empire Club Speeches* 1 (1903-04): 84-94.

–. 'The Pacific Cable.' *Queen's Quarterly* 5, 3 (1898): 225-35.
–. 'Presidential Address: The Unit Measure of Time.' *Transactions of the Royal Society of Canada* 8, 3 (1890): 3-6.
–. 'A Problem in Political Science.' *Transactions of the Royal Society of Canada* 7, 3 (1889): 33-40.
–. *Terrestrial Time: A Memoir*. [London?], 1876.
Franklin, Ursula M. *The Real World of Technology*. [1992]. Rev. ed. Toronto: Anansi, 1999.
–. 'Will Women Change Technology or Will Technology Change Women?' In *Knowledge Reconsidered: A Feminist Overview*. Ed. Ursula M. Franklin et al., 81-90. Ottawa: Canadian Research Institute for the Advancement of Women, 1984.
Frye, Northrop. *Anatomy of Criticism: Four Essays*. Princeton: Princeton University Press, 1957.
–. 'Conclusion to a *Literary History of Canada*.' In *The Bush Garden: Essays on the Canadian Imagination*. Ed. Northrop Frye, 213-52. Toronto: Anansi, 1971.
–. 'Conclusion to a *Literary History of Canada*.' 2nd ed. In *Divisions on a Ground: Essays on Canadian Culture*. Ed. Northrop Frye, 71-88. Toronto: Anansi, 1982.
–. *Creation and Recreation*. Toronto: University of Toronto Press, 1980.
–. *The Critical Path: An Essay on the Social Context of Literary Criticism*. Toronto: Harvester Press, 1983.
–. *The Double Vision: Language and Meaning in Religion*. Toronto: University of Toronto Press, 1991.
–. *The Educated Imagination*. Massey Lectures: Second Series. Toronto: CBC, 1963.
–. *Fables of Identity: Studies in Poetic Mythology*. New York: Harcourt, 1963.
–. *The Great Code: The Bible and Literature*. Toronto: Academic Press, 1982.
–. *The Modern Century*. The Whidden Lectures. Toronto: Oxford University Press, 1967.
–. *Words with Power: Being a Second Study of 'The Bible and Literature.'* Toronto: Viking, 1990.
Frye, Northrop, ed. *The Bush Garden: Essays on the Canadian Imagination*. Toronto: Anansi, 1971.
–. *Division on a Ground: Essays on a Canadian Culture*. Toronto: Anansi, 1982.
Frye, Northrop, narrator. *Journey without Arrival*. Toronto: National Film Board, 1964.
Galbraith, John. 'The Function of the School of Applied Science in the Education of the Engineer.' *University of Toronto Monthly* 1, 5 (1901): 150-57.
–. *In the New Capital; or, The City of Ottawa in 1999*. [1897]. Introduction by R. Douglas Francis. Ottawa: Penumbra Press, 2000.
–. *Technical Education: Address Delivered by Professor Galbraith at the Opening of the Engineering Laboratory of the School of Practical Science, Toronto, February 24, 1892*. Toronto, 1892.
Galbraith, John Kenneth. *Affluent Society*. Boston: Houghton Mifflin, 1960.
–. *The New Industrial State*. 3rd ed. Boston: Houghton Mifflin, 1978.
–. *The Socially Concerned Today*. Toronto: University of Toronto Press, 1998.
Giedion, Siegfried. *Mechanization Takes Command: A Contribution to Anonymous History*. New York: Oxford University Press, 1948.
Gill, L.W. 'Wireless Telegraphy.' *Queen's Quarterly* 10, 3 (1903): 268-73.
Gingell, Susan, ed. *E.J. Pratt on His Life and Poetry*. Toronto: University of Toronto Press, 1983.
Grant, George. *Collected Works of George Grant*. 3 vols. Toronto: University of Toronto Press, 2005.
–. *English-Speaking Justice*. Toronto: Anansi, 1974.
–. 'An Ethic of Community.' In *Social Purpose for Canada*. Ed. Michael Oliver, 3-26. Toronto: University of Toronto Press, 1961.
–. 'Knowing and Making.' *Transactions of the Royal Society of Canada* 4, 12 (1974): 59-67.
–. *Lament for a Nation: The Defeat of Canadian Nationalism*. Toronto: McClelland and Stewart, 1965.
–. *Philosophy in the Mass Age*. Toronto: Copp Clark, 1966.
–. *Technology and Empire: Perspectives on North America*. Toronto: Anansi, 1969.
–. *Technology and Justice*. Toronto: Anansi, 1986.
–. *Time as History*. Massey Lectures. Toronto: CBC, 1969.

–. 'The Uses of Freedom: A Word and Our World.' *Queen's Quarterly* 62, 4 (1955-56): 515-27.

Grove, Frederick Philip. 'Apologia Pro Vita Et Opere Suo.' *Canadian Forum* 11, 131 (1931): 420-22.

–. *In Search of Myself.* [1946]. Reprinted with an Introduction by D.O. Spettigue. New Canadian Library No. 94. Toronto: McClelland and Stewart, 1974.

–. *It Needs to Be Said.* [1929]. Introduction by W.J. Keith. Ottawa: Tecumseh Press, 1982.

–. *The Master of the Mill.* [1944]. Introduction by R.E. Walters. New Canadian Library No. 19. Toronto: McClelland and Stewart, 1967.

Guthrie, Norman Gregor, ed. *The Poetry of Archibald Lampman.* Toronto: Musson Books, 1927.

Haliburton, T.C. *The Letter-Bag of the Great Western, or, Life in a Steamer.* [1840]. New York: Routledge, 1987.

–. *The Season Ticket.* [1860]. Toronto: University of Toronto Press, 1973.

Havelock, Eric A. 'The Crucifixion of Intellectual Man.' Introduction to *Prometheus, with a Translation of Aeschylus'* Prometheus Bound [1951]. Seattle: University of Washington Press, 1968, 3-109.

–. *Harold A. Innis: A Memoir.* Preface by H. Marshall McLuhan. Toronto: Harold Innis Foundation, Innis College, University of Toronto, 1982.

–. *The Liberal Temper in Greek Thought.* New Haven: Yale University Press, 1957.

–. *The Literate Revolution in Greece and Its Cultural Consequences.* Princeton: Princeton University Press, 1982.

–. *The Muse Learns to Write: Reflections on Orality and Literacy from Antiquity to the Present.* New Haven: Yale University Press, 1986.

Heidegger, Martin. *The Question Concerning Technology and Other Essays.* Trans. with an Introduction by William Lovitt. New York: Harper and Row, 1977.

Horowitz, Gad, and George Grant. 'A Conversation on Technology and Man.' *Journal of Canadian Studies* 4, 3 (1969): 3-6.

Hume, J.G. *Political Economy and Ethics.* Toronto: J.E. Bryant, 1892.

Innis, Harold A. *The Bias of Communication.* [1951]. Introduction by Paul Heyer and David Crowley. Toronto: University of Toronto Press, 1991.

–. *The Cod Fisheries: The History of an International Economy.* [1940]. Rev. ed. Toronto: University of Toronto Press, 1954.

–. *Empire and Communications.* [1950]. Rev. ed. Foreword by Marshall McLuhan. Toronto: University of Toronto Press, 1972.

–. *Essays in Canadian Economic History.* Ed. Mary Q. Innis. Toronto: University of Toronto Press, 1956.

–. *The Fur Trade in Canada: An Introduction to Canadian Economic History.* [1930]. Rev. ed. Toronto: University of Toronto Press, 1979.

–. *A History of the Canadian Pacific Railway.* [1923]. Foreword by Peter George. Toronto: University of Toronto Press, 1971.

–. *Political Economy in the Modern State.* Toronto: Ryerson, 1946.

–. *Settlement in the Mining Frontier.* Canadian Frontiers of Settlement Series, Vol. 9. Toronto: Macmillan, 1936.

–. *The Strategy of Culture.* Toronto: University of Toronto Press, 1952.

–. 'This Has Killed That: An Unpublished Paper.' *Journal of Canadian Studies* 12, 5 (1977): 3-5.

–. 'A Trip through the Mackenzie River Basin.' *University of Toronto Monthly* 25, 4 (1925): 151-53.

–. 'The Work of Thorstein Veblen.' In *Essays in Canadian Economic History.* Ed. Mary Q. Innis, 17-26. Toronto: University of Toronto Press, 1956.

Innis, Harold A., ed. *Select Documents in Canadian Economic History, 1497-1783.* Toronto: University of Toronto Press, 1929.

James, William. *Pragmatism.* New York: Longmans, Green, 1907.

–. *Talks to Teachers on Psychology and to Students on Some of Life's Ideals.* New York: Henry Holt, 1899.

Keefer, T.C. 'The Canals of Canada.' *Transactions of the Royal Society of Canada* 2, 3 (1893): 25-50.

–. *Handbook and Official Catalogue of the Canadian Section, Paris Universal Exhibition, 1878*. London, 1878.

–. *Philosophy of Railroads and Other Essays*. Ed. H.V. Nelles. Toronto: University of Toronto Press, 1972.

–. 'Presidential Address: Canadian Water Power and Its Electrical Product in Relation to the Undeveloped Resources of the Dominion.' *Transactions of the Royal Society of Canada* 2, 5 (1899): 3-40.

–. 'President's Address to Canadian Society of Civil Engineers,' January 12, 1880. *Transactions of the Society* 2, 1 (1888): 3-36.

–. 'Travel and Transportation.' In *Eighty Years' Progress of British North America, Showing the Wonderful Development of Its Natural Resources, by the Unbounded Energy and Enterprise of Its Inhabitants*. Ed. Henry Youle Hind, 99-265. Toronto: L. Stebbins, 1864.

King, William Lyon Mackenzie. 'The Four Parties to Industry.' Empire Club of Canada. *Addresses* 17 (1919): 160-80.

–. *Industry and Humanity: A Study in the Principles Underlying Industrial Reconstruction*. [1918]. Introduction by David Jay Bercuson. Toronto: University of Toronto Press, 1973.

–. *The Secret of Heroism: A Memoir of Henry Albert Harper*. [1906]. Toronto: Hunter-Ross, 1919.

Kingwell, Mark. *Better Living: Report from a Culture on the Brink*. Boston: Faber and Faber, 1996.

–. *Dreams of Millennium: Report from a Culture on the Brink*. Boston: Faber and Faber, 1996.

–. *Practical Judgments: Essays in Culture, Politics, and Interpretation*. Toronto: University of Toronto Press, 2002.

–. *The World We Want: Restoring Citizenship in a Fractured Age*. Boston: Rowman and Littlefield, 2000.

Kingwell, Mark, ed. *Marginalia: A Cultural Reader*. Toronto: Penguin Books, 1999.

Kroker, Arthur, and Michael A. Weinstein. *Data Trash: The Theory of the Virtual Class*. New York: St. Martin's Press, 1994.

Lampman, Archibald. 'At the Ferry,' 'The City,' and 'The Railway Station.' In *The Poems of Archibald Lampman (Including 'At the Long Sault')*. Ed. Margaret Coulby Whitridge, 150-53; 116; 118. Toronto: University of Toronto Press, 1974.

–. 'The City of the End of Things,' 'The Land of Pallas,' and 'To a Millionaire.' In *The Poems of Archibald Lampman*. 3rd ed. Ed. Duncan Campbell Scott, 179-82; 201-10; 276-77. Toronto: Morang and Co., 1905.

–. 'Socialism.' In *The Essays and Reviews of Archibald Lampman*. Ed. D.M.R. Bentley, 186-90. London, ON: Canadian Poetry Press, 1996.

Leacock, Stephen. 'The Apology of a Professor: An Essay on Modern Learning.' [1910]. In *The Social Criticism of Stephen Leacock*. Ed. Alan Bowker, 28-39. Toronto: University of Toronto Press, 1973.

–. *Arcadian Adventures with the Idle Rich*. [1914]. Toronto: McClelland and Stewart, 1959.

–. *The Boy I Left behind Me*. London: Bodley Head, 1947.

–. *Canada: The Foundations of Its Future*. Montreal: Privately printed, 1941.

–. 'Democracy and Social Progress' and 'Our National Organization for the War.' In *The New Era in Canada: Essays Dealing with the Building of the Canadian Commonwealth*. Ed. J.O. Miller, 13-33, 409-21. Toronto: J.M. Dent, 1917.

–. *Elements of Political Science*. Boston: Houghton Mifflin, 1921.

–. *Last Leaves*. Toronto: McClelland and Stewart, 1945.

–. 'Literature and Education in America.' [1909]. In *The Social Criticism of Stephen Leacock*. Ed. Alan Bowker, 13-26. Toronto: University of Toronto Press, 1973.

–. 'The Man in Asbestos: An Allegory of the Future.' *Nonsense Novels,* 159-76. New York: John Lane Company, 1921.

–. *My Discovery of the West: A Discussion of East and West in Canada*. Toronto: Allen, 1937.

–. *My Recollection of Chicago* and *The Doctrine of Laissez-Faire*. Introduction by Carl Spadoni. Toronto: University of Toronto Press, 1998.

–. *Nonsense Novels*. New York: John Lane, 1921.

–. *Our Heritage of Liberty: Its Origins, Its Achievement, Its Crisis: A Book for War Time*. London: Bodley Head, 1942.
–. 'The Riddle of the Depression.' Empire Club of Canada. *Addresses* 31 (1933-34): 70-83.
–. *Sunshine Sketches of a Little Town*. [1912]. Toronto: McClelland and Stewart, 1960.
–. 'The Unsolved Riddle of Social Justice.' [1920]. In *The Social Criticism of Stephen Leacock*. Ed. Alan Bowker, 71-145. Toronto: University of Toronto Press, 1973.
–. "What Is Left of Adam Smith?" *Canadian Journal of Economics and Political Science* 1, 1 (1935): 41-51.
Le Bon, Gustave. *The Psychology of Peoples: Its Influence on Their Evolution*. London: T. Fisher Unwin, 1899.
Le Play, Frédéric. *On Family, Work, and Social Change*. Ed. and trans. Catherine Bodard Silver. Chicago: University of Chicago Press, 1982.
Lee, Dennis. *Civil Elegies and Other Poems*. Toronto: Anansi, 1972.
Loudon, James. 'The Universities in Relation to Research.' *University of Toronto Monthly* 2, 9 (1902): 234-44.
MacCallum, A.B. 'The Old Knowledge and the New.' Presidential Address. *Proceedings and Transactions of the Royal Society of Canada* 3, 11 (1917): Appendix A, 59-73.
Mackenzie, A. Stanley. 'The War and Science.' Presidential Address. *Transactions of the Royal Society of Canada* 3, 12 (1918): 1-6.
Marx, Karl. *Critique of Political Economy*. Vol. 1. [1867]. New York: Random House, 1976.
–. *Grundrisse: Foundations of a Critique of Political Economy*. [1857-58]. Trans. M. Nicolaus. London: Pelican Books. 1973.
Marx, Karl, and Frederick Engels. *The Communist Manifesto*. New York: Pathfinder Press, 1987.
McLuhan, Eric, and Frank Zingrone, eds. *Essential McLuhan*. Toronto: Anansi, 1995.
McLuhan, Marshall. *Counterblast*. Toronto: McClelland and Stewart, 1969.
–. 'The Future of Man in the Electric Age.' [1965]. In *Understanding Me: Lectures and Interviews*. Ed. Stephanie McLuhan and David Staines, 56-75. Toronto: McClelland and Stewart, 2003.
–. *The Gutenberg Galaxy: The Making of Typographic Man*. Toronto: University of Toronto Press, 1962.
–. *The Mechanical Bride: The Folklore of Industrial Man*. New York: Vanguard Press, 1951.
–. *Understanding Media: The Extensions of Man*. New York: McGraw-Hill, 1965.
–. Introduction. In *The Bias of Communication*. By Harold Innis, vii-xvi. Toronto: University of Toronto Press, 1964.
–. 'The Later Innis.' *Queen's Quarterly* 60, 3 (1953): 385-94.
McLuhan, Marshall, with Quentin Fiore. *The Medium Is the Massage: An Invention of Effects*. New York: Bantam Books, 1967.
McLuhan, Marshall, and Eric McLuhan. *Laws of Media: The New Science*. Toronto: University of Toronto Press, 1988.
McLuhan, Marshall, and Bruce R. Powers. *The Global Village: Transformations in World Life and Media in the 21st Century*. Oxford: Oxford University Press, 1989.
Menzies, Heather. *Computers on the Job: Surviving Canada's Microcomputer Revolution*. Toronto: James Lorimer, 1982.
–. *Fast Forward and Out of Control: How Technology Is Changing Your Life*. Toronto: Macmillan, 1989.
–. 'In His Image: Science and Technology.' In *Twist and Shout: A Decade of Feminist Writing in* This Magazine. Ed. Susan Crean, 156-65. Toronto: Second Story Press, 1992.
–. *No Time: Stress and the Crisis of Modern Life*. Vancouver: Douglas and McIntyre, 2005.
–. *Whose Brave New World? The Information Highway and the New Economy*. Toronto: Between the Lines, 1996.
–. *Women and the Chip: Case Studies of the Effects of Informatics on Employment in Canada*. Montreal: Institute for Research on Public Policy, 1981.
Mumford, Lewis. *The Myth of the Machine: Technics and Human Development*. New York: Harcourt, 1966.
–. *Technics and Civilization*. New York: Harcourt, 1934.
Murray, J. Clark. 'Pragmatism.' *University Magazine* 14, 1 (1915): 102-14.
Peters, R.S., ed. *Brett's History of Psychology*. London: George Allen and Unwin, 1953.

Pitt, David, G., ed. *E.J. Pratt*. Toronto: Ryerson Press, 1969.

Scott, Duncan Campbell, ed. *The Poems of Archibald Lampman*. 3rd ed. Toronto: Morang, 1905.

Scott, M.O. 'Marconi in Canada.' *Canadian Magazine* 18 (1902): 338-40.

Shelley, Mary. *Frankenstein, or, The Modern Prometheus*. [1818]. New York: Penguin Books, 1985.

Slick, Sam [T.C. Haliburton]. *The Clockmaker, Series One* [1836]. Repr. *The Clockmaker, Series One, Two, and Three*. Ed. George L. Parker. Ottawa: Carleton University Press, 1995.

Thoreau, Henry David. *Walden*. [1854]. Repr. and abrid. Ed. J. Lyndon Shanley. Princeton: Princeton University Press, 1971.

Toynbee, Arnold. *Lectures on the Industrial Revolution of the Eighteenth Century in England*. [1884]. London: Longmans, Green, 1928.

Veblen, Thorstein. *The Theory of the Leisure Class*. [1899]. New York: Dover Publications, 1994.

Whitridge, Margaret Coulby, ed. *The Poems of Archibald Lampman (Including 'At the Long Sault')*. Toronto: University of Toronto Press, 1974.

Wiener, Norbert. *Cybernetics, or, Control and Communication in the Animal and the Machine*. New York: Technology Press, 1948.

–. *Human Use of Human Beings: Cybernetics and Society*. 2nd ed. New York: Doubleday, 1954.

Secondary Sources

Adas, Michael. *Machines as the Measure of Men: Science, Technology, and Ideologies of Western Dominance*. Ithaca, NY: Cornell University Press, 1989.

Akin, William E. *Technocracy and the American Dream: The Technocrat Movement, 1900-1941*. Berkeley: University of California Press, 1977.

Ayre, John. *Northrop Frye: A Biography*. Toronto: Random House, 1989.

Babe, Robert E. *Canadian Communication Thought: Ten Foundational Writers*. Toronto: University of Toronto Press, 2000.

Baragar, Fletcher. 'Influence of Veblen on Harold Innis.' *Journal of Economic Issues* 30 (1996): 667-83.

Barber, William J. 'Political Economy in an Atmosphere of Academic Entrepreneurship: The University of Chicago.' In *Breaking the Academic Mould: Economists and American Higher Learning in the Nineteenth Century*. Ed. William J. Barber, 3-14. Middletown: Wesleyan University Press.

Barrett, William. *The Illusion of Technique*. New York: Doubleday, 1979.

Beard, Charles A., ed. *Whither Mankind: A Panorama of Modern Civilization*. Westport: Greenwood Press, 1928.

Beattie, Munro. 'Archibald Lampman.' In *Our Living Tradition: Seven Canadians*. Ed. Claude T. Bissell, 63-88. Toronto: University of Toronto Press, 1957.

Berger, Carl. 'The Other Mr. Leacock.' *Canadian Literature* 55 (Winter 1973): 23-40.

–. *Science, God, and Nature in Victorian Canada*. Toronto: University of Toronto Press, 1983.

–. 'Sir Daniel Wilson.' *Dictionary of Canadian Biography*. Vol. 12, 1891-1900. Toronto: University of Toronto Press, 1990, 1109-14.

–. *The Writing of Canadian History: Aspects of English-Canadian Historical Writing since 1900*. 2nd ed. Toronto: University of Toronto Press, 1986.

Berman, Marshall. *All That Is Solid Melts into Air: The Experience of Modernity*. New York: Penguin Books, 1988.

Birney, Earle, 'E.J. Pratt and His Critics.' In *Our Living Tradition: Seven Canadians*. Ed. Claude T. Bissell, 123-47. Toronto: University of Toronto Press, 1957.

Bissell, Claude T. 'Haliburton, Leacock, and the American Humorous Tradition.' *Canadian Literature* 39 (Winter 1969): 5-18.

–. 'Herbert Marshall McLuhan.' In *Marshall McLuhan: The Man and His Message*. Ed. George Sanderson and Frank Macdonald, 5-11. Golden, CO: Fulcrum, 1989.

Blaise, Clark. *Time Lord: The Remarkable Canadian Who Missed His Train and Changed the World*. Toronto: Knopf, 2000.

Bonnett, John. 'Communication, Complexity, and Empire: The Systematic Thought of H.A. Innis.' PhD diss., University of Ottawa, 2001.

Bowker, Alan. Introduction. *The Social Criticism of Stephen Leacock*. Ed. Alan Bowker, ix-xliii. Toronto: University of Toronto Press, 1973.

Box, Ian. 'George Grant and the Embrace of Technology.' *Canadian Journal of Political Science* 15, 3 (1982): 503-15.

Briggs, Asa. 'The Pleasure Telephone: A Chapter in the Prehistory of the Media.' In *The Social Impact of the Telephone*. Ed. Ithiel de Sola Pool, 40-68. Cambridge, MA: MIT Press, 1977.

Brooke, Michael Z. *Le Play: Engineer and Social Scientist*. London: Longman, 1970.

Bruce, Robert V. 'Alexander Graham Bell and the Conquest of Solitude.' In *Technology in America: A History of Individuals and Ideas*. 2nd ed. Ed. Carroll W. Pursell Jr., 105-16. Cambridge, MA: MIT Press, 1990.

Burpee, Lawrence. *Sandford Fleming: Empire Builder*. Toronto: Oxford University Press, 1915.

Carey, James. 'Harold Adams Innis and Marshall McLuhan.' *Antioch Review* 27 (1967): 5-39.

–. 'Innis "in" Chicago.' In *Harold Innis in the New Century*. Ed. Charles R. Acland and William J. Buxton, 81-104. Montreal and Kingston: McGill-Queen's University Press, 1999.

–. 'The Language of Technology: Talk, Text, and Template as Metaphors for Communication.' In *Communication and the Culture of Technology*. Ed. Martin J. Medhurst et al., 19-39. Pullman: Washington State University Press, 1990.

–. 'McLuhan and Mumford: The Roots of Modern Media Analysis.' *Journal of Communication* 31, 3 (1981): 162-78.

–. 'Space, Time, and Communications: A Tribute to Harold Innis.' In *Communication as Culture: Essays on Media and Society*, 142-72. Boston: Unwin Hyman, 1989.

–. 'Technology and Ideology: The Case of the Telegraph.' *Prospects: The Annual of American Cultural Studies* 8 (1983): 303-25.

Carey, James, and John J. Quirk. 'The Mythos of the Electronic Revolution.' *The American Scholar* Part 1, 39, 2 (1970): 219-41; Part 2, 39, 3 (1970): 395-425.

Cashman, Sean Dennis. *America in the Gilded Age: From the Death of Lincoln to the Rise of Theodore Roosevelt*. 3rd ed. New York: New York University Press, 1984.

Channell, David F. *The Vital Machine: A Study of Technology and Organic Life*. New York: Oxford University Press, 1991.

Charland, Maurice. 'Technological Nationalism.' *Canadian Journal of Political and Social Theory* 10, 1-2 (1986): 196-220.

Chittick, V.L.O. *Thomas Chandler Haliburton ('Sam Slick'): A Study in Provincial Toryism*. New York: AMS Press, 1966.

Christian, William. *George Grant: A Biography*. Toronto: Anansi, 1995.

–. 'George Grant and Love: A Comment on Ian Box's "George Grant and the Embrace of Technology."' *Canadian Journal of Political Science* 16, 2 (1983): 349-54.

–. *Harold Innis as Economist and Moralist*. Guelph, ON: Department of Political Science, University of Guelph, 1981.

Christians, Clifford G., and Jay M. Van Hook, eds. *Jacques Ellul: Interpretative Essays*. Urbana: University of Illinois Press, 1981.

Cogswell, Fred. 'Thomas Chandler Haliburton.' *Dictionary of Canadian Biography*. Vol. 9, 1861-1970. Toronto: University of Toronto Press, 1976, 348-57.

Collins, W.E. 'Archibald Lampman.' In *Archibald Lampman*. Ed. Michael Gnarowski, 125-42. Toronto: Ryerson Press, 1970.

Cook, David. *Northrop Frye: A Vision of the New World*. Montreal: New World Perspectives, 1985.

Cook, Ramsay. *The Regenerators: Social Criticism in Late Victorian English Canada*. Toronto: University of Toronto Press, 1985.

–. 'Stephen Leacock and the Age of Plutocracy, 1913-1921.' In *Character and Circumstance: Essays in Honour of Donald Grant Creighton*. Ed. John Moir, 163-81. Toronto: Macmillan, 1970.

Cooper, Barry. 'On Reading *Industry and Humanity*: A Study in the Rhetoric Underlying Liberal Management.' *Journal of Canadian Studies* 13, 4 (1978-79): 28-39.

Cox, Robert W. 'Civilizations: Encounters and Transformations.' *Studies in Political Economy* 47 (1995): 7-31.
Craven, Paul. *An Impartial Umpire: Industrial Relations and the Canadian State 1900-1911.* Toronto: University of Toronto Press, 1980.
Creet, Mario. 'Sir Sandford Fleming.' *Dictionary of Canadian Biography.* Vol. 14, 1911-1920. Toronto: University of Toronto Press, 1998, 359-62.
Creighton, Donald. *Harold Adams Innis: Portrait of a Scholar.* [1957]. Toronto: University of Toronto Press, 1978.
Crowley, David, and Paul Heyer, eds. *Communication in History: Technology, Culture, Society.* New York: Longman, 1991.
Curry, Ralph. *Stephen Leacock and His Works.* Toronto: ECW Press, n.d.
Czitrom, Daniel. *Media and the American Mind: From Morse to McLuhan.* Chapel Hill: University of North Carolina Press, 1982.
Davies, Robertson. 'Stephen Leacock.' In *Our Living Tradition: Seven Canadians.* Ed. Claude T. Bissell, 128-49. Toronto: University of Toronto Press, 1957.
Davis, Arthur, ed. *George Grant and the Subversion of Modernity.* Toronto: University of Toronto Press, 1996.
Den Otter, A.A. *The Philosophy of Railways: The Transcontinental Railway Idea in British North America.* Toronto: University of Toronto Press, 1997.
Diggins, John P. *The Bard of Savagery: Thorstein Veblen and Modern Social Theory.* New York: Seabury Press, 1978.
Di Norcia, Vincent. 'Communications, Time, and Power: An Innisian View.' *Canadian Journal of Political Science,* 23, 2 (1990): 335-57.
Djwa, Sandra. *E.J. Pratt: The Evolutionary Vision.* Toronto: Copp Clark, 1974.
Dorfman, Joseph. *The Economic Mind in American Civilization.* Vols. 3 and 4. New York: Viking Press, 1949, 1959.
–. *Thorstein Veblen and His America.* New York: Viking Press, 1961.
Dudley, Leonard M. 'Space, Time, Number: Harold A. Innis as Evolutionary Theorist.' *Canadian Journal of Economics* 28, 4 (1995): 754-69.
Duffy, Dennis. *Marshall McLuhan.* Toronto: McClelland and Stewart, 1969.
Dupuis, N.F. 'Sir Sandford Fleming.' *Queen's Quarterly* 23, 2 (1915): 124-27.
Eggleston, Wilfrid. 'Frederick Philip Grove.' In *Our Living Tradition: Seven Canadians.* Ed. Claude T. Bissell, 105-27. Toronto: University of Toronto Press, 1957.
Eksteins, Modris. *Rites of Spring: The Great War and the Birth of the Modern Age.* Boston: Houghton Mifflin, 1989.
Emberley, Peter C. *By Loving Our Own: George Grant and the Legacy of* Lament for a Nation. Ottawa: Carleton University Press, 1990.
Farrington, Benjamin. *Francis Bacon: Philosopher of Industrial Science.* [1961]. New York: Octagon Books, 1979.
Feenberg, Andrew. *A Critical Theory of Technology.* New York: Oxford University Press, 1991.
Ferkiss, Victor. *Nature, Technology, and Society: Cultural Roots of the Current Environmental Crisis.* New York: New York University Press, 1993.
Ferns, Henry, and Bernard Ostry. *The Age of Mackenzie King.* Toronto: James Lorimer, 1976.
Fleming, James Rodger. 'Science and Technology in the Second Half of the Nineteenth Century.' In *The Gilded Age: Essays on the Origins of Modern America.* Ed. Charles W. Calhoun. Washington, DC: Scholarly Resources, 1996.
Forbes, R.J. *The Conquest of Nature: Technology and Its Consequences.* New York: Frederick A. Praeger, 1968.
Fortner, Robert S. 'The Canadian Search for Identity, 1846-1914: Communication in an Imperial Context.' *Canadian Journal of Communication* 6 (1979): 24-31.
–. 'Communication and Canadian National Destiny.' *Canadian Journal of Communication* 6 (1979): 43-57.
–. 'Communication and Regional/Provincial Imperatives.' *Canadian Journal of Communication* 6 (1979): 32-46.

–. 'Communications and Canadian-American Relations.' *Canadian Journal of Communication* 7 (1980): 37-52.

Francis, R. Douglas. 'The Anatomy of Power: A Theme in the Writings of Harold Innis.' In *Nation, Ideas, Identities: Essays in Honour of Ramsay Cook*. Ed. Michael D. Behiels, 26-40. Toronto: Oxford University Press, 2000.

–. *Frank H. Underhill: Intellectual Provocateur*. Toronto: University of Toronto Press, 1986.

–. 'Modernity and Canadian Civilization: The Ideas of Harold Innis.' In *Globality and Multiple Modernities: Comparative North American and Latin American Perspectives*. Ed. Luis Roninger and Carlos H. Waisman, 213-29. Brighton: Sussex Academic Press, 2002.

–. 'Stephen Leacock and Technology in an Age of Reorientations.' In *Canada and the Nordic Countries in Times of Reorientation: Literature and Criticism*. Vol. 12. Ed. Jørn Carlsen, 73-86. Aarhus, Denmark: Nordic Association for Canadian Studies, 1998.

–. 'Technology and Empire: The Ideas of Harold Innis and George P. Grant.' In *Canada and the End of Empire*. Ed. Philip Buckner, 285-98. Vancouver: UBC Press, 2004.

Frankl, George. *The Social History of the Unconscious*. London: Open Gate Press, 1989.

Frankman, Myron J. 'Stephen Leacock, Economist: An Owl among the Parrots.' In *Stephen Leacock: A Reappraisal*. Ed. David Staines, 51-58. Ottawa: University of Ottawa Press, 1986.

Gauvreau, Michael. 'Baptist Religion and the Social Science of Harold Innis.' *Canadian Historical Review* 76, 2 (1995): 161-204.

–. 'Philosophy, Psychology, and History: George Sidney Brett and the Quest for a Social Science at the University of Toronto, 1910-1940.' Canadian Historical Association. *Historical Papers* (1988): 209-36.

Gerrie, James. 'Canada's Lost Tradition of Technological Criticism.' In *The River of History: Trans-National and Trans-Disciplinary Perspectives on the Immanence of the Past*. Ed. Peter Farrugia, 217-45. Calgary: University of Calgary Press, 2005.

Gidney, R.D., and W.P.J. Millar. *Inventing Secondary Education: The Rise of the High School in Nineteenth-Century Ontario*. Montreal and Kingston: McGill-Queen's University Press, 1990.

Gnarowski, Michael, ed. *Archibald Lampman*. Toronto: Ryerson Press, 1970.

Goodwin, Craufurd D.W. *Canadian Economic Thought: The Political Economy of a Developing Nation, 1814-1914*. Durham, NC: Duke University Press, 1961.

Gordon, Daniel M. 'Our Late Chancellor [Sir Sandford Fleming].' *Queen's Quarterly* 23, 2 (1915): 111-23.

Gordon, W. Terrence. *Marshall McLuhan: Escape into Understanding: A Biography*. Toronto: Stoddart, 1989.

Goyder, John. *Technology and Society: A Canadian Perspective*. Peterborough, ON: Broadview Press, 1997.

Gray, Charlotte. *Reluctant Genius: The Passionate Life and Inventive Mind of Alexander Graham Bell*. Toronto: HarperCollins, 2006.

Green, Lorne Edmond. *Chief Engineer: Life of a Nation-Builder – Sandford Fleming*. Toronto: Dundurn Press, 1993.

–. *Sandford Fleming*. Toronto: Fitzhenry and Whiteside, 1980.

Hamilton, A.C. *Northrop Frye: Anatomy of His Criticism*. Toronto: University of Toronto Press, 1990.

Haraway, Donna J. *Simians, Cyborgs, and Women: The Reinvention of Nature*. New York: Routledge, 1991.

Hård, Mikael, and Andrew Jamison, eds. *The Intellectual Appropriation of Technology: Discourses on Modernity*. Cambridge, MA: MIT Press, 1998.

Hardison Jr., O.B. *Disappearing through the Skylight: Culture and Technology in the Twentieth Century*. New York: Viking, 1989.

Hart, Jonathan. *Northrop Frye: The Theoretical Imagination*. London: Routledge, 1994.

Herbertson, Dorothy. *The Life of Frederic Le Play*. Ledbury, UK: Le Play House Press, 1950.

Houghton, Walter E. *The Victorian Frame of Mind, 1830-1870*. New Haven, CT: Yale University Press, 1957.

Hudson, Randolph, ed. *Technology, Culture, and Language: Selected Essays*. Boston: D.C. Heath, 1966.

Hughes, Thomas P. *American Genesis: A Century of Invention and Technological Enthusiasm, 1870-1970*. New York: Penguin Books, 1990.

–. 'A Technological Frontier: The Railway.' In *The Railroad and the Space Age: An Exploration in Historical Analogy*. Ed. Bruce Mazlish, 53-73. Cambridge, MA: MIT Press, 1965.

Irving, John A. 'The Achievement of George Sidney Brett (1879-1944).' *University of Toronto Quarterly* 14, 4 (1945): 329-65.

James, Cathy. 'Reforming Reform: Toronto's Settlement House Movement, 1900-20.' *Canadian Historical Review* 82, 1 (2001): 55-90.

Kasson, John F. *Civilizing the Machine: Technology and Republican Values in America, 1776-1900*. New York: Grossman, 1976.

Kern, Stephen. *The Culture of Time and Space, 1880-1918*. Cambridge, MA: Harvard University Press, 1983.

Kroker, Arthur. *Technology and the Canadian Mind: Innis/McLuhan/Grant*. Montreal: New World Perspectives, 1984.

–. *The Will to Technology and the Culture of Nihilism: Heidegger, Nietzsche, and Marx*. Toronto: University of Toronto Press, 2004.

Krutch, Joseph Wood. *The Modern Temper: A Study and a Confession*. New York: Harcourt, 1929.

Kuffert, L.B. *A Great Duty: Canadian Responses to Modern Life and Mass Culture, 1939-1967*. Carleton Library Series No. 199. Montreal and Kingston: McGill-Queen's University Press, 2003.

Kuhns, William. *Post-Industrial Prophets: Interpretations of Technology*. New York: Weybright and Talley, 1971.

Kusher, J., and R.D. MacDonald. 'Leacock: Economist/Satirist in *Arcadian Adventures and Sunshine Sketches*.' *Dalhousie Review* 56, 3 (1976): 493-509.

Landes, David S. *The Unbound Prometheus: Technological Change and Industrial Development in Western Europe from 1750 to the Present*. Cambridge: Cambridge University Press, 1969.

Lapham, Lewis H. Introduction. *Understanding Media: The Extensions of Man*. By Marshall McLuhan. Cambridge, MA: MIT Press, 1994, ix-xxiii.

Layton Jr., Edwin T., ed. *Technology and Social Change in America*. New York: Harper and Row, 1973.

Lee, Alvin A., and Robert D. Denham, eds. *The Legacy of Northrop Frye*. Toronto: University of Toronto Press, 1994.

Legate, David M. *Stephen Leacock: A Biography*. Toronto: Doubleday, 1970.

Leiss, William. *The Domination of Nature*. New York: George Braziller, 1972.

–. *Under Technology's Thumb*. Montreal and Kingston: McGill-Queen's University Press, 1990.

MacDonald, R.D. 'Lee's *Civil Elegies* in Relation to Grant's "Lament for a Nation."' *Canadian Literature* 98 (1983): 10-30.

MacLean, Hugh. *Man of Steel: The Story of Sir Sandford Fleming*. Toronto: Ryerson, 1969.

Macphail, Andrew. 'Sir Sandford Fleming.' *Queen's Quarterly* 36, 2 (1929): 185-204.

Marcell, David W. *Progress and Pragmatism: James, Dewey, Beard, and the American Idea of Progress*. Westport, CT: Greenwood Press, 1974.

Marchand, Philip. *Marshall McLuhan: The Medium and the Messenger*. Toronto: Random House, 1989.

Marcus, Alan I., and Howard P. Segal. *Technology in America: A Brief History*. San Diego: Harcourt, 1989.

Marx, Leo. 'The Impact of the Railroad on the "American Imagination" as a Possible Comparison for the Space Impact.' In *The Railroad and the Space Program: An Exploration in Historical Analogy*. Ed. Bruce Mazlish, 202-16. Cambridge, MA: MIT Press, 1965.

–. *The Machine in the Garden: Technology and the Pastoral Ideal in America*. London: Oxford University Press, 1964.

–. 'On Heidegger's Conception of "Technology" and Its Historical Validity.' *Massachusetts Review* 25, 4: 638-79.

–. 'The Railroad-in-the-Landscape: An Iconological Reading of a Theme in American Art.' In *The Railroad in American Art: Representations of Technological Change*. Ed. Susan Danly and Leo Marx, 183-209. Cambridge, MA: MIT Press, 1988.

Massolin, Philip. *Canadian Intellectuals, the Tory Tradition, and the Challenge of Modernity, 1939-1970*. Toronto: University of Toronto Press, 2001.
Mazlish, Bruce. 'Historical Analogy: The Railroad and the Space Program and Their Impact on Society.' In *The Railroad and the Space Program*. Cambridge, MA: MIT Press, 1965, 1-52.
McDougall, Robert L. 'Thomas Chandler Haliburton.' In *Our Living Tradition: Seven Canadians*. Ed. Claude. T. Bissell, 13-30. Toronto: University of Toronto Press, 1959.
McKillop, A.B. *Contours of Canadian Thought*. Toronto: University of Toronto Press, 1987.
–. *A Disciplined Intelligence: Critical Inquiry and Canadian Thought in the Victorian Era*. [1979]. Montreal and Kingston: McGill-Queen's University Press, 2001.
McMaster, R.D. 'Criticism of Civilization in the Structure of *Sartor Resartus*.' *University of Toronto Quarterly* 37, 3 (1968): 268-80.
Melody, William H., et al., eds. *Culture, Communication, and Dependency: The Tradition of H.A. Innis*. Norwood, NJ: Ablex Publishing, 1981.
Meynaud, Jean. *Technocracy*. Trans. Paul Barnes. London: Faber and Faber, 1968.
Millard, Rodney. *The Master Spirit of the Age: Canadian Engineers and the Politics of Professionalism*. Toronto: University of Toronto Press, 1988.
Mitcham, Carl. 'Philosophy of Technology.' In *A Guide to the Culture of Science, Technology, and Medicine*. Ed. Paul T. Durbin, 282-363. New York: Free Press, 1980.
–. *Thinking through Technology: The Path between Engineering and Philosophy*. Chicago: University of Chicago Press, 1994.
Moriarty, Catherine. *John Galbraith, 1846-1914: Engineer and Educator: A Portrait*. Toronto: Faculty of Applied Science and Engineering, University of Toronto, 1989.
Mueller, Herbert J. *The Children of Frankenstein: A Primer on Modern Technology and Human Values*. Bloomington: Indiana University Press, 1970.
Neill, Robin. *A New Theory of Value: The Canadian Economics of H.A. Innis*. Toronto: University of Toronto Press, 1972.
Nelles, H.V. 'Thomas Coltrin Keefer.' *Dictionary of Canadian Biography*. Vol. 14, 1911-1920. Toronto: University of Toronto Press, 1998, 552-55.
O'Brien, A.H. *Haliburton: A Sketch and Bibliography*. 2nd ed. Montreal, 1909.
Ostrander, Gilman M. *American Civilization in the First Machine Age: 1890-1940*. New York: Harper and Row, 1970.
Pacey, Arnold. *The Maze of Ingenuity: Ideas and Idealism in the Development of Technology*. 2nd ed. Cambridge, MA: MIT Press, 1993.
Parker, Ian, et al., eds. *The Strategy of Canadian Culture in the 21st Century*. Toronto: TopCat Communications, 1988.
Patterson, Graeme. *History and Communications: Harold Innis, Marshall McLuhan, the Interpretation of History*. Toronto: University of Toronto Press, 1990.
Postman, Neil. *Technolopy: The Surrender of Culture to Technology*. New York: Vintage, 1993.
Purcell, Caroll W. *Technology in America: A History of Individuals and Ideas*. 2nd ed. Cambridge, MA: MIT Press, 1990.
Quentin, Anthony. *Thoughts and Thinkers*. London: Duckworth, 1982.
Reaney, James. '*Towards the Last Spike*: The Treatment of a Western Subject.' In *E.J. Pratt*. Ed. David G. Pitt, 73-82. Toronto: Ryerson Press, 1969.
Robertson, Ian Ross. 'The Historical Leacock.' In *Stephen Leacock: A Reappraisal*. Ed. David Staines, 33-49. Ottawa: University of Ottawa Press, 1986.
Robinson, Gertrude J., ed. 'The Medium's Messenger: Understanding McLuhan.' Special issue, *Canadian Journal of Communication* (Fall 1989): 1-160.
Sanderson, George, and Frank Macdonald, eds. *Marshall McLuhan: The Man and His Message*. Golden, CO: Fulcrum, 1989.
Schivelbusch, Wolfgang. *The Railway Journey: The Industrialization of Time and Space in the 19th Century*. Berkeley: University of California Press, 1986.
Schmidt, Larry, ed. *George Grant in Process: Essays and Conversations*. Toronto: Anansi, 1978.
Segal, Howard P. *Technological Utopianism in American Culture*. Chicago: University of Chicago Press, 1985.
Semmel, Bernard. *Imperialism and Social Reform: English Social-Imperial Thought, 1895-1914*. Cambridge, MA: Harvard University Press, 1960.

Shore, Marlene. *The Science of Social Redemption: McGill, the Chicago School, and the Origins of Social Research in Canada*. Toronto: University of Toronto Press, 1987.

Shortt, S.E.D. *The Search for an Ideal: Six Canadian Intellectuals and Their Convictions in an Age of Transition*. Toronto: University of Toronto Press, 1976.

Simpson, Lorenzo C. *Technology, Time, and the Conversations of Modernity*. New York: Routledge, 1995.

Smith, Gregory Bruce. 'Heidegger, Technology, and Postmodernity.' *Social Science Journal* 28, 3 (1991): 369-90.

Spadoni, Carl, ed. Introduction. *My Recollection of Chicago* and *The Doctrine of Laissez-Faire*. By Stephen Leacock. Toronto: University of Toronto Press, 1998, vi-xxxix.

Stacey, C.P. *A Very Double Life: The Private Life of Mackenzie King*. Toronto: Macmillan of Canada, 1976.

Stamps, Judith. 'Innis in the Canadian Dialectical Tradition.' In *Harold Innis in the New Century: Reflections and Refractions*. Montreal and Kingston: McGill-Queen's University Press, 1999, 46-66.

–. *Unthinking Modernity: Innis, McLuhan, and the Frankfurt School*. Montreal and Kingston: McGill-Queen's University Press, 1995.

Stamp, Robert M. 'Teaching Girls Their "God Given Place in Life."' *Atlantis* 2, 2 (1977): 18-34.

Sussman, Herbert L. *Victorians and the Machine: The Literary Response to Technology*. Cambridge, MA: Harvard University Press, 1968.

Surtees, Lawrence. 'Alexander Graham Bell.' *Dictionary of Canadian Biography*. Vol. 15, 1921-1930. Toronto: University of Toronto Press, 2005, 78-88.

Sutherland, Ronald. *Frederick Philip Grove*. Toronto: McClelland and Stewart, 1969.

Taylor, M. Brook. 'Haliburton as a Historian.' In *The Thomas Chandler Haliburton Symposium*. Ed. Frank M. Tierney, 103-22. Ottawa: University of Ottawa Press, 1985.

Theall, Donald F. *Beyond the Word: Reconstructing Sense in the Joyce Era of Technology, Culture, and Communication*. Toronto: University of Toronto Press, 1995.

–. *Understanding McLuhan: The Medium Is the Rear View Mirror*. Montreal and Kingston: McGill-Queen's University Press, 1971.

Thompson, Graham M. 'Sandford Fleming and the Pacific Cable: The Institutional Politics of Nineteenth-Century Imperial Communications.' *Canadian Journal of Communication* 15, 2 (1990): 64-75.

Tichi, Cecelia. *Shifting Gears: Technology, Literature, Culture in Modernist America*. Chapel Hill: University of North Carolina Press, 1987.

Trachtenberg, Alan. *The Incorporation of America: Culture and Society in the Gilded Age*. New York: Hill and Wang, 1982.

Umar, Yusuf K., ed. *George Grant and the Future of Canada*. Calgary: University of Calgary Press, 1992.

Vanderburg, Willem H. *Labyrinth of Technology*. Toronto: University of Toronto Press, 2000.

–. *Living in the Labyrinth of Technology*. Toronto: University of Toronto Press, 2005.

–. ed., *Perspectives on Our Age: Jacques Ellul Speaks on His Life and Work*. Rev. ed. Toronto: Anansi, 2004.

Wajcman, Judy. *Feminism Confronts Technology*. University Park, PA: Pennsylvania State University Press, 1991.

Ward, James. *Railroads and the Character of America, 1820-1887*. Knoxville: University of Tennessee Press, 1986.

Watson, A. John. 'Harold Innis and Classical Scholarship.' *Journal of Canadian Studies* 12, 5 (1977): 45-61.

–. *Marginal Man: The Dark Vision of Harold Innis*. Toronto: University of Toronto Press, 2006.

Watt, F.W. 'Critic or Entertainer? Stephen Leacock and the Growth of Materialism.' *Canadian Literature* 5 (1960): 33-42.

Wernick, Andrew. 'The Post-Innisian Significance of Innis.' *Canadian Journal of Political and Social Theory* 10, 1-2 (1986): 128-50.

Whitney, Charles. *Francis Bacon and Modernity*. New Haven: Yale University Press, 1986.

Wiebe, Robert H. *The Search for Order 1877-1920*. New York: Hill and Wang, 1976.

Whitaker, Reginald. 'The Liberal Corporatist Ideas of Mackenzie King.' *Labour/Le Travail* 2 (1977): 137-69.

–. '"To Have Insight into Much and Power over Nothing": The Political Ideas of Harold Innis.' *Queen's Quarterly* 90, 3 (1983): 818-31.

Willmott, Glenn. *McLuhan, or Modernism in Reverse*. Toronto: University of Toronto Press, 1996.

Winner, Langdon. *Autonomous Technology: Technics-out-of-Control as a Theme in Political Thought*. Cambridge, MA: MIT Press, 1977.

Young, C.R. *Early Engineering Education at Toronto, 1851-1919*. Toronto: University of Toronto Press, 1958.

Zeller, Suzanne. '"Merchant of Light": The Culture of Science in Daniel Wilson's Ontario, 1853-1892.' In *Thinking with Both Hands: Sir Daniel Wilson in the Old World and the New*. Ed. Marinell Ash and Elizabeth Hulse, 115-38. Toronto: University of Toronto Press, 1999.

Zimmerman, Michael E. 'Beyond "Humanism": Heidegger's Understanding of Technology.' In *Heidegger: The Man and the Thinker*. Ed. Thomas Sheehan, 219-27. Chicago: Precedent Publishing, 1981.

Index